AIR CRASHES
and
MIRACLE LANDINGS

AIR CRASHES
and
MIRACLE LANDINGS
85 CASES - How and Why
Second Edition

by

Christopher Bartlett

Second Edition
ISBN 978-0-9560723-6-8
January 2018
Updated September 2018

First edition
published April 12, 2010,
revised first edition September 2012.

by
OpenHatch Books, Gillingham, Dorset, UK
openhatchbooks.com
chrisbart.com

Cover
by
Lysa Bartlett

Introduction

This expanded second edition *of Air Crashes and Miracle Landings* stresses the human aspect even more than the first, though time for reflection means views can be more nuanced. Advances in technology and the lessons learnt from the various accidents mean most will never happen again.

As before, each chapter covers a specific type of incident in chronological order. While showing the evolution of accidents over time, this means those of the greatest interest may well be nested within the chapter.

With the Kindle you can click on an entry in the Table of Contents or a word (e.g. flight number or location) in the "Search" function and go straight to it.

To avoid the printed book becoming impossibly long, the glossary has been placed in the second part of our new book *Plane Clever: Booking Strategies and All About Flying*, available in print and on the Kindle.

For
all those who died
and
all those who survived.

Lessons
were learned
for the benefit
of us all.

ToC/Sub-Chapters !

FIGURES

Diagrams and images are included only where essential for explanation, since so many can be easily found on the Internet.

On the Kindle, images often appear small but can usually be expanded to show finer detail.

Chapter 1
LOSS OF POWER
OVER WATER

The Amelia Earhart Mystery (Howland Island, 1937)
America's "favorite missing person"

> Aviatrix Amelia Earhart fully deserved her celebrity status, even though
> it initially came about fortuitously when in 1928 she was offered the
> opportunity to be the first woman to fly across the Atlantic after the one
> initially chosen pulled out fearing the flight was too perilous.
> Her life up until then had been no bed of roses. She fell seriously ill
> during the 1918 Spanish flu pandemic, her family had financial
> difficulties, and she strove hard in many domains, including aviation,
> where, as always, she tried to further the position of women.
> On what was to be her last major exploit, an equatorial round-the-world
> flight, in 1938, she got two-thirds of the way but then failed to reach a
> minuscule island in the middle of the Pacific Ocean.
> The president of the United States ordered a large-scale search to no
> avail, and ever since people have been wondering what happened to her
> and her navigator.

Earhart later commented she had been little more than a sack of potatoes on that 1928 transatlantic flight. However, the sack must have had grit and courage, as just crossing the Atlantic was a difficult and extremely risky thing to do at the time. Also, the aircraft had to be flown by instruments, something for which she had not at the time been trained. Even so, on their return to Manhattan, she and the two male pilots received a ticker-tape welcome, with her seen as a plucky young woman and getting top billing.

Her fame further increased when, in 1932, she became the second person after Charles Lindberg to fly solo across the Atlantic. Though her fifteen-hour flight from Newfoundland to Ireland was much shorter than Lindberg's historic thirty-three-hour flight in 1927 from Long Island to Paris, it proved a considerable ordeal, with icing problems and leaking fuel splashing onto her face.

With the help of her husband, publisher G. P. Putnam, she had stepped on a publicity treadmill, involving a grueling lecture schedule and a constant need to find trailblazing feats to stay in the headlines and keep the money coming in. In the summer of 1937, aged forty and probably somewhat jaded, she was going to embark on what should have been her last major flying exploit, an eastward around-the-world flight along the equator in a Lockheed Electra twin-engine airliner fitted with extra fuel tanks in place of the passenger seats.

She and Putnam were mortgaged to the hilt despite financing from Purdue University. An earlier mishap when taking off in the Electra from Honolulu on their initial attempt to make the around-the-world trip in a westerly direction had added to their costs. On their second attempt, a change in the trade winds forced them to make it in an easterly direction.

Accompanying her on the trip in the key role of navigator was ex–Pan Am navigator Fred Noonan. While reports that he was fired from the airline for a drink problem are disputed, he was an exceptionally capable navigator and an expert at celestial navigation.

Starting from Oakland, California, Earhart and Noonan flew to Miami and then on to Brazil before crossing the narrowest part of the Atlantic to Dakar. After crossing Africa, they flew on to Pakistan, Burma, Singapore, and Darwin in Northern Australia. At times Earhart was quite sick with dysentery, and she must have been exhausted when she arrived at Lae, Papua New Guinea, after the short hop from Darwin. Lae was the departure point for the long transpacific legs, first to tiny Howland Island, and from there on to Hawaii.

Noonan was none too happy, and after an argument with Amelia at Lae apparently went on a drinking spree. When her husband, back in the States, heard about it, he reportedly urged her to call off the venture. She was reluctant to do so, as they had successfully flown twenty-two thousand miles and only had seven thousand to go, and she thought the hardest part had been accomplished.

Whatever his condition, Noonan must have been quite daunted by the prospect of navigating over water with no landmarks to an island only 1.6 square kilometers in area, rising only a foot or two above the sea.

Interestingly, it was the introduction of flying boats in the early thirties that enabled the United States to take over a number of essentially uninhabited islands from the British, which, in the case of Howland, it did only in 1935.

Having jurisdiction at this convenient midway point between Lae and Hawaii in 1937, the US government had prepared a landing strip on the island especially for Amelia's benefit. A US coast guard cutter was also ordered to stand by nearby to render assistance.

Unfortunately, the published coordinates for Howland were out by some five or six miles.

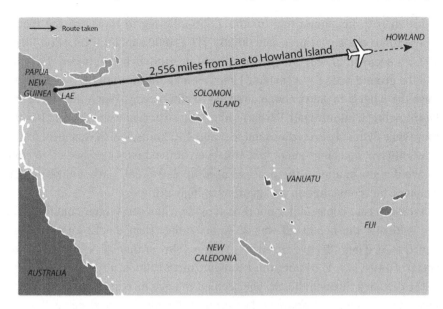

From Lae, Papua New Guinea, to minuscule Howland Island.

For navigational reasons, Amelia and Noonan timed their departure from Lae so they would fly over land and various identifiable islands in daylight, and then fly overnight over the ocean to Howland, arriving there just after dawn. This meant that while over the ocean and far from land, Noonan would be able to check their position from the stars, provided the weather was not too cloudy. Even after dawn, with no stars to get a fix, Noonan would be able to determine longitude but not latitude by how far the sun was above the horizon for the time of day.

He would also be using dead reckoning, whereby one's position is calculated according to one's speed, direction of travel, and elapsed time. As one's actual speed and direction over the ground, as opposed to through the air, depend on the direction and speed of the wind, the method can introduce considerable errors over long distances. Of course, for dead reckoning, correct identification of the waypoints is essential, and as mentioned below it seems Amelia and Noonan misidentified a ship as one that was meant to serve as a waypoint, partly due to their inability to communicate by radio.

The classic way to compensate for the limitations of dead reckoning is to purposely aim for a point well to the left or right of one's destination and, on completing the required distance, make the ninety-degree turn to the right or left to find it. Provided the offset is enough to allow for the navigational error, this avoids the risk of arriving just off to the side of one's destination and flying farther and farther away from it in that direction.

Laden with maximum fuel for the 2,556-mile leg to Howland, the Electra labored into the air[1] at precisely 00:00 GMT (10:00 a.m. local time) on July 2, 1937. It seems that only the ground effect (air sandwiched between the wings and the ground acting as a cushion[2]) kept them in the air until they were able to raise the wheels to gain enough airspeed to climb away. When they were four hours eighteen minutes out, Earhart radioed an airport manager back at Lae that they were flying at seven thousand feet and 140 knots, which was ideal for the most efficient consumption of fuel. However, at five hours out she had to climb to avoid a storm and, in the process, used up extra fuel, since, unlike modern airliners, fuel consumption was greater[3] at high altitudes.

Worse was to come. In her hourly call made when seven hours out from Lae, she said that the headwind was 25 knots rather than the 12 knots she had anticipated. If her effective speed was 12 knots slower than her expected ground speed of 140 − 12 = 128 knots, this headwind meant almost a 10 percent increase in the distance through the air. She seemed to have increased engine power to increase airspeed and optimize her ground distance covered per gallon, but there was certainly still a very significant fuel penalty due to the headwind. Though she was able to call Lae for the first part of the trip, there is no indication she was able to receive any message from Lae in return.

The US government had placed a ship, the *Ontario*,[4] at the halfway point, but it seems that though Earhart and Noonan saw a ship, it was another one forty miles farther north. This may have resulted in them thinking they were on the correct course when in fact they were heading to an area north of Howland Island. (A message she had wanted to send to the *Ontario* before leaving Lae regarding the radio signals the ship should emit had not reached it.)

As the Electra got farther from Lae and nearer Howland, the coast guard cutter *Itasca*, waiting just off the island, began to receive some messages but could not seem to contact them. Finally, at 7:42 a.m., nineteen and a half hours after their departure from Lae, the *Itasca* received the following message from the Electra:

> *KHAQQ calling Itasca.*
>
> *We must be on you but cannot see you . . . Gas is running low . . . Been unable to reach you by radio. We are flying at one thousand feet.*

Indeed, the signal was so loud the radio operator expected to see them overhead. However, there was no sign of them. The transmission had not lasted long, and the *Itasca* could not get a fix—some say because its direction finder was not working. The *Itasca* emitted smoke, and had there been no cloud, the Electra could have flown higher and seen them from fifty miles.

An hour later, at 8:44 a.m., there was a message in which, according to Doris L. Rich's biography[5], "Amelia's voice—shrill and breathless, her words tumbling over one another—came in."

We are on the line of position 157-337[6]. Will repeat message... We are running north and south.

When nothing more was heard, it became increasingly obvious that the Electra must have come down somewhere. The crew on board the *Itasca* searched the area where they thought they might have been, but there was no sign of them, not even wreckage. The nearest land to Howland Island was 325 miles away, so if Earhart had indeed been as near the island as she thought she was, there was little chance of her having reached any other land. Some later alleged she had left their life raft behind to save weight.

On President Roosevelt's orders, a major search was launched, involving nine US Navy ships and sixty-six aircraft, at a cost of $4 million or more. They searched a vast area of the Pacific, including the various islands where some people suggested they might have come down. When the authorities finally called off the search after two weeks, the rumors began.

One rumor was that Earhart was a spy for the US government. Hardly anyone believes this to be true now. In the author's view, she would have been much better prepared, especially regarding the use of the radio, had she been a spy. One theory with much traction in the form of witness statements is that she came down on Mili Atoll, to the northwest of Howland, where she and Nolan, suspected as spies, were captured by the Japanese and taken, with their aircraft, to Saipan, where they died. According to witnesses, the embarrassed US government engaged in a major cover-up operation.

During the search, radio messages seeming to emanate from Earhart, some of which caused the US Navy to change the focus of its searches, were received. While many were obviously hoaxes, the navy could not be certain that some were not genuine. Subsequently, a woman's shoe was found on an island she might possibly have reached, only for it to be found not to be her size.

There are fifty or more books, any number of magazine and newspaper articles and TV documentaries, not to mention online forums, covering her disappearance. The 2009 movie *Amelia* put some meat on the bones of her personal life and her affair with Gene Vidal. The dashing Gene Vidal was a West Point[7] quarterback who had also come seventh in the 1924 Olympic Games pentathlon and was a prominent figure in aviation. He also happened to be the father of the late Gore Vidal.

In an interview[8] with Robert Chalmers, writing for the UK's *The Independent* newspaper, Gore Vidal mentioned that he once asked his father why he had not married Earhart, to which his father replied, "I have never really wanted to marry another boy." Gore Vidal went on to say, "And she was like a boy." He then drew Chalmers's attention to a photo of her in a prominent place on a dresser, remarking, "You see how one could see courage in those eyes," and made it clear how much he admired her and even felt for her.

Searchers have spent tremendous sums trying to locate her aircraft on relatively close islands, such as Nikumaroro, and underwater near Howland. Recently, more credence has been given to some of the radio messages received after her disappearance purporting to be from her. Though searchers are truly fascinated, there is a profit motive. Anyone who succeeds in finding the lost Electra could put the artifacts on show and write the definitive book about her disappearance.

Tom Crouch, senior curator of aeronautics at the National Air and Space Museum, in Washington, who perhaps like others feels he has devoted more than enough time to the case, has said, "I'm convinced that the mystery is part of what keeps us interested. In part, we remember her because she's our favorite missing person." One might push curator Crouch's argument further by saying that people even enjoy arguing about the clues, some of which may have been spun to show the parties, and notably officialdom, in a better light.

Furthermore, some dispute facts once taken as gospel,[9] such as that navigator Noonan was a true alcoholic fired from Pan Am, it being pointed out that he seemed in good form in pictures of him helping Amelia into the aircraft at Lae. Even the idea that the coast guard cutter *Itasca* could have produced a meaningful column of smoke is disputed, since doing so for any significant time would have compromised one of the vessel's boilers, and any smoke would have quickly dissipated in the prevailing wind.

That said, any reader hooked on the subject should note that local time at Howland is GMT plus ten and a half hours, and that this half-hour step can be confusing when comparing times at other places, such as Lae. What is somewhat unsettling is that the great strength of the transmission "We must be on you but cannot see you . . . Gas is running low . . . " gave the impression they were very close.

As mentioned, it even made the *Itasca*'s wireless operator think she might be right overhead and certainly indicated that Noonan had brought her quite close to the minuscule, low-profile Howland Island (whose published coordinates, as said, were out by five or six miles). What's more, her final transmission, at 8:44 a.m., was at maximum strength.

This signal strength, and the later statement "We are running north and south," suggest that she was so determined to reach Howland that she kept trying to get there by going back and forth rather than at some point continuing southward on that line of position, which would very likely have taken them over other islands. Of course, if she waited too long, reaching any other island would be impossible.

Here and elsewhere, commentators have emphasized the general use of the already-mentioned *offset navigation technique*, whereby one aims to one side of one's destination. On reflection, there is a problem with this in the Earhart context. Firstly, when Amelia Earhart says at the outset "We must be on you but cannot see you," it seems to imply they had flown directly to where they thought Howland was. The final message had been

> We are on the line of position 157-337. Will repeat message... We are running north and south.

Noonan would have first determined the position of this line when the sun had risen some two hours earlier (see endnote reference to an Internet explanation of the significance of 157-337 degrees) and added to the extra time required so that the line would traverse Howland. If the unexpected headwinds had not delayed them, the dead reckoning element would have been much reduced.

One interesting theory[10] about what could have gone wrong navigationally is that Noonan got the date wrong for his calculations on approaching Howland Island after crossing the International Date Line. Just before they crossed it at 6 a.m. local time, when he would have been exhausted tired after almost eighteen hours flying, he would have been feverishly trying to get his last star fixes before dawn using the figures in the tables for July 3. Then, just after crossing it he would have been anxious not to miss getting that essential fix on the rising sun.

Could fatigue, stress and fixation on the various tasks have made him forget to switch back to July 2? If so, their position would have been out by one degree of longitude or sixty nautical miles; and with Howland mapped 6 miles out of position, that would have made them think they were at Howland when 66 miles away—too far to see but near enough for the radio signal to have been so strong.

What Earhart perhaps had not realized was that the chances of finding Howland visually in less-than-perfect weather would anyway be slim. In fact, the weather was far from ideal that day, and the only sure way of finding the island would have been by the radio direction finder. In addition, her lack of fuel might have tempted her to try to fly straight to Howland rather than safely to one side to avoid flying away from it. Everything thus depended on the ability to communicate by radio and the transmission of radio signals on frequencies that respective radio direction finders could pick up. Apparently, there was a lack of coordination between her and the coast guard prior to the flight. The coast guard later tried to portray her as a prima donna unwilling to listen to reason, leading some to interpret this to mean she was a spy unwilling to obey orders from the conventional military.

Though some contemporary female pilots were highly critical of her flying abilities, General Leigh Wade, who had flown with her, claimed she was a born flier "with a delicate touch on the stick." The movie clip—the real one, not the simulated Amelia movie one— showing how she managed the difficult takeoff from Lae, with the aircraft so heavy with extra fuel that it could hardly get off the ground, seems to bear that out. Many a pilot would have pushed the nose up too far and stalled the aircraft. However, there was an early incident in her flying career where her judgment was in question in that she dropped down through cloud not realizing what would have happened had the cloud extended right to the ground.

Anyway, she would have been very reluctant to give up, as she and husband Putnam would be very unlikely to get their money back after a failed exploit. From a modern airline safety point of view, she at first seems to have been silly not to turn back to Lae when she found the climb to a higher altitude and the unexpected increase in headwind meant she would have little fuel in reserve even if she made a perfect course to Howland. However, going back to Lae would have involved going again over nine-thousand-foot peaks on Bougainville Island, in the Solomon Islands.

What is more, the takeoff from Lae, even though helped by the headwind that later caused her to use more fuel than expected, had been a touch-and-go affair, and she might not have been so lucky a second time. Lack of high-octane aviation fuel on the island would mean waiting until further supplies arrived and paying for their shipment. Weather conditions might be even worse the next time, and there was no guarantee the US Navy would keep the Itasca on station at Howland. Finally, she could not be sure the Electra could withstand another takeoff at one and half times its normal maximum takeoff weight.

On her other flights toward large landmasses, having ample reserves of fuel was not so critical—for instance, on her transatlantic solo flight she was said to have been hoping to emulate Charles Lindbergh's 1927 historic flight from New York's Long Island to Le Bourget Field, Paris, and yet ended up in Ireland, both large landmasses.

Her, some might say, forgotten navigator, the recently married Noonan, must also have had considerable courage to embark on the venture. Were the fatigue and stress getting to her and making her disapprove of his conduct, or was he unhappy about preparations for the particularly difficult flight ahead?

View of A Modern Air Accident Inquiry

A modern air accident inquiry might say the direct causes of the tragedy, apart from the inherent risk, included:

1. Failure to master the use of radio communication and particularly radio direction-finding technology (RDF), failure to ensure it was set up and tested properly, and failure to check the compatibility of the radio communication and direction-finding equipment.

2. Failure to employ backup Morse code radio equipment (and associated direction-finding modes), perhaps on account of failure to master Morse code.

Lack of funds probably meant the venture was not professionally managed, resulting in a lack of liaison between the parties especially regarding use of the radio equipment. Even had the *Itasca* been able to get an RDF fix, it would not have enabled it to give her the vector to reach Howland, as she was not receiving its transmissions. It would, though, have given some general idea of where to look for her—then and even now.

The Search Goes On

For eighty years people in the United States and all over the world have been speculating on what happened to Amelia and Noonan, but nowadays the main impetus for endeavors to find her aircraft comes from an organization called TIGHAR (pronounced "tiger"), standing for The International Group for Historic Aircraft Recovery. Its members donate funds for searches, with them now focusing on Nikumaroro Atoll (also called Gardner Island) on the line of position 157-337, passing through Howland but two hours' flying time to the south.

They are not the first to look for Amelia-related items there, and previous seekers have found artifacts made from aluminum, some from a B-24. In 1940, a partial skeleton was discovered but dismissed when a British doctor said it was of a male. At the time, the island was part of the British Gilbert and Ellice Islands Crown Colony.

The bones disappeared, but the doctor had taken notes, and TIGHAR recently got two forensic anthropologists—Karen Burns, PhD, and Richard Jantz, PhD—to review them. They concluded that "the morphology of the recovered bones, insofar as we can tell by applying contemporary forensic methods to measurements taken at the time, appears consistent with a female of Earhart's height and ethnic origin."

TIGHAR also are giving increased credence to radio transmissions, purportedly from Amelia after she had come down, that were picked up by a sixteen-year-old girl radio ham in Texas and a woman in Melbourne, listening on a harmonic of the frequency Amelia used.

For TIGHAR to keep the pot boiling and funds coming in for further missions with a submersible to explore around Nikumaroro, it is only natural for them to constantly come up with—as they always do—new reasons to hope Amelia's aircraft wreckage will soon be found. Its pronouncements are picked up by the world's media and published almost verbatim for a fascinated public. However doubtful one may be, one still gets hooked.

But see https://earharttruth.wordpress.com/tag/nikumaroro/ for one highly dubious skeptic's correspondence with Gillespie from TIGHAR.

In 2017, the National Geographic Channel had a program about Amelia Earhart showing a photo purportedly of her and Nolan on a pier, implying she had been captured by the Japanese. However, a researcher in Japan found the photo in question in a book published two years before her disappearance!

Until hard evidence or the wreck of her Electra is found, people will always wonder.

As curator Tom Crouch said, "She is our favorite missing person."

---

[1] http://www.tighar.org/amelia_3.rm is a 30-second clip of the takeoff. It shows Earhart and Noonan apparently looking fit when boarding.
Also, see www.tighar.org.

[2] To describe the ground effect as simply a cushioning effect is an oversimplification, as the aerodynamics are much more complicated.

[3] *Amelia Earhart: The Mystery Solved*, Elgen M. Long, Marie K. Long.

[4] http://www.thehistorynet.com/ahi/blearhart/index3.html.

[5] *Amelia Earhart: A Biography*, by Doris L. Rich.

[6] Best explanation of significance of "157-337."
http://tighar.org/forum/FAQs/navigation.html

[7] US military academy

[8] Interview by Robert Chalmers, *The Independent*, 25 May, 2008.

[9] See http://www.tighar.org/forum/FAQs/Forumfaq.html

[10] See datelinetheory.com

Captain Coolly Announces No Engines Working (Jakarta, 1982)
Out-of-this-world phenomenon

In 1982, a British Airways 747 encountered a strange, almost mystical phenomenon high up in the lonely sky off Indonesia. With the aircrew not knowing why, all four engines failed, leaving the aircraft gliding through the night sky, with high mountains between them and the nearest diversion airport.

British Airways Flight 9, June 24, 1982

Many of the passengers—and particularly those who had boarded in London and endured many stops and delays on what was then the longest scheduled five-stopover flight, from London to New Zealand—were dozing off after the evening meal. Everything had been going smoothly on that leg from Kuala Lumpur to Perth, Australia, with only one more stopover to go.

Then some two and a half hours into the leg, passengers on the Boeing 747 started complaining that there were too many smokers. On seeing the haze at the back of the economy-class cabin, the cabin crew wondered why so many were lighting up when they would normally be trying to get some sleep. As the smoke and acrid smell of burning increased, cabin staff surreptitiously went around ensuring a smoldering cigarette was not about to start a fire.

Meanwhile, profiting from a quiet moment on the flight deck, the captain had taken the opportunity to go to the toilet and stretch his legs. The first officer and flight engineer left in charge on the flight deck had the autopilot handling the aircraft and only needed to keep a lookout and be ready for anything unexpected.

Having passed right over Indonesia's capital, Jakarta, the 747 was about to cross the mountains and head out over the Indian Ocean, stretching from India to Antarctica. Cruising at thirty-seven thousand feet, with bright stars overhead and little cloud below, the first officer and flight engineer began to see odd light effects ahead, which they later said were like Saint Elmo's fire, a phenomenon in which lightning dances around in the sky. Yet their weather radar gave no indication of storm clouds.

Soon, with flashes of light and tiny balls of fire rushing at them and exploding on the windscreen, the crew's amazement switched to concern, and the first officer called the captain, who was relaxing just below. Hurrying up the stairs, Captain Moody could hardly believe what he was seeing. No one could work out what it was; it seemed out of this world. High up in the sky, with no light pollution, anything producing some degree of light is immediately obvious, and a strange glow seemed to be enveloping the leading edge of the wings and the nacelles of the engines. Worse still, the engines themselves appeared to be illuminated from inside. Passengers seated behind the engines were startled to see bright particles issuing from them. Their rough running began to rouse the few passengers still asleep. A pungent smell of smoke prompted the flight engineer to check for fire and consider shutting down the air-conditioning, even though he could not find any reason to do so.

Two minutes after entering the zone where the strange phenomenon was occurring, the instruments indicated a pneumatic valve pressure problem for the number four (outside right) engine, which afterward surged and flamed out. A minute later the number two engine failed likewise, and failure of engines number one and three soon followed. They had become a glider with 15 crew members and 248 passengers, including children and babies.

Though airliners do not make good gliders, their great cruising height—in this case thirty-seven thousand feet—means they usually have some time in which to restart the engines or select somewhere they might be able to land in the most unlikely event of the engines failing at cruising height. However, in this case their predicament was particularly serious in that the nearest diversion airport (Jakarta-Halim) had high mountains on its approach, and it would be quite impossible to keep above the 11,500-foot minimum height needed to reach it safely.

Ditching a 747 in the sea in daytime when able to judge the height and direction of the swell would be difficult enough, but it was 10 p.m., and the chances of a successful ditching in the dark would be virtually zero, especially as unknown to them at the time their landing lights had been sandblasted and would have been useless. While any survivors might find the warm waters of the Indian Ocean more hospitable than, say, the cold Atlantic, the likelihood of someone bleeding would mean it would not be long before sharks came for a late supper.

The flight engineer made repeated attempts to restart the engines, much to the consternation of the passengers at the rear of the aircraft, who saw burning fuel mixed with whatever had accumulated in the engines spewing out explosively each time.

With no power for cabin pressurization, the air pressure inside the aircraft was gradually dropping. After about five minutes it fell so much that the crew had to don their oxygen masks, not only because of their physical exertion but because they needed all their mental faculties, and the human brain uses a surprising amount of oxygen. It was only then that the first officer found his oxygen mask had been stowed incorrectly and was unusable, forcing them to drop down from 26,000 feet to 20,000 feet for his benefit.

As they lost further valuable height and sank to 17,500 feet, with the cabin pressure still falling, the passengers' oxygen masks descended.

The situation seemed to be becoming more and more critical. Little did they know their imposed loss of height would be their salvation. Almost twelve minutes after losing all power, they found themselves nosing into normal air below the unnatural zone. Shortly afterward, the first engine they had shut down, and the least subjected to whatever the cause was, sprang into life.

One engine alone would not provide enough power for them to climb but would keep them aloft for much longer and give them some measure of control over their destiny. Subsequently, the other three engines also came back to life in quick succession.

Rather prematurely they reported this reprieve to the tower controller at Jakarta, who told them to climb to fifteen thousand feet, as the mountains between them and the airport were making them invisible to the Jakarta radar. In complying, they found themselves yet again in the abnormal zone, and one of the resurrected engines begun to run so roughly that they had to shut it down. They lost no time in dropping down into the clear air again.

With careful nursing, the three other engines seemed to be sustaining their power. Thinking their tribulations were over, they approached the airport at Jakarta in high spirits, only to find the "sandblasted" windscreen had frosted over so much that everything was a blur.

They even had to request that the tower decrease the intensity of the runway lights because of the glare. Fortunately, each windscreen panel had a narrow strip of clear glass down each outboard side, allowing them to see just enough to land safely. The captain performed well, making a very good landing with help from the first officer reading out the heights.

One passenger was violently sick due to mental stress as they came in, but all things considered it was surprising more passengers were not similarly indisposed.

The 747's landing at Jakarta had in the end been normal, but in view of the doubtful state of the engines, Moody asked that they be towed to the terminal, with the passengers thanking God and above all the crew, personified by the captain, for their deliverance.

The odds against all four engines of an aircraft failing for reasons other than fuel depletion are so great that it is—or was—thought to be virtually impossible. However, the aircrew had trained for such an eventuality and performed very competently, as did the cabin crew. The constantly repeated efforts to restart the engines, probably stopped them clogging up completely.

They later learned that what they had encountered had been ash from a volcano, which explained why being forced to lose height due to the first officer's faulty oxygen mask had been such a blessing. Had they spent longer in the ash-laden environment, it is conceivable that the engines would have silted up too much to be relit. Some would argue the aircrew should have put on their oxygen masks earlier and thereby solved the problem, with the first officer's mask before being obliged to descend. However, not doing so happened to work in their favor.

The authorities closed the skies around the volcano (Mount Galunggung) to aircraft for a week. Just four days after they were reopened, a Singapore Airlines 747 encountered conditions like those that had caught the BA crew by surprise. Forewarned, the SIA crew took evasive action, but even so had to divert to Jakarta-Halim, as three of their engines had overheated. There was an almost identical incident in 1989, when a KLM flight from Amsterdam to Anchorage, Alaska, encountered an ash cloud from Mount Redoubt and all four engines flamed out. As in the BA case, the pilots were able to restart them on exiting the plume of ash.

In a commendation ceremony for both the aircrew and cabin crew, BA was happy to bathe in the reflected glory. Captain Eric Moody, later left the airline to become a TV pundit on aviation matters and a much-in-demand after-dinner speaker.

In a deftly written review of the TV documentary *All Engines Failed*, the UK's *Daily Mail* said Moody "displayed the stiff-upper-lip spirit that built an empire" when he uttered the words that are every air passenger's worst nightmare: "Ladies and gentlemen, this is your captain speaking. We have a small problem. All four engines have stopped. We are doing our damnedest to get it under control. I trust you are not in too much distress."

The crisis resolved itself thanks to difficulties with the first officer's oxygen mask forcing Moody to bring the aircraft down to uncontaminated air where the engines could restart before silting up completely. When all engines flame out (stop), pilots are loath to gratuitously cede height, as that lessens the time available to restart them.

To be fair, the captain's calm demeanor and theatrically delivered words not only reassured the passengers but must have helped the crew perform optimally in the testing circumstances.

The strain Moody was under in the sky above the Indian Ocean and the ramifications career-wise are perhaps evident from the fact that a year later his hair had turned completely white. How pilots cope with the instant notoriety and PTSD arising from such events is often overlooked amidst the adulation and awe. Relations with colleagues can become strained.

Nowadays, more help is at hand. For example, Sullenberger who famously ditched his Airbus A320 in the Hudson River tells how he could not sleep or think normally for quite some time afterward but got advice about what would happen mentally that proved correct. He said talking about the incident was a great help—something that might well apply to Richard de Crespigny who wrote a book about QF32 (see Chapter 10) and gives lectures and talks.

Pilot Struggling with Hijacker Ditches off Beach (Comoros, 1996)
Within Sight of Bathers; Video Seen around the World

> The media often unfavorably compare this ditching with the "Miracle on the Hudson," described later in the chapter. They fail to point out that the Ethiopian pilot was fighting off a hijacker, had insufficient electrical power and thus only limited control of the aircraft, and had to cope with swell of the sea.
>
> *Ethiopian Airlines Flight 961, November 23, 1996*

In 1996, an Ethiopian Airlines Boeing 767 was seized by three hijackers, who thought that the range quoted in the in-flight magazine meant it could fly all the way to Australia, when in fact the pilots had only loaded enough fuel for the leg in question, which was a quarter of the distance.

They would not believe this was not possible and insisted on going on. In an impossible situation, the captain, Leul Abate, tried to keep within sight of the African coast, but the hijackers noticed this. Pretending he was complying and going out toward the open sea, the captain flew toward the Comoros Islands. The fuel was running out when they neared them, but this did not prevent the hijackers remonstrating with the pilots and preventing them from landing at the main airport.

Finally, with no fuel left and no engine power, they had to ditch in the sea just off a beach. The left wing and its engine touched the water first, and as a result the aircraft started to spin leftward and broke up. Of the 175 crew and passengers, only 48 survived. Premature inflation[1] of life jackets, despite a warning not to do so, was the cause of a number of fatalities.

Wingtip touches the water, then the engine. Aircraft spins round and breaks up.

Aircraft should stay afloat for quite a while, provided they do not break up. Even with a fair amount of fuel on board, they will not sink like a stone, since the fuel (kerosene) is less dense than water and provides some buoyancy, and if the fuel has run out, the tanks will be full of air. That said, the high stalling speed of modern jetliners makes a successful ditching very tricky. Having the flaps fully extended in the extreme downward position to give more lift and a lower landing speed as on land may not be desirable, as the resistance generated when they hit the water at speed risks tearing the wings off.

Letting the landing gear down[2] to reduce the speed is not an option either, since the aircraft would topple forward and cartwheel the moment it broke the surface of the water. On the other hand, the engines are designed to break off when striking an obstacle and would normally do so, except that in the Ethiopian ditching the aircraft struck the water while banking, thus causing it to spin around rather than slowing with both engines breaking off.[3]

Though Sully who ditched his A320 in New York's Hudson River faced the risk of coming down on buildings, the actual ditching was much more difficult in Abate's case:

1. He was coming down in the sea with a certain swell;

2. All the while fighting off one of the hijackers;

3. Because there was no fuel left he did not have the auxiliary power unit (APU) to generate electrical power. He only had minimal power for the most basic instruments and controls generated by the ram air turbine (RAT). The aircraft was therefore difficult to control, and an untoward bank caused by interference by the hijackers could not be easily corrected.

However, Abate did succeed in bringing the aircraft down by a beach with tourists, from which rescue attempts could be made.

---

[1] Besides their bulk making it more difficult for people to get out, prematurely inflated life jackets can cause the wearer to float upwards if the cabin fills with water resulting in their being trapped against the ceiling (or the floor if the aircraft is upside-down).

[2] The undercarriage can be allowed to drop down through gravity if there is no hydraulic power to operate it. However, many accounts report that reluctant pilots are to do this if there is any chance of having to ditch as it is impossible to retract it thereafter.

[3] Unlike the flaps, which have to withstand strong forces pushing them backwards, the main purpose of the engine mountings is to withstand the thrust of the engines pushing forwards, and they are designed to break off backwards if they hit a solid object—water being like a brick wall when struck at speed.

A330 Glides Eighty Miles Over Atlantic to an Island (Azores, 2001)
A remarkable feat; though pilot should not have transferred fuel

> Switching to today's ETOPS[1] era, with twin-engine aircraft allowed to fly far from possible landing places, we have an instance where an airliner ran out of fuel over the Atlantic eighty miles from the nearest landing place, a US military airfield on an island in the Azores. Reaching it was one thing; adjusting height and speed so as not to overshoot or undershoot was quite another.
>
> *Air Transat Flight 236, August 24, 2001*

The nearest possible landing place was a US air force base on an island in the Azores eighty miles away.

Carefully nursing his Airbus A330, captain Piché managed to glide there and come in at four hundred miles per hour rather than risk falling out of the sky by attempting an extra circuit to lose more height and speed. With no air brakes or flaps, and with only the landing gear to slow him, he raised the nose to decrease the sink rate and increase forward resistance.

Hitting hard near the threshold, he burst eight tires—because the antilocking system was not functioning and with the wheels locked by the brakes, the tires just skidded over the runway without rotating. The aircraft with its terrified passengers shuddered to a halt almost three-quarters of the way down the runway. Waiting fire service appliances quickly ensured the sparks resulting from friction between exposed metal wheel rims and the runway did not start a fire, which was unlikely in view of the empty, unruptured fuel tanks and their being empty, though there would have been hydraulic fluid.

All on board evacuated in less than ninety seconds thanks to the excellent work of the cabin crew, whom some passengers later accused of panicking, as they were shouting so loudly—the airline retorted that the crew had to shout to ensure everyone heard them. Only twelve passengers sustained injuries in the course of the evacuation, and those were minor. Some were vomiting beside the runway after the tension of the long glide.

On Piché's return to Canada, people showered praise on him, but to one journalist he obliquely replied, "I don't consider myself a hero, sir. I could have done without this." This probably rightly conveyed his sentiment of just having done his job and having been lucky but could also have intimated his fear that details of his prison sentence in the United States for allegedly smuggling marijuana in a light plane in his younger days would leak out.

Before readers jump to conclusions regarding his suitability to be a captain, they should note that in applying for a job with the airline he did not hide that fact. Indeed, in an interview quite some time after his great feat, he claimed the experience of facing death several times as the only French-speaking inmate in an American jail was what had given him the fortitude to make the successful landing after enduring the stress of the long glide.

Investigations subsequently showed the cause of the incident was faulty maintenance by the airline. However, there remained the question as to whether Piché and his first officer aggravated the situation by poor judgment. Were they largely responsible for their own misfortune? Alternatively, had the sophisticated computer systems and displays on the aircraft been at fault?

The Canadian authorities' imposition of a $250,000 fine on Air Transat and the reduction of their ETOPS rating to a maximum diversion time of only ninety minutes attest to the seriousness of the maintenance lapses. They discovered that on finding metal filings in the oil, engineers temporarily installed the only replacement Rolls-Royce engine they had available on site for loan in such situations—one that happened to be a slightly older version, and one not equipped with a fuel pump. With no pump, they decided to use the one fitted to their own newer engine that they were removing, despite the concern expressed by one member of the team.

The airliner had crossed the Atlantic thirteen times since the change to the older model engine without anyone realizing that using parts designed for the newer engine on the older version would allow the fuel line to chafe. On that fourteenth crossing the fuel line ruptured and broke. The fuel kept pouring out of the wing, and even though Captain Piché asked a crew member to check, they were unable to see it in the darkness.

The inquiry's findings were as follows:

1. The main cause was irresponsible maintenance;

2. Piché and his first officer had possibly aggravated the fuel leak situation by making wrong decisions regarding the transfer of fuel between tanks.

Having reviewed the aircraft's computer systems and displays, Airbus took measures to make pilots aware of abnormal fuel consumption even without doing the calculations themselves. They revised the manual to stress the need even further to avoid transferring fuel where any suspicion of a fuel leak existed. Rolls-Royce was the subject of some criticism for putting Air Transat in a difficult situation by not providing the pump together with the engine.

The Pilots' Performance
Being able to glide so far was due not only to good calculation of the optimum glide slope and avoiding overcontrolling, but also to the aerodynamics of the A330 and the great height at which they were initially flying. Perhaps the key to this success was the way Piché eased the aircraft along and risked coming in too

fast rather than losing more height by circling, as it was suggested the first officer wanted to do, and the possibility of stalling away from the runway.

Some commentators have remarked that the aircraft would have toppled over the cliff at the far end of the runway and into the sea had Piché not landed so adroitly. In fact, the "cliff" is an optical illusion when looking along the runway toward the end, as the ground merely slopes downward beyond the runway, giving the impression that it is a sheer drop to the sea. Even had the aircraft been about to go over the so-called cliff, all would not have been lost, as Lajes is a military field, with arresting chains at either end of the runway. Controllers can raise these if an aircraft (with possibly a weapons payload) looks as though it will overrun.

Had there been enough electrical power to keep the flight recorders functioning, investigators would have been able to learn more about the glide and final control inputs. With the batteries drained in the course of the long glide, the ram air turbine[2] (RAT) was providing essential power for instruments and controls. However, electricity for the flight recorders was not considered essential at the time, though the NTSB now says it should be maintained in all circumstances.[3]

Piché, with his charming, unconventional personality, became a hero, especially in Canada, with hundreds of people wanting to interview him. He was asked to open various events, almost like royalty. He had a biography written about him and spent some time on sick leave, apparently with alcohol problems brought about by the unwanted roller-coaster ride of fame and shame. He then gave up drinking entirely and returned to work as a pilot at the airline.

It is a very human story, made more dramatic by the perception that the landing was on a remote island, which consequently must have poor facilities. As mentioned above, this was not the case. Piché was the first to admit that luck played a part. The weather was almost perfect, unlike the day before and the day after. Air traffic control (ATC) had by chance allotted him a route sixty miles farther south than usual, thus putting him just within gliding range of the Azores when the fuel ran out.

Other pilots have successfully made dead-stick landings, and most of them, including Piché, deserve praise. The question is, to what extent was Piché responsible for his own misfortune?

Culture of the Airline

The culture of the airline surely played a part. When the fuel imbalance became serious enough to be worrying, it seems the pilots hesitated to divert immediately, as they were wondering what the airline would say if they landed in the Azores with full tanks. Pilots working for a major carrier might not be under such pressure, as their diversion, if finally proven unnecessary, would only be a mere blip compared with the airline's vast scenario of operations.

For Air Transat, the cost of the diversion and schedule disruption would have been of concern, but the fear of losing its ETOPS classification through too many diversions for mechanical reasons would have worried it even more. As already mentioned, a regulation designed to ensure safety sometimes has the opposite result.

Lawyers

If the pilot actually saves the aircraft and its occupants, some passengers, egged on by lawyers, are prone to exploit the situation. For instance, when a Boeing 777 crash-landed miraculously just inside London Heathrow's perimeter fence, with everyone evacuating virtually unharmed, the passengers were full of praise. However, after a while some began murmuring about compensation, leading the airline to treat them with kid gloves, with the chief executive calling them personally on the phone to show his concern for their well-being.

In this Azores incident, a lawyer immediately contacted the airline (Air Transat) to arrange a deal whereby Air Transat would pay out a once-only sum, rumored to be $10,000 per passenger, in compensation. Apparently, some hundred passengers signed up for that. Another law firm was soon establishing a class action lawsuit.

---

[1] Explanations of technical terms, such as ETOPS, can be found in *PLANE CLEVER*, Christopher Bartlett, OpenHatch Books.

[2] The Ram Air Turbine (RAT) is a "windmill" that drops down to generate emergency power for key instruments.

[3] The NTSB has stressed the need to ensure CVRs and FDRs have the required electrical power in all circumstances—for instance when a Swissair airliner lost electrical power in a fire, the recorders stopped working and investigators had great difficulty in determining subsequent events.

Was There Really a Miracle on the Hudson? (NYC, 2009)

On a talk show, Sully said, "Everyone is looking for a hero."

> In awe of the fairy tale outcome, we almost all saw this simply in terms of
> the captain's amazing achievement: ditching his Airbus A320 in New
> York City's Hudson River with not a single life lost. Looking more closely,
> one can see that it was not that simple.
>
> *US Airways Flight 1549, January 15, 2009*

Cactus is the call sign air traffic controllers use for US Airways. This seems bizarre until one realizes that was the call sign used by American West Airlines (AWE), with which US Airways merged and which indeed flies to many locations where cacti grow. For air traffic control, the merged airline also kept the American West identification code, AWE. But for the public, the name US Airways was retained, allegedly because it sounded more prestigious, and doing so had the advantage of keeping both parties to the merger happy. Also, there was a lot of confusion both with ATC and passengers over differentiating American and American West. For ATC, it was much better and safer to use Cactus. USAir was always an awkward mouthful anyway for controllers.

After lifting off from LaGuardia's Runway 4 at 3:25:33 p.m. on January 15, 2009, and bound for Charlotte, two hours' flying time away, Cactus 1549 (US Airways Flight 1549) was handed over to departure (control) with clearance to climb to five thousand feet. Including the two pilots, Captain Chesley Sullenberger ("Sully") and First Officer Jeffrey Skiles, and three cabin crew, there were a hundred and fifty-five people on board the Airbus 320. It was a beautiful day but rather cold.

Exchanges with Air Traffic Control (ATC)

To simplify the following transcriptions of exchanges between the aircraft and ATC, we refer to US Airways flights as "Cactus" rather than "AWE," and use "Departure" when in fact it is "New York TRACOM[1] LaGuardia Departure."

Departure was the controller handling flights from climb-out (in other words, after they had taken off) to the time they reached approximately eighteen thousand feet and were then handed over to the ATC center (in this case New York Center) covering the upper skies for the whole area.

After two or three routine exchanges between Departure and Cactus 1549, Departure gave instructions to turn left as a first step on the two-hour flight to Charlotte.

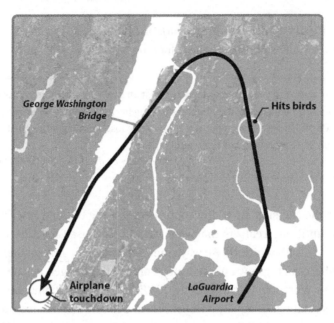

Route followed by the A320 (based on image courtesy of NTSB)

3:27:32 p.m. Departure:

Cactus 1549 turn left heading two seven zero.

There was no confirmation from the aircraft. Then four seconds later, with a slight error in the call sign indicative of the stressful situation:

3:27:36 p.m. Cactus 1549:

Ah, this is Cactus 1539 [sic] . . . Hit birds. We lost thrust in both engines. We're turning back toward LaGuardia.

Departure:

Okay. Yeah. You need to return to LaGuardia. . . Turn left, heading of, uh, two two zero.

Cactus 1549:

Two two zero.

Departure called the LaGuardia tower, telling them to stop all departures, that the aircraft in trouble was in fact Cactus 1549, and that they had lost all engines. The tower (local controller) found this difficult to grasp, perhaps because the simultaneous loss of all engines is almost unheard of.

Departure clarified:

Cactus 1549 has lost thrust in both engines.

Tower:

Got it.

Departure to Cactus 1549:

Cactus 1549, if we can get it to you, do you want to try to land Runway 13?

Runway 13 was perpendicular to the one from which Sully had just taken off and had the advantage that aligning with it would be quicker and involve less turning. In a turn an aircraft loses lift and the sink rate increases, and normally extra power has to be applied to compensate.

Cactus 1549:

We're unable. We may end up in the Hudson.

Departure:

All right, Cactus 1549, it's going to be left traffic to Runway 31.

Cactus 1549:

Unable.

Departure:

Okay, what do you need to land?

Departure:

Cactus 1549, Runway 4 is available if you want to make left traffic to Runway 4.

Cactus 1549:

I am not sure if we can make any runway. Oh, what's over to our right? Anything in New Jersey, maybe Teterboro?

(Departure called Teterboro airport and told them that LaGuardia Departure had an "emergency inbound" and asked if their Runway 1 would be okay. Approval was given.)

Departure to Cactus 1549:

Cactus 1529 turn right two eight zero. You can land Runway 1 at Teterboro.

Cactus 1549:

We can't do it.

Departure:

Okay. Which runway would you like at Teterboro?

3:29:28 p.m. Last transmission from Cactus 1549:

We're going in the Hudson.

Departure:

I'm sorry, say again, Cactus.

Departure quickly dealt with another aircraft before saying:

Cactus, ah, Cactus 1549, radar contact is lost. You also got Newark airport off your two o'clock and about seven miles.

Having overheard the earlier exchange, the pilot of Eagle 4718 called Departure:

Eagle 4718, I don't know, I think he said he was going in the Hudson.

On the outside chance that it might be useful, Departure transmitted information about the runway available at Newark, seven miles away from Cactus 1549's presumed position, which was no longer visible on the radar due to the tall buildings.

Events in the Air

Just before Sullenberger (nickname Sully) reported the bird strike and announced his intention to return to LaGuardia, the Airbus had been climbing through 3,200 feet under takeoff power, with everything quite normal.

The flock of birds appeared from almost nowhere. The first officer, flying the aircraft at the time, saw them first. Sully saw them just as they were about to hit and later said he wanted to duck. They struck the windscreen and other parts of the aircraft with unusually loud thuds, as they were so large. A passenger in first class later said he saw a "gray shape" shoot past his window and go into the engine and that he knew it had done so because of the colossal *bam*. The engines powered down, with another passenger later saying the engines "sounded like a spin dryer with a tennis shoe rolling around in it," according to *The Miracle of the Hudson Plane Crash*, Channel 4 (UK), February 19, 2009.

Sully took over control from his first officer, Jeffrey Skiles, with the words "My aircraft," and he or the A320's computer, or both, pushed the nose down so the aircraft would not stall. Interviewed later on *60 Minutes*, Sully said, "It was the worst, sickening, pit-of-the-stomach, falling-through-the-floor feeling I've ever felt in my life. The physiological reaction I had of this was strong, and I had to force myself to use my training and to force calm on the situation."

As detailed in the transcription above, Sully informed air traffic control (ATC) of the situation and indicated his intention to turn back to LaGuardia, but after dismissing the options of returning to LaGuardia or trying Teterboro, he said he would have to come down in the Hudson. He therefore continued his turn to the left until he was flying along the left bank of the wide river.

He skirted the George Washington Bridge, with its 604-foot (184-meter) twin towers. Manhattan's Central Park was ahead on the left. With the first officer, Skiles, vainly trying to restart an engine using a checklist conceived for dual engine failure at a very much greater height, Sully brought the aircraft over the river itself. He made a single announcement for the benefit of the cabin crew, as well as the passengers:

This is the captain; brace for impact!

The aircraft was at five hundred feet, and the time between the captain's announcement and the actual impact seemed inordinately long, perhaps because the great height of the buildings in Manhattan gave passengers the impression they were lower than they really were. In addition, the A320 was leveling out and losing speed, so it was not traveling so fast on nearing the water.

The flight attendants were following their training, shouting "Brace! . . . Brace! . . . Heads down! . . . Stay down!" in unison at intervals. One flight attendant later mentioned her concern that some passengers were raising their heads to look out of the windows. This was understandable, as some ninety seconds elapsed between the captain saying, "Brace for impact" and the actual impact. Not realizing "impact" meant one with water, many passengers feared they might do a 9/11—that is, fly straight into a building and end up inside a ball of fire.

Lest we forget the short time frame, Sully only had about three minutes from the time the engines failed to the moment they would inevitably come down, either under control, with the possibility of survival if a suitable spot could be found, or out of control following a stall, with very little hope of but a few surviving.

Also, while Sully was concentrating on flying the aircraft, First Officer Jeffrey Skiles wasted valuable time on hopeless attempts to extract some meaningful thrust from the engines, which were actually still running, but at very low power and were useless. The NTSB investigators later recommended engine makers try to develop some system whereby pilots could assess the nature and extent of damage via their instruments.

The Impact

Sully had successfully restarted the auxiliary power unit (APU) to generate electricity; in addition, the ram air turbine (RAT) had apparently deployed automatically. This meant he had more than adequate power to run his cockpit displays and control systems. For the Airbus A320, the manufacturer recommends Flaps 3 and a minimum approach airspeed of 150 knots, with eleven degrees of pitch at touchdown. According to FDR data, the airplane touched down on the Hudson River at a calibrated airspeed of 125 knots with a pitch angle of 9.5 degrees and a right roll angle of 0.4 degrees.

Calculations indicated that the airplane ditched with a descent rate of 12.5 feet per second, a flightpath angle of –3.4 degrees, an angle of attack between 13 and 14 degrees, and a side slip angle of 2.2 degrees. The lack of airspeed meant Sully could not break the too-rapid sink rate with his flare.

What is certain is that the bottom of the fuselage at the rear struck the water first and sufficiently hard to damage some panels. In consequence, the experiences of those on board differed considerably, with the female flight attendant at the rear possibly getting the worst of it, and the pilots at the front getting the best. Apparently, the two pilots were surprised at how smooth their touchdown had been.

On coming to a halt, the aircraft slewed around, no doubt because only one engine had sheared off. (Aircraft engines slung under the wings are attached in such a way that they will break off under an extreme rearward force, such as in the case of hitting an object when crash-landing, or the water when ditching. This prevents the wing(s) being ripped off.)

Events on the Water

Thankful for how well it had appeared—at least at the front—to have gone, the two pilots knew that only half the battle had been won. Everything depended now not only on getting everyone out safely but also on their being picked up before the icy water (36°F, or 2°C) of the Hudson River took its toll. Pictures of the passengers standing on the wings so happy to be still alive give the impression the evacuation was something of a picnic. This is far from the truth, particularly for the passengers and the flight attendant at the very rear of the aircraft.

As mentioned, the impact with the water damaged fuselage panels at the rear. Water seeped mainly in from there and not, as some accounts state, from a rear door that had somehow cracked open. A female flight attendant tried to shut it without success. Fit people at the back clambered over seatbacks to escape the influx of water, not knowing whether the situation might suddenly worsen. As they moved forward toward relative safety, the flight attendant who had been at the rear realized for the first time that she might survive.

The first officer had been using a ditching checklist written on the assumption that the decision to ditch would be made at a great height, with plenty of time to prepare. This meant that he did not get as far as the line at the end where it said, "Press the ditching button" (at two thousand feet), which would have closed various vents and the like. Thus, more water than otherwise would have been the case was coming in. Even so, this did not mean the aircraft would immediately sink completely.

Sully had brought the aircraft down between two ferry terminals, and help was soon on its way. Before leaving the aircraft, he walked up and down the aisle twice to make sure no one remained. He then stepped out and told rescuers to save people on the wings first, as he knew the life rafts could stay afloat almost indefinitely. The first ferry had arrived on scene within about three minutes of the accident, and the other six within ten minutes. A New York Fire Department rescue boat arrived within eight minutes, followed by two coast guard craft nine minutes later. Video footage showed all of the occupants were rescued within about twenty minutes of the ditching. Early on, a helicopter arrived on the scene but pulled back so that its downdraft would not blow people off the wings—consequently, a frogman had to jump bravely from a much greater height than usual to help a woman in difficulty in the water.

Finally, the rescuers saved everyone, with only five people having what are classed as serious injuries: the flight attendant who had been at the rear and who did not notice a gash in her leg extending down to the muscle until rescued; a passenger with a fracture of the small ossified extension to the lower part of the sternum; a passenger with hypothermia, kept in hospital for two days; and two passengers with a fractured shoulder who did not initially go to hospital but who in post-accident interviews had medical records proving it.

The Investigation

The investigators were bound to review Sully's decisions and carry out simulator what-if tests to see whether he could have made it back to LaGuardia. In fact, some pilots reenacting the event on the simulator succeeded in doing so by immediately turning back, but none could when thinking time was added in. The idea of factoring in the thirty-five seconds the pilots needed to assess the situation and work out the best course of action came from the investigators and not from Sully.

Even if it had been just about possible to make it back to the airport, would it have been sensible to take the risk when unsure? The investigators concluded Sully made the right decision.

One of the best books on the whole affair is *Fly by Wire: The Geese, The Glide, the "Miracle" on The Hudson*, by William Langewiesche, who not only writes beautifully but, even more importantly, researches his topics in depth. He describes that public hearing at the NTSB offices at L'Enfant Plaza, in Washington, DC, six months after the event as a missed opportunity, with many questioners either pulling their punches or basking in the glory of the hero, asking leading questions to show themselves or their organizations in the best possible light. Representatives from Airbus were, on the other hand, chomping at the bit, frustrated that no attempt was made to highlight the fact that their sophisticated Airbus flight control system, which did most of the actual flying and did not allow the A320 to stall, deserved some of the praise.

Richard de Crespigny, the captain of a Qantas Airbus 380 that he brought safely back to Singapore after it was crippled by an uncontained engine disintegration, was full of praise for his aircraft, unlike Sully, who said nothing, perhaps because it was America and he had been told by his various handlers to be guarded in his remarks.

The NTSB made many recommendations, most of which have been virtually ignored, perhaps because such events are a rarity, though the NTSB points out that many airports have stretches of water close by.

Sully the Movie and TV Documentaries

The movie *Sully*, with Tom Hanks as a very credible and likeable Sully, proved a big box office hit in the US and abroad, in great part by making the hero look even more heroic, at the expense of the NTSB investigators.

Unlike the gripping movie *Apollo 13*, where the high drama lasted three days as the capsule with the three astronauts circled the moon, this incident lasted a mere three minutes from the moment of the bird strike to coming down safely on the water. Therefore, to hold cinemagoers' interest, the filmmakers introduced scenes showing Sully imagining his aircraft hitting Midtown buildings in 9/11 style, and more disconcertingly created a false scenario in which Sully and Skiles are portrayed as victims of a drawn-out NTSB witch hunt, the dramatic finale being that public hearing at the NTSB, at which Sully pulls a cat out of a bag to dramatically prove for all to see that the callous investigators were totally wrong because they had forgotten to factor in "thinking time." In fact, as said, it was the NTSB's idea to factor it in and say he could not have got back to LaGuardia and had indeed acted correctly.

For balance, one should perhaps see the gripping *Air Crash Investigation* TV series, where the honorable NTSB investigators figure personally. Also, the excellent documentary *Miracle of the Hudson Plane Crash*, broadcast in the UK by Channel 4 in February 2009, has good explanations, actual footage, and interviews with the passengers and rescuers, and shows just how difficult the rescue from the icy water was, since the ferries first on the scene were not designed to pluck people out of water or from life rafts, nor retrieve them from aircraft wings—Sully told them to concentrate on those on the wings, as the life rafts could stay afloat indefinitely. Also, everyone survived despite the floating escape chutes not automatically separating from the fuselage and potentially being dragged down by the sinking A320—one of the crew on a rescue craft provided a knife to cut one free.

An Imaginary Father's Response

Let us suppose the incident had occurred other than in full view in New York City and the public had not yet been told the details. A proud father not wanting to boast of his son's unseen exploit might well have replied to a journalist, "Yes, the boy did well. He not only got them down safely but went through the cabin making quite sure no one remained. I do wonder, though, how many other pilots could have done that."

Patrick Smith, in his bestseller *Pilot Confidential*, says that in many respects landing on the calm Hudson was akin to landing on a twelve-mile-long runway and did not require exceptional piloting abilities.

Great Competency Plus Fortunate Circumstances

Competency includes judgment. Of the utmost importance was Sully's good judgment, firstly, in restarting the auxiliary power unit so the Airbus computer could fly the aircraft optimally, all the while providing full instrument displays, and secondly in deciding not to attempt to reach an airport, which could have meant coming down on buildings, killing all on board together with many on the ground.

Even so, the "miracle" of saving every single soul on board was contingent on there being a long straight stretch of smooth water within gliding range, rescue craft three minutes away from the spot where they would logically touch down, and—what was most unusual for an overland domestic flight—the aircraft having life rafts and life vests.

There has been argument between the two esteemed authors already mentioned in this piece, with William Langewiesche saying the A320's computer was the real hero, and Patrick Smith endorsed by others critical of fly-by-wire systems, such as on the A320, where the pilots tell the computer what they want to do, and the computer decides whether it is possible. The latter say the computer has a mind of its own, and that in this case it prevented the aircraft flaring, which resulted in it hitting the water much too hard. They even cite the A320 that came down on trees at Habsheim Air Show, as described in Chapter 17, allegedly because the computer would not let it climb. They fail to point out that in both cases the computer would not let it climb because the pilots were flying it too slowly to do so, and that a stall could have been worse than hitting too hard.

Sully's actual flying skill—the thing for which understandably the public hold him in such high esteem—was the area where he performed creditably but not exceptionally. He let his airspeed fall too low and thus was unable to make an effective flare to break the excessive sink rate, which ultimately resulted in the panels under the tail section breaking and letting in water and the shock injuring the flight attendant sitting there. The NTSB investigators attributed this failure to maintain airspeed to tunnel vision brought about by lack of time, stress, work overload, the distractions of various alarms, the off-putting sight of skyscrapers alongside, and concentrating on maneuvering the aircraft, and other audible warnings being prioritized over airspeed warnings at that low height.

They pointed out that without thrust from the engines, it would be extremely difficult for any pilot to achieve the recommended ditching speeds, assuming he or she knew them. However, Sully did strike the water with the wings virtually level, and at such an angle the aircraft did not spin, cartwheel, or break up. He did well enough, but other than keeping his nerve while performing that maneuver, he probably did not need all the talents developed from the age of three onwards extolled in his autobiography.

With the rear of the aircraft damaged and icy water coming in, perhaps the real "miracle" was the rescue, with not a single fatality thanks to the diligence of so many—boatmen, helicopter men, not forgetting Sully, who checked so thoroughly that everyone was out and gave advice and assistance to those in difficulty. This was also thanks to the cabin crew, all with twenty if not thirty years' experience, showing there is something to be said for that. Admittedly, an average domestic flight might well have had more passengers with impaired mobility and made their task more difficult.

One of the many lessons noted by the NTSB investigators is a little chilling for those of us confined in seats so close to the one in front that in adopting the brace position we cannot bend fully over and grab our ankles, and instead have to press on the top of the back of the seat in front with our hands or arms. Two female passengers so doing suffered a shoulder fracture.

A man generously took charge of the baby a mother was carrying on her lap, but had the deceleration been far greater he might not have been able to save it as he did. Babies and infants in arms should ideally have their own seats and restraints; however, that might mean people unable to afford the extra seat would travel by road, which is far more dangerous.

One gratifying endnote is that the passengers who risked the lives of their fellow passengers and cabin crew by bringing their luggage with them lost it in the river, while those that did not got it returned.

One cannot take away Sully's achievement. Just imagine the gut-wrenching feeling he had to overcome on finding himself engineless over a packed city with no height to play with. Interestingly, in Richard de Crespigny's book *FLY*, Sully is quoted as saying he would have been better prepared in his younger days flying as a fighter pilot, regularly having to cope with emergencies, rather than as an airline pilot (fortunately) not experiencing anything really challenging or unexpected over many years.

---

[1] Aircraft are handled by the "tower" (local controller (LC)) at takeoff. Once aloft, they are soon handed over to Terminal Radar Approach Control (TRACOM), which controls flights up to a height of 18,000 feet and a range of say 50 miles. Very often TRACOM is called Approach or Departure (control). Here we are talking about "New York TRACOM LaGuardia Departure."

Chapter 2
LOSS OF POWER
OVER LAND

Conscientious Crew Forget How Much Fuel Left (Portland, 1978)
Classic case that heralded CRM (crew resource management)

> This absurd United Airlines DC-8 accident led to the introduction of a
> new way for pilots to work together called CRM.
> *United Airlines Flight 173, December 28, 1978*

On December 28, 1978, a United Airlines McDonnell Douglas DC-8 with the flight number 173 was approaching Portland Airport in Oregon. The 189 people on board included the captain, the first officer, a flight engineer, a deadheading captain two weeks from retirement, and four cabin attendants. There were six infants. The ETA was 17:15, and so far everything was on schedule.

On leaving the gate at the previous stopover at Denver, there had been enough fuel for the two-hour, twenty-six-minute leg to Portland, with an additional forty-five-minute fuel reserve to meet FAA requirements and a further approximately twenty-minute fuel reserve to meet company contingency requirements. Portland Approach Control had given them clearance to make a straight-in landing on Runway 28, so they should have been landing with their fuel reserve of just over an hour intact.

As they descended through six thousand feet with the runway in sight, the first officer, who was the pilot flying (PF), requested fifteen degrees of flap and the lowering of the landing gear. The captain extended the flaps to fifteen degrees and initiated the lowering of the landing gear, which to his surprise dropped down with an unusual thump and more quickly than usual. The captain later said the first officer had noticed a simultaneous yaw to the right.

The jolt was even more noticeable to the flight attendants and passengers farther back in the aircraft. More disconcertingly, one of the landing gear indicator lights failed to show green, presenting the crew with the worrying possibility that the right-hand landing gear assembly was not properly down and locked.

Unaware of any such problem, Portland Approach told them to switch to the Portland Tower frequency for the actual landing. UA173 declined, saying they had a "gear problem" and wanted to stay with approach, maintaining a height of five thousand feet and a speed of 170 knots. According to AVweb.com, one of the approach controllers in question said as recently as 1998/9 that the controller gave the captain the option of holding at six thousand feet over the Laker outer compass locator until he sorted out the problem. From there he could have made a dead-stick landing at any time onto either runway, but instead he opted to orbit twenty miles or so southeast of the airport.

Approach agreed to this request to orbit, telling them to turn left onto a heading of two hundred degrees that would take them away from the airport and out of the path of any other incoming aircraft.

They carried out various checks, even contacting the company's San Francisco Maintenance Control Center, who could not think of anything further.

The captain then briefed the senior flight attendant about getting ready for an emergency landing and possible evacuation but did not give her a deadline or suggest any state of urgency. (He later said he assumed it would take ten or fifteen minutes, and that final preparations could be carried out as they came in to land.)

At that time, all three aircrew were fully aware of the fuel situation, since at 17:46:52 the flight engineer replied to the first officer that they had five thousand pounds of fuel left. Not only did the first officer acknowledge this, he immediately asked the captain,

What's the fuel show now?

The captain replied,

Five.

The first officer duly repeated,

Five.

Further confirmation that there was only about five thousand pounds of fuel left was given by the fact that the inboard fuel pump lamps started to blink, as they are meant to do if the fuel level falls below five thousand pounds. At that point, the aircraft was about thirteen nautical miles south of the airport and heading away from it.

With not much time before the fuel would run out, they yet again discussed the status of the landing gear, before being interrupted by a course change and advisory from approach control.

17:50:20 Captain to flight engineer:

Give us a current card on weight. Figure about another fifteen minutes.

Flight engineer:

Fifteen minutes?

Captain:

Yeah, give us three or four thousand pounds on top of zero fuel weight.

Flight engineer:

> *Not enough. Fifteen minutes is gonna . . . really run us low on fuel here.*

The two pilots then busied themselves with preparations for landing, while the flight engineer replied to questions from the company representative at Portland about the fuel load on landing and the number of passengers. Finally, the flight engineer asked the captain whether he could inform the company representative that they would be landing at "five after." The captain said, "Yeah."

At 17:55:04 the flight engineer reported completion of the approach descent check. When the first officer, at 17:56:53, asked about how much fuel he had, the flight engineer replied that four thousand pounds remained—one thousand pounds in each tank. The captain then sent the flight engineer to the passenger cabin "to kinda see how things are going." In his absence, the captain and first officer discussed the importance of giving the flight attendants time to prepare, what cockpit procedures would be required in the event of an emergency evacuation, and finally whether the brake antiskid devices would be working.

At 18:01:12, Portland Approach put them on another orbit.

> *United 173 heavy turn left, heading one niner five.*

By complying with those instructions to orbit instead of requesting immediate clearance to proceed with the landing in accordance with the initial timetable, the first officer—with the captain's apparent acquiescence—sealed the fate of the aircraft.

At 18:01:34, the flight engineer returned to the flight deck with the news that everything would be ready in another two or three minutes. He and the captain briefly discussed the mood and state of the passengers.

18:02:22 Flight engineer:

> *We got about three on the fuel and that's it.*

Captain:

> *Okay. On touchdown, if the gear folds or something really jumps the track, get those boost pumps off so that . . . you might even get the valves open.*

At 18:02:44, approach control, no doubt surprised that the whole affair was taking so long, asked to be appraised of the situation and was told by the first officer that they still had an indication of a landing gear abnormality, they expected to land in about five minutes, and would need emergency services standing by. At 18:03:14, with the aircraft about eight nautical miles south of the airport and heading away from it, Portland Approach told them to advise when they would like to begin their approach.

The captain replied,

> *They've about finished in the cabin. I'd guess about another three, four, five minutes.*

As is customary for a possible emergency landing, Portland Approach asked for a final confirmation as to the number of souls on board and the amount of fuel remaining.

The flight crew spent the next two and half minutes discussing technical matters. They wondered whether the horn that warns pilots if they are landing without the landing gear locked down might not be a helpful indication of its true status. They also wondered whether the tripping of the landing gear circuit breakers might mean the spoilers and antiskid brakes would not deploy automatically.

The senior flight attendant came back to the flight deck to report that she thought they were ready, with the aircraft by then about seventeen nautical miles south of the airport and still heading away from it. After talking to her for about a half minute and going into detail about whether able-bodied people were by the exits, and noting the fact that the off-duty captain would be in the first row of coach behind the galley, the captain said to her:

Okay. We're going to go in now. We should be landing in about five minutes.

It was 18:06:40, and they were nineteen nautical miles south-southwest of the airport and still heading away from it.

Before the flight attendant could answer, the flight engineer or the first officer interjected:

I think you lost number four, buddy, you . . .

Flight attendant:

Okay, I'll make the five-minute announcement. I'll go. I'm sitting down now.

First officer to flight engineer:

Better get some cross feeds open there or something!

Flight engineer:

Okay!

"All righty," said the flight attendant on starting to make her way back to the passenger cabins.

At 18:06:46, the first officer twice told the captain they were going to lose an engine. The captain twice replied, "Why?" to which the first officer each time replied, "Fuel." They both told the flight engineer to open the cross feeds, and there seemed to be some confusion as to how much fuel they had.

At 18:07:06, the first officer declared the engine had flamed out. At 18:07:12, almost an hour after declining their initial clearance to make a landing approach, the captain finally asked for it again. The controller accordingly gave him an initial heading vector of 010 degrees, to which they turned, and they were again nineteen nautical miles south-southwest of the airport but at least heading for it.

18:07:27 Flight engineer

We're going to lose number three in a minute too.

The crew tried to see if they could keep the engines going by opening cross feeds.

18:10:47 Approach:

Eighteen flying miles from the field.

18:12:43 Approach:

Twelve flying miles from the field.

The flight engineer declared they had just lost two engines, "One and two."

18:13:29 Approach:

Eight or nine flying miles from the field.

18:13:38 Captain:

They're all going. We can't make Troutdale [a nearer general aviation airport].

18:13:43 First officer:

We can't make anything!

The captain then told him to declare a Mayday. The first officer did so, ending his transmission to the tower with the words:

We're going down. We're not going to be able to reach the airport.

The captain later said that in the minute or so he could remain in the air there were few options, and no good ones. Congestion due to rush-hour traffic would make landing on the highway along the bank of the Columbia River a disastrous proposition, and icy water in the fast-moving river itself would certainly result in many fatalities, even should he be successful in ditching there in the darkness. He therefore opted to bring the aircraft down in a wooded area with no visible lights about six nautical miles east-southeast of the airport.

His skill no doubt helped keep casualties to a minimum. Initial impacts were with trees, before the aircraft destroyed two fortunately unoccupied homes. Just before the aircraft came to a halt almost five hundred meters from the point of initial impact with the first tree, the vertical stabilizer on the tail snagged some high-tension electric transmission lines, and this encounter no doubt helped bring the aircraft to a final halt. An outbreak of fire, earlier so much feared by the crew, did not occur, since there was no fuel left. A tree penetrating the fuselage near the flight engineer's position caused some fatalities back from there.

Remarkably, "only" 8 passengers and 2 crew (the flight engineer and the senior flight attendant, who was said to have saved lives by preparing so well for the possible emergency evacuation) were killed. A further 21 passengers and 2 crew members incurred serious injuries. Out of the 189 persons on board, 158 escaped relatively unscathed.

The passengers did not receive any warning from the aircrew that the aircraft was coming down and initially thought they were landing at the airport. After they started striking the trees, a flight attendant shouted out, "Hold your ankles!"

The Inquiry

The inquiry revealed that a fault due to corrosion in part of the landing gear mechanism caused one set of wheels to free-fall down into place rather than sink down gently. The temporary imbalance in drag (one set of wheels came down earlier than the other) produced the yaw noticed by the first officer. Furthermore, in the process of dropping down into place, the gear damaged the micro-switches used to show whether the gear was fully down and locked. In fact, the aircraft could have landed safely at any time.

The inquiry concluded that the crew, having become so absorbed in trying to solve the landing gear problem and thinking about the possible consequences, failed properly to consider the fuel consumption rate. However, they did note that when asked about the fuel they would have if they landed as planned (at five past the hour), the flight engineer had said, "Fifteen minutes is gonna . . . really run us low on fuel here!"

For the hour they were orbiting, the captain was considering every eventuality, including the mental state of the passengers, in detail. He even sought to arrange for the San Francisco Maintenance Office to handle certain procedures to avoid the risk of bad local publicity. At one point the off-duty, about-to-retire captain, who later returned to the passenger cabin to better help with a possible evacuation, suggested the flight crew put on their jackets so as to be more recognizable and to expose less bare flesh in the event of fire.

Much of what the captain did and thought about was very commendable, but one must wonder whether the drip-drip effect of spending so much time thinking about what might happen in the unlikely worst possible scenario did not affect the crew members' grasp of the overall situation.

Cockpit Resource Management (CRM)

The incident revealed the need for formal policies and programs to ensure aircrew function well as a team, with each having defined complementary duties rather than everyone focusing on a single matter. United Airlines took the experience to heart and became the first to initiate such a program, using some of the techniques already used by business and management consultants.

As mentioned at the beginning of this chapter, they called this cockpit resource management (CRM), now changed to crew resource management, to reflect the vital role of others, such as cabin crew, ground staff, and maintenance staff. This has prevented numerous accidents.

Miscalculation Meant Only Half the Fuel Loaded (Gimli, 1983)
Superlative piloting technique saved the day

> Considering there are relatively few instances of aircraft, such as Piché's, gliding long distances to a safe landing, it is surprising that another Canada-registered aircraft had eighteen years before featured in a similar feat, but over land.
>
> *Air Canada Flight 143, July 23, 1983*

Captain Bob Pearson and First Officer Quintal were flying their sophisticated Air Canada Boeing 767 between Ottawa and Edmonton when it ran out of fuel out of gliding range to an airport.

They had set off with their fuel gauges not working, and the flight management computer (FMC) was calculating the amount of fuel in the tanks by deducting the amount consumed by the engines from the amount measured by dipsticks on departure. Thus, the amount shown depended on correctly knowing the amount of fuel initially in the tanks.

Regardless of whatever the FMC said, they should have had no need to worry, as the dipstick checks at Montreal and then Ottawa had shown they had ample fuel for the 3,574-kilometer leg to Edmonton and would still have the required reserves.

Just over halfway, with the readout from the FMC indicating several tons of fuel left, a beeping sound drew the relaxing pilots' attention to low pressure in one of the two fuel pumps feeding the left engine. Soon afterward the instruments indicated low pressure in the other pump. Was it a computer glitch or faulty sensors? However, the aircraft was virtually new, and to have two pumps, or their pressure sensors, fail was troubling enough for Pearson to decide to divert to the nearest appropriate airport, which was Winnipeg. Still at forty-one thousand feet, they throttled back the engines to begin descent into Winnipeg, some 120 nautical miles away.

Shortly afterward the displays warned of a similar low fuel-pump pressure situation for both pumps feeding the engine on the opposite side. For so many pumps to be affected simultaneously was a sure sign of a fuel problem, and their fears were confirmed when a few minutes later the number one engine flamed out, followed three minutes later by the number two engine. Normally, the fuel measurement system would have given a warning once the fuel level fell to two tons, giving them more than adequate time (over land) to make an emergency landing under power. But of course, that was not working.

They were by then down to twenty-five thousand feet, with still sixty-five nautical miles to go to Winnipeg. As in the case of Piché's aircraft, a small

RAT (ram air turbine) had dropped down from the underside of the aircraft to provide just enough power for essential flying controls. The once-sophisticated screens were blank, and all they had was the artificial horizon, airspeed indicator, altimeter, and a magnetic compass, which was difficult to read because, unlike the usual gyrocompass, it was not stable.

With no vertical speed indicator, judging the optimum glide path was exceedingly difficult, and amid his other tasks, the first officer had to get the Winnipeg controller to constantly give the distance remaining to Winnipeg and try to work it out from tables.

With them down to 10,000 feet and descending to 9,500 feet, the distance remaining was forty-five nautical miles. They were losing height faster than expected, and at that rate all hope of making Winnipeg had gone. The situation was desperate.

Just as the captain was about to ask Winnipeg for anything nearer, the first officer suggested they try Gimli Air Force Base, where he had been temporarily stationed during his military service. He knew it had two runways of sufficient length. Reassuringly, air traffic control informed them Gimli was only twelve nautical miles from their location.

The air traffic controller guided them toward Gimli but with some difficulty, as not being operational, the field itself was not shown on the charts. The captain was informed of this but told light aircraft were using the right-hand runway. The controller could not guarantee the runway would be clear.

They did not want to lose too much height before being sure they could identify the Gimli base, and the trouble was that although the first officer knew the general area he could not be sure exactly where it would be. When he finally did sight it, they were much too high, posing the problem of how to lose both height and speed in the distance that remained without the help of flaps or air brakes.

By a lucky coincidence, Captain Pearson happened to be an experienced glider pilot. With no air brakes, he had to resort to a series of difficult sideslips and yaws, with the ailerons sending the aircraft one way and the rudder the opposite way to provide a braking effect that only a glider pilot could execute, he brought the speed and height down to a reasonable degree; and with the runway ahead, told the first officer to lower the landing gear.

In the dusk, they could only see one clearly defined, whitish runway ahead and assumed incorrectly it must be the (actually darker) right-hand one on which they planned to land. Thus, they were unknowingly lining up with the disused left-hand runway, which that weekend was being used as a racing car circuit, with a strip down the middle and the cars going down one side and up the other.

The first officer duly selected "Landing Gear Down," but nothing happened. Frantically, he searched in the manual to see how to let the wheels fall just by gravity but could not find the explanation—actually, and logically, in the hydraulics section—because the index failed to mention it. He then tried the alternative gear extension switch and, with a sigh of relief, heard the wheels drop down. To his consternation there was no green light to confirm the nose wheel had locked.

To slow the aircraft and lose more height the captain had to bank the aircraft steeply and apply rudder to make it fly crabwise, frightening the passengers looking straight down at astonished people on a golf course. Even so, coming in fifty knots faster than normal, he made a perfect landing on the wrong runway, only to find there were people, including children, on it.

Luckily, the last race of the day had just finished, and all were congregated at the far end. Also, having caught sight of the silent aircraft coming in, drivers rushed to move their cars, while others screamed at a boy riding up and down the runway, quite oblivious to what was happening. Luckily, the aircraft missed him. The captain applied the brakes as hard as possible, and as the aircraft slowed, the unlocked nose-wheel assembly collapsed to produce a shower of sparks. The failure of the front wheel was a blessing in disguise, as it helped stop the aircraft just short of a group of very surprised adults and children.

With the collapse of the nose-wheel assembly, the aircraft had tilted forward with the tail high up in the air, making the chutes at the rear too steep to use safely. However, with relatively few passengers on board, all were able to exit easily and safely from the front. Again, with no fuel there was little likelihood of a major fire, and fire extinguishers from the race organizers ensured smoke coming from the front did not develop into a serious fire.

As usual a series of failures and misunderstandings were responsible for this ridiculous near disaster.

Multiple Causes

There are so many little factors that not every reader will want to read this next part in full:

1. Noncompliance with minimum equipment list (MEL)
   For safety reasons, essential systems in aircraft are in duplicate, triplicate, and sometimes quadruplicate. In addition, some items are not safety critical, in the sense that they do not significantly affect safety. Therefore, for every aircraft there is a MEL showing what systems or items must be working. The pilots run through this list before every flight and check their aircraft meets minimum requirements before setting out, even though the maintenance personnel will have verified this before handing over.

Sometimes there may be a special-case MEL with conditions attached to it. For instance, when maintenance staff in Tokyo found a faulty engine on a 747 that could not be easily or cheaply repaired there, the airline, British Airways, decided to have it repaired in London. It was able to fly the aircraft back to London on three engines on the condition that there were no passengers and it had an extra pilot, who would have to come from London to assist. A three-engine ferry flight must have a specially trained captain, as improper application of asymmetric power can be disastrous.

Many of the provisions in the MEL are obvious and apply to similar aircraft, but others are specific to a particular aircraft type. When a new aircraft is developed, the aircraft manufacturer issues a MEL, which the airline and manufacturer update. The newly developed and sophisticated Boeing 767 (with screens replacing many traditional dials) was a completely new aircraft for Air Canada, so the MEL was not definitive. With so many revisions, pilots often had to check with maintenance management to be sure of what was permissible.

When Pearson arrived at Montreal Airport with First Officer Quintal, they met the incoming captain, Captain Weir, and naturally discussed the condition of the aircraft. Weir said there was a problem with the fuel gauges and went on to mention the fact that they had once blanked out. Weir himself had misunderstood the technician in Edmonton, who had said they had blanked out coming in from Toronto several weeks earlier; Weir thought he meant that they had blanked on the flight that had just come in from Toronto. Anyway, Captain Pearson believed the aircraft he was to take over had come in with all the fuel gauges blanked out and that it would be reasonable to carry on likewise, especially as the new part was waiting in Edmonton, their destination.

It would have been difficult for Pearson to refuse to fly an aircraft that he thought other captains had deemed acceptable, and especially so when maintenance had confirmed that complying with the MEL did not pose a problem. In fact, the other captains had flown the aircraft with the fuel gauges working with input from just one processor channel, which satisfied the MEL, provided ground staff carried out a dipstick check as well. Maintenance was wrong to say departing with none of the fuel gauges working complied with the MEL.

2. The Technical Fault

Much earlier the technician in Edmonton had found that the entire fuel gauge system blanked out when one of the digital processor channels failed. Failure of one element in a two-channel fuel indication system should not have crippled the entire system. However, poor design (not by Boeing) meant there was not enough power when one side failed, and this accounted for the whole system going blank from time to time, because of a fickle connection due to cold soldering in the right-hand channel.

Any car owner knows how difficult it is to find an intermittent fault, as it is never there when you are looking for it. It can also lure the user into a sense of false security, as things may revert to normal for a long time, only for a failure to occur when there is a temperature or pressure change or when there is turbulence. This is why the faulty part in Pearson's aircraft was not repaired earlier.

3. Misunderstandings: Language

The qualified technician in Edmonton found he could get the fuel gauges to work by isolating the defective processor channel. He therefore switched off that channel and agreed with maintenance that the aircraft could fly like that while waiting for the spare part, provided a dipstick check was carried out. He left a note on the tripped circuit breaker saying the channel was inoperative. In the maintenance log, he wrote:

SERVICE CHK—FOUND FUEL QTY IND BLANK—FUEL QTY #2 C/B PULLED & TAGGED—FUEL DIP REQD PRIOR TO DEP. SEE MEL.

4. Too Many Cooks

Had no one else stuck their oar in, all would have been well, at least as regards keeping the fuel gauges functioning. Unfortunately, while waiting to dipstick the tanks after refueling in Montreal, a less-qualified technician noticed the pulled circuit breaker with the tag saying, "Not Operational." Though not properly qualified to test the system, he tried resetting the breaker, thus making the fuel gauges go blank. Before he had time to re-pull the breaker and see if that made the gauges work, he was called away to check the tanks using a dipstick. He forgot about it but did remember to put a note in the log:

FUEL QTY IND U/S. SUSPECT PROCESSOR UNIT AT FAULT. NIL STOCK.

He then signed off the maintenance log as satisfactory.

5. Metric Conversion Error

When dealing with fuel volumes, such as when refueling or using a dipstick to check the amount of fuel in a tank, the unit used is the liter. However, weight is what the pilots want to know, as it is so important when flying. For instance, the V takeoff speeds vary according to the gross weight of the aircraft.

As Stanley Stewart says in his excellent account in *Emergency: Crisis on The Flightdeck*, the technician was referring to the fuel indicators going blank during the check, but that would not be clear from the text. Of course, the technician's "SEE MEL" supposedly, but not obviously, restated the need to have a dipstick check, as well as one processor channel working.

Up to that time the unit used to express weight at the airline had been pounds. Like many countries other than the United States, Canada was changing over to the metric system, and the 767 was the first aircraft type in the Air Canada fleet to have the fuel weight expressed in kilograms rather than pounds.

Everyone dealing with that 767 knew the weight of the fuel had to be expressed in kilograms, but in converting liters to kilograms they all used or copied the wrong conversion factor. Any kid educated at school in the metric system would have known that one liter of water weighs one kilogram and would realize that with kerosene being slightly lighter than water, one would have to use a conversion factor of slightly less than one (actually 0.8). Instead, they used the conversion factor for converting liters to pounds, which is 1.77. As a result, they only had some ten tonnes (metric tons) of fuel in the tanks and imagined they had twenty-two, and that was the figure programmed into the FMC which was calculating the amount of fuel remaining by deducting the amount burned from that.

On checking the calculations, Captain Pearson, not expecting an error of that nature, failed to notice the use of the wrong conversion factor.

Conclusion

This incident was the result of many errors and failures. As renowned safety academic, Professor Reason, says, with the best will in the world it is impossible to prevent all errors, and the important thing is finding ways to cope with them.

Failure to follow the MEL and the bad design of the overall fuel processor system, which allowed failure of one part to cripple the entire system, would seem to be key errors. Badly relayed information and cryptic technical logs also played a significant part. Unfamiliarity with weight and volume conversions was also a factor, meaning the obvious mistake was not realized.

A Piloting Feat

Captain Pearson's landing on that short landing strip with the need to use his glider-flying skills to lose so much height and attenuate airspeed represented a remarkable feat.

TACA Deadstick Landing on Grass Levee (New Orleans, 1988)

A great feat of airmanship by a captain with one eye

> The captain was going to ditch in the water but at the last moment saw a relatively flat grass levee on the bank. He landed there with no loss of life despite having only one eye.
>
> *TACA Flight 110, May 24, 1988*

TACA, a small airline based in El Salvador and founded in 1931, was so proud of its brand-new Boeing 737-3 that when Flight 110 encountered severe hail, the pilots' first concern was that it might damage the paintwork.

The flight had taken off in fine weather from Belize City with thirty-eight passengers and seven crew, but on crossing over the Gulf of Mexico, known for its fickle weather, the weather radar showed storms ahead. The pilots tried to skirt the parts shown in red, with the heaviest rain, and keep to the areas shown in green and yellow, where the rain would be less intense.

Though the captain, Carlos Dardano, and second officer, Dionisio Lopez, were relatively young, they were very experienced. Sitting in the jump seat was a senior training pilot, who had come along get some experience of being on a new 737. As they approached their destination, New Orleans, they commenced their descent from thirty-five thousand feet. They saw a gap, shown in green and yellow areas, with red ones on either side ahead, and entered the cloud at thirty thousand feet, having turned on the engine anti-icing and engine auto-ignition so that an engine would immediately restart should it flame out.

As they descended lower and lower, they were surprised at the amount of rain, turbulence, and especially hail, despite it not having been displayed as red on their radar. That was because since hail is dry, it does not show up properly on radar. At sixteen thousand feet, both engines flamed out almost simultaneously, leaving them gliding without even electrical power for their instruments, other than a few key ones with battery backup. Their radio was out, so they could not communicate with air traffic control (ATC).

They started the auxiliary power unit (APU) in the tail, but that took time to power up, at which point, with electricity for their communications, they could discuss possible landing places with ATC. They were low in dense air and had the engines windmilling at the right speed for restarting but were unsuccessful Getting desperate, they finally managed to restart them using the engine starters with electric power supplied by the APU. Sighing with relief, they prematurely informed ATC they no longer needed to come down immediately and would continue to New Orleans. ATC gave them the course for that.

However, the engines were not producing any meaningful thrust. Worse, when the pilots opened the throttles to push more fuel in, the engines overheated to the point where it was obvious they would catch fire or explode, which meant they had to be shut down immediately and definitively. (The overheating was because it was a hot start—that is, excess fuel in the engine was catching fire. The pilots cannot be blamed, because at such a low height they could not take their time and allow it to bleed away.)

They were at three thousand feet with just three minutes left. They had dismissed the suggestion by ATC of coming down on a highway, for, with vehicles on it, the maneuver would be fraught with danger, both for those in the aircraft and on the ground. The only serious option was to ditch in one of the wide waterways. One lay straight ahead, and the captain aimed to ditch near the bank. Then, as they came down below two thousand feet, copilot Lopez pointed out the grass-covered levee to the side on the right. Perhaps to get a clearer view and make sure the ground there was as flat and obstacle-free as it appeared, Captain Dardano may have continued to aim for the water before doing a deft sideslip to line up with the levee.

Passing over the wall at the near end, he touched down perfectly and with no reverse thrust to help bring the thirty-seven-ton airliner to a halt by delicate application of the brakes to prevent it skidding out of control. There was no fire and all on board evacuated safely via the slides, with only one very minor injury. Truly remarkable. With one engine completely replaced and the other repaired, two test pilots flew the aircraft, with just enough fuel to reach the airport, off the levee.

CFM56 Engine Used Worldwide

There was great concern, for the CFM56 engine was used in many aircraft worldwide in addition to the ubiquitous 737. The fact that such a reliable engine could in certain circumstances flame out in a storm, with subsequent problems when restarted, was very troubling.

The investigators had the engine maker redo the acceptance tests, but no matter how much water they injected, they could not get the engines to stop. They then studied the readings on the TACA cockpit data recorder and noticed that when the engines flamed out the aircraft was beginning its descent into New Orleans under low power. The test was therefore repeated at low power settings and showed that a large amount of water in the core—simulating the presence of hail—did make it flame out.

Minor modifications were made to CFM56 engines, including better pathways to bleed away water, reshaping fan blades to help divert small hail particles from the core, and a sensor to make the ignitors fire automatically in heavy rain or hail.

Avianca 52 Copilot Failed to Say "Emergency" (New York, 1990)
Also, did not inform the controllers it was their last chance to land

> The survivors and relatives of those who died when Avianca Flight 52
> ran out of fuel while attempting to land at New York's JFK airport were
> incensed when reminded the official inquiry attributed the accident
> almost entirely to the first officer's failure to use the term "emergency" in
> his radio transmissions to air traffic control.
>
> *Avianca Flight 52, January 25, 1990*

The lights in the passenger cabin of the Colombian Avianca Boeing 707 flickered as the fuel supply to the engines became erratic. With so little fuel left, no measure could save them other than coming down on a runway or flat, open space. However, JFK airport was fifteen miles away, and the hilly ground of the affluent residential district of Cove Neck, on Long Island, lay ahead.

A few seconds later the engines fell silent, leaving only the rustle of the wind against the fuselage, soon to be drowned out by the screams and exclamations of the passengers realizing they might be facing their maker.

How, in what one would imagine to be one of the most sophisticated air traffic control (ATC) zones in the world, could the pilots and passengers of Avianca Flight 52 find themselves in such a predicament? It was due to what, with hindsight, was a whole series of missed opportunities to avoid disaster.

The first of these was not diverting to their alternate, Boston, when, on approaching the New York control zone an hour and a half earlier, controllers informed them their wait in the holding pattern would be at least forty-five minutes. The pilots possibly thought the controller was being careful and that the wait would not be very much longer. In fact, they had to hold for seventy-seven minutes.

Then, as the aircraft was subsequently handed over from one controller to another, the first officer, who was handling radio communications, used phrases such as "We're running out of fuel." He evidently thought this clearly indicated their fuel predicament, but he failed to convey the true situation to the controllers, who had perhaps fifty aircraft in the sky, all in a sense running out of fuel and all wanting priority. If they started to let aircraft that had not declared an emergency jump the queue, a traffic jam would develop over the airport, perhaps compromising the safety of other aircraft also low on fuel.

Another factor explaining the controllers' apparent lack of probing into Avianca 52's status was that, with the aircraft being handed over successively from controller to controller, none had the time to build up a detailed picture.

Aircraft have to be pigeonholed in the controller's mind, and this is particularly so at busy times; for them it is either a normal flight or declared emergency.

When after seventy-seven minutes Flight 52 was allowed to exit the holding pattern (after the crew were asked how much longer they could hold), it was passed on to the approach controller, who, unaware of their predicament, greeted them as follows:

21:03:11 Approach:

> Avianca zero five two heavy, New York Approach, good evening. Fly heading zero six zero.

After acknowledging this, the Avianca flight crew, consisting of the captain, first officer, and flight engineer, agreed on the need, when less than a thousand pounds of fuel remains in any tank, to avoid doing anything, such as raise the nose too much or accelerate violently, that might cause it to slosh to one side, leaving the outlet uncovered.

As the controllers brought them in and gave them course changes, the first officer and flight engineer surmised they were being accommodated and that the controllers were aware of their situation. At no point did they tell the approach controller they were low on fuel, no doubt assuming that the previous controller told him. Apart from the controller telling them to make their speed 160 knots if practical, there is nothing of note from the controller before he hands them over to the tower controller, who greets them:

21:15:23 Tower:

> Avianca zero five two heavy, Kennedy Tower, runway two two left. You're number three following seven two seven traffic on a, ah, niner mile final.

The tower, finding the more modern aircraft following behind was in danger of catching up with the old Boeing 707, asked Avianca 52 for their airspeed (140 knots) and asked them to increase it by 10 knots, impatiently telling them "Increase! Increase!" Avianca 52's captain, who was flying the aircraft, seemed to be having some difficulty hearing these exchanges and what the first officer and flight engineer were saying.

They proceeded with the standard prelanding checks and the lowering of the landing gear. Duly cleared to land, they asked for a wind check and were told it was 190 degrees at 20 knots. (The wind speed at their location was apparently of the order of 60 knots, with the difference between that and the 20 knots given to them for the airport representing considerable wind shear.)

The tower, still concerned about the separation from the TWA aircraft behind them, asked for their airspeed again, and on being told it was one four five, asked the TWA aircraft behind if they could match it.

The TWA pilot said:

> Okay, we'll do our best.

The Avianca flight was all set for landing but sank a little below the glide slope. The tower, increasingly concerned about the separation, asked the TWA craft to reduce its final airspeed, if feasible. With the TWA crew saying they could not go slower, the tower asked Avianca 52 to increase theirs by ten knots, but finding they were getting too close, ordered the TWA heavy to turn off left and maintain two thousand feet. The tower then informed American Airlines Flight 40 they had become number two in the landing sequence, behind a 707 (Avianca 52).

It was then, with everything seemingly fine for the landing, that Avianca 52 encountered wind shear two and half nautical miles from the runway. The aircraft sank, with the "Whoop! Whoop! Pull up!" from the ground proximity warning system (GPWS) telling the crew they were in danger of hitting the ground. To recover, the captain pushed the throttles forward, thus using up much of the remaining fuel. After sinking to the dangerously low height of two hundred feet two miles from the runway, the aircraft finally pulled out of its descent.

Captain:

Where is the runway?

The GPWS repeated "Whoop! Whoop! Pull up!" three more times.

Captain:

The runway! Where is it?

The automatic "Glide slope!" warning sounded twice.

First officer:

I don't see it! I don't see it!

The captain ordered the raising of the landing gear as they aborted the landing.

The glide slope warning sounded twice again, presumably because they were by then above it. The first officer then informed the tower they were executing a missed approach. It is very likely that the pilots failed to see the runway in the poor visual conditions due to the nose-up attitude of the aircraft at the critical moment as they recovered from the perilous sink rate brought about by the wind shear.

The tower told them to climb and maintain two thousand feet and subsequently asked them to confirm they were making a left turn, according to the standard missed landing procedure, exactly as the TWA craft had done just before. The captain then specifically told the first officer to tell the controllers it was an emergency. Instead, the first officer simply confirmed to the controller they were executing the left turn as instructed, adding that they were running out of fuel.

First officer to ATC:

That's right, to one eight zero on the heading—and, ah, we'll try once again. We're running out of fuel.

The tower simply said "Okay" and gave the next aircraft, American Airlines Flight 40, clearance to land, adding that a DC-9 had reported wind shear, with a gain and loss of ten knots, from seven hundred feet down to the surface. The Avianca captain once again told his first officer to tell the tower it was an emergency, adding, "Did you tell him?"

First officer replied:

Yes, sir. I already advised him.

This was not strictly true, as the first officer had not used the term "emergency." In addition, as pointed out by the NTSB investigators, the flight engineer had failed to remind the pilots that to all intents and purposes that would represent their one and only chance to land. This fact should also have been made clear to approach control and the tower lest they order a go-around, such as the one they ordered the TWA aircraft to execute for lack of separation.

Also, even without using the term *"emergency,"* it is difficult to understand why, in the even more desperate situation following the missed approach, the first officer failed to inform the tower they had under ten minutes of fuel left. Some commentators have suggested it was because they were unable to work out a precise figure!

Whether it would have been possible to free up either of JFK's very long 31L or 31R runways and get the Avianca flight far out enough to line up and come in with sufficient fuel remaining is open to question. Performing flying club antics with an airliner would have been difficult enough even in good visibility.

Thus, not realizing the severity of the situation, the tower controller, who was about to hand over to a colleague at the end of his shift, simply handed them over to the approach controller.

The captain told the first officer to tell approach they didn't have fuel, but the first officer, after automatically acknowledging the order to climb and maintain three thousand feet, reverted to saying, "We are running out of fuel, sir." The controller replied "Okay" and gave them a new heading.

Again, the captain asked the first officer if he had advised ATC they didn't have fuel. He confirmed that he had, adding optimistically, "And he's going to get us back."

The approach controller then gave instructions to two other aircraft. After giving Avianca 52 a new heading, he showed his concern as one can see from the following exchange.

21:26:35 Approach control:

And Avianca zero five two heavy, ah, I'm going to bring you fifteen miles northeast and then bring you back onto the approach. Is that fine with you and your fuel?

21:26:43 First officer:

I guess so. Tha [sic] you very much.

The captain asked what the controller said, but before the first officer could tell him, the flight engineer bizarrely said, "The guy is angry."

Approach control continued to give instructions both to them and to other aircraft as if things were normal. At one point, on being asked to climb, they replied in the negative, saying they were running out of fuel. Approach replied, "Okay" and gave a slightly different heading. The controller, knowing they did not have much fuel, evidently wanted to avoid them creeping up on the aircraft in front and having to go around again.

21:31:01 Approach control:

Okay, and you're number two for the approach. I just have to give you enough room so you can make it without, ah, having to come out again.

The first officer acknowledged, and the controller replied, "Thank you, sir." This was hardly the sign of an angry controller, unless said sarcastically, and presumably the controller would be far too busy for such niceties. The controller then dealt with a couple of other aircraft before giving Avianca 52 a thirty-degree change of heading to the left to bring it nearer the heading for the outer marker.

21:32:38

[CVR anomaly—hiccup due to fluctuating power supply, no doubt corresponding to flickering of cabin lights.]

21:32:39 Flight engineer:

Flame out! Flame out on engine number four.

21:32:49 Captain:

Show me the runway.

21:32:49 First officer to controller:

Avianca zero five two. We just, ah, lost two engines and, ah, we need priority, please.

The controller then gave them a new heading to intercept the localizer more quickly.

21:32:56:

[Sound of engine(s?) spooling down]

The captain and first officer then talked about setting the ILS (instrument landing system). The captain said, "Set the ILS. Let's see."

21:33:04 Approach control:

> *Avianca zero five two heavy, you're one five miles from the outer marker; maintain two thousand until established on the localizer. Cleared for ILS two two left.*

Avianca 52 acknowledged this.

21:33:22 Captain:

> *Did you select the ILS?*

21:33:22 First officer:

> *It is ready on two.*

21:33.24 [End of CVR recording.]

In view of the CVR hiccups when the power supply fluctuated, it is evident that there was no battery backup, and the end of the recording did not correspond with the impact with the ground.

The flight data recorder provided no evidence, because someone had rendered it inoperable by taping back the foil inside. However, examination of the engines at the crash site immediately revealed they had not been rotating under power when the aircraft struck the ground.

Surviving passengers and the only surviving crew member, the leading flight attendant, were able to describe the last moments. Radar records gave useful information about heights and tracks.

According to a witness on the ground, the aircraft dropped silently out of the sky. Without evidence from the flight data recorder, ascertaining the precise sink rate and forward speed prior to impact was impossible.

From the distribution of the debris and the injuries to passengers, it was possible to deduce that the forward speed on impact had not been so great. Although shattered, most parts, relatively speaking, were in the right places, with the wings sticking out from the fuselage in the normal place and not lying somewhere else.

The aircraft apparently belly flopped into a gully, hitting the odd tree and slithering up the higher far side. The fuselage snapped in at least two places, with one break right behind the flight deck, so that the nose, with the flight crew inside, flipped over the brow to land near a house.

Lack of fuel meant there was no fire, but the great g-forces of the impact meant all 85 survivors were injured in some way. Of the 158 persons on board, 73 died.

Most deaths were due to head and upper-body injuries, and three doctors involved in the treatment of the injured wrote a paper analyzing the injuries, suggesting it might provide valuable lessons regarding better constraints for passengers.[1] Cost-benefit considerations meant that not all suggestions would be implemented.

As regards passenger survival, this accident was somewhat atypical in that the aircraft had completely empty tanks and getting out before succumbing to the effects of smoke inhalation was not the key to survival.

As usual, a whole series of factors were responsible for the disaster:

1. Lack of assertiveness on the part of the first officer, perhaps explained by an inferiority complex when dealing with the—in his mind— "superior" American controllers. At one point the flight engineer even remarked that the controller sounded angry, when this does not seem to be the case.

2. Hoping for the optimum scenario, though the controller at the outset might have done better to suggest a more probable hold time rather than saying "at least forty-five minutes."

3. The creeping up of events, in that they remained on hold for so long that the option of diverting to their alternate, Boston, was lost. This could be classed as indecision.

4. The first officer handling the radio communications did not even once use the word "emergency," though he was told to do so by the captain.

5. Neither did he inform the tower that their attempt to land was in fact their one and only chance, in which case the controller using radar could have tried to talk them down until they could see the runway.
    However, they were expecting it to be a routine landing, but even so, by informing the controller it was their last chance would have meant he or she would have done their utmost to avoid ordering a go-around.

6. Finally, there was a fatal dose of bad luck in that just as they were coming down toward the runway, expecting everything to be finally all right, they had to raise the nose and add power to regain height lost due to the wind shear, and probably thereby failed to pick out the runway in the murk.

ATC was under great pressure, having to cope with so many aircraft holding in the difficult weather conditions, and having to order some to go around because of lack of separation. Had this not been so, the controllers might have felt able to devote time to exploring what the first officer meant when he repeatedly told them—but different controllers—that the aircraft was running out of fuel, something that would have been true to a greater or lesser extent for many of the aircraft they were handling at the time.

However, the courts subsequently made the FAA, as the controllers' employer, liable for 40 percent of the $200 million awarded as compensation.

Unnecessary Fatalities—frail seat attachments

Because the broken off parts of the fuselage were mostly intact, the deceleration had not been too brutal and there was no fuel to produce a fire, more people should have survived. That they did not was because the frail seat attachments on the cabin floor had hardly been improved since the sixties, and passengers were hurtled forward in their seats with them pilling up on each other. Some of those weak seats had become contorted with the passengers strapped in them suffering hip and spinal fractures.

The passengers had not been warned of the impending impact and the need to assume the brace position which might have saved some despite the flimsy seats and fixations.

BA 777 from Beijing Loses Thrust on Late Final (Heathrow, 2008)

Captain saves situation by not taking over at last minute but accused of freezing at the controls.

> With only seconds remaining, the captain did not take over but decreased the flap setting to reduce drag just enough for them to get over the perimeter fence and crash down short of the runway.
>
> It took a considerable time for the investigators to demonstrate how and why the interruption of the fuel supply to the engines occurred.
>
> *British Airways Flight 38, January 17, 2008*

After a long flight from Beijing, which only differed from many others in that the aircraft flew through some exceptionally cold air early on, the British Airways Boeing 777 was coming in quite normally to land at London's Heathrow's Runway 27L. As is usual after what is virtually a long, steady glide under the new arrangement to save fuel, a little extra thrust was required at the last minute to prevent the aircraft losing too much airspeed and sinking below the glide path.

To the pilots' dismay, this was not forthcoming, and it looked as if the aircraft was destined to touch down just before reaching the airport. The first officer was flying the aircraft at the time, and although normally the captain takes over in a crisis, he let him continue—an apparently sensible decision, as the first officer had the feel of the aircraft and there was so little time left. The captain reduced the amount of flap to 25 percent, thus reducing the drag and allowing the aircraft to fly farther. The aircraft staggered over the perimeter fence and came down heavily on the grass just beyond some thousand feet short of the paved runway.

The right main landing gear broke off, while the force of the impact forced the landing gear on the left into the wing. The first officer managed to keep the aircraft in a straight line and thus prevent it from cartwheeling. After skidding across the grass, it ended up just at the beginning of the runway paving. Despite an escape of fuel due to the pilots' failure to switch off the fuel supply to the engines correctly, there was no fire and there were no fatalities. The 136 passengers and 16 crew members evacuated the aircraft via the chutes, with one passenger suffering a broken leg. The airline and no doubt its insurers soon deemed the aircraft not worth repairing and classed it as a write-off.

Initially, the media reported many passengers considered it a nonevent, with some only thinking it had been a hard landing. Their main gripe seemed to be their insensitive treatment on reaching the terminal. Yet some days later, with the arrival of lawyers on the scene, some were talking of the great distress they had suffered as justification for suing the airline.

This in turn led to the CEO of BA personally contacting passengers to head off legal action by showing personal concern. In this context, one might mention that some at BA think some business- and first-class passengers regard the airline as a soft touch in that they greatly exaggerate their suffering when things go wrong in order to obtain free flights and/or upgrades. Passengers in economy are not treated so benignly by the airline.

That said, there were injuries and one Australian man sitting in seat 30K had his leg crushed when a 12-inch cylindrical beam of the front wheel assembly pierced the skin of the fuselage. Luckily, he was not in the brace position with his head near his knees.

In their preliminary report the UK's Air Accident Investigation Board (AAIB) concluded that ice in the fuel was the probable cause and recommended modifications to the fuel supply system on 777s fitted with Rolls-Royce engines. While it was not possible at that time to find cast-iron proof that ice in the fuel was the cause, concern increased, as there had been another case where an engine of a 777 flying in cold conditions had temporarily lost power.

The NTSB, the US equivalent of the AAIB, was more forceful in its recommendations, announced simultaneously with those issued by the AAIB, which the UK investigators regarded as bad form. The NTSB maintained that interim precautionary measures, such as coming down to warmer air rather than staying high up, could expose aircraft to greater risk than usual.

Conspiracy Theory

As mentioned elsewhere, when people from the same country as the airline and the manufacturer of the aircraft or engines in question investigate an accident, someone will say there has been a cover-up. Even in this case, people claiming to have "contacts airside at Heathrow" were alleging the aircraft had run out of fuel. As evacuating passengers could smell the leaking fuel, one can discount this, but it does show how easily misinformation can be spread.

Cause Finally Proven

The UK investigators looked at every conceivable reason for the engines to not respond but failed to find anything wrong with the computers or the programming. Indeed, the valves supplying fuel to the engines had opened fully in response to the demand for more thrust. The quality of the fuel itself was checked and found to be above average. It was from South Korea and had been shipped to a Chinese port and sent directly by pipeline to Beijing Airport. Investigators found some matter in the fuel tanks, no doubt left there at the time of manufacture.

The investigators concluded that ice must have been the cause, but despite numerous attempts over many months, they were unable to replicate a situation where ice formed on the inside of the fuel pipes and broke off to block the fuel/lubricating oil heat exchanger.

They compared the flight in question with thousands of other flights and found that although the weather over Russia had been exceptionally cold, the pilots had constantly checked that the temperature of the fuel never fell to the level where it would become waxy. There was only one case where a Rolls-Royce engine on a 777 had behaved similarly. However, that had happened to a Delta aircraft at cruising height and only involved a single engine and had resolved itself when the pilot throttled back before reapplying power, something the BA pilots obviously could not have done in their predicament.

Finally, investigators noted again that the aircraft had flown for hours so precisely on autothrottle that there had never been a demand for a surge in power. The simple conclusion was that this allowed ice to build up inside the fuel pipes and remain there until the sudden demand for power prior to landing. They noted that this had been true in the case of the Delta 777 too.

Replicating the situation in tests, they found ice built up on the inside of the pipes, and when a sudden demand for high power was invoked, the ice broke off in such quantities that the heat exchanger was blocked. The cause had been proven conclusively, much to the relief of the many users of the Boeing 777.

The problem only pertained to 777s fitted with Rolls-Royce engines, because the tubes in their heat exchangers carrying the fuel through the hot oil protruded a few millimeters from the part carrying the oil and could not melt large quantities of ice deposited on them when cold fuel was flowing. Incidentally, the pilot of the Delta 777 unblocked the exchanger by throttling back and giving the ice time to melt. The temporary solution was to power up the engines from time to time to prevent the buildup of ice inside the pipes. The long-term solution was to redesign the heat exchanger.

Sour Aftertaste

An unwarranted whispering campaign largely among cabin crew at the airline that Captain Burkill had frozen at the controls finally led to him resigning in disgust, only for him to find no other serious airline would employ him.

Even though Captain Burkill was blameless, any pilot involved in a serious incident can have difficulty finding employment elsewhere, and in his case the rumors did not help. Falling on relatively hard times, he wrote a well-received book, *Thirty Seconds to Impact*, in conjunction with his wife about the incident and, notably, its aftermath.

Finally, British Airways reinstated him.

Poisoned Chalice

Captain Burkill makes a very interesting point about how accolades can be a poisoned chalice and cites a case at BA during the writing of the book where the pilots in control of an aircraft had very probably saved the lives of all on board through fantastic airmanship. There had been talk of giving them a medal, but they refused to receive it, preferring to stay anonymous, having seen for themselves how such recognition can lead to problems—perhaps particularly at BA, contemporarily and historically.

Chapter 3
RUNWAY OVERRUNS

"Safest Airline's" 100 mph Overrun (Bangkok, 1999)

747 aquaplaned after captain cancelled go-around at last moment

> Australia's Qantas airline mostly flies long-haul routes to the world's
> major airports, where there is relatively little risk. With good pilots and
> an absence of bad luck, it could boast of never having suffered a hull loss
> (in the jet age). The airline came close to blotting its copybook in stormy
> weather in Bangkok, and indeed only avoided doing so by carrying out
> the most expensive repairs ever made to a civilian aircraft.
>
> *Qantas Flight 001, September 23, 1999*

L ooking out from the twenty-eighth-floor balcony of a tower apartment on
the bank of Bangkok's Chao Phraya River, the author watched torrential rain
of an intensity he had never witnessed. He wondered how pilots of incoming
aircraft could cope with it. As if in answer to that question, the next day's
Bangkok Post had a few lines saying a Qantas 747 had been involved in some
trouble at the airport, with no injuries. The incident was termed a mere mishap.
However, when more details became available, perhaps through disaffected
Qantas staff, the "mishap" became headline news in Thailand.

Apparently, the jumbo was still traveling at 100 miles an hour (160
kilometers an hour) when it ran onto the grass at the end of the paved runway
overrun area after coming in to land. The *Bangkok Post* was later to comment,
"It was a miracle a fire leading to many deaths had not occurred," adding that
the landing had also been a "fiasco" in that, as explained later, some systems on
the aircraft "assumed" it was about to take off, which for a moment it was.

The Thai authorities were miffed to discover Qantas staff had removed the
quick access data recorder but finally asked the then highly respected Australian
Transport Safety Bureau (ATSB) to investigate the incident. After all, in the
absence of significant injuries, and with damage limited to Australian property,
the Thais preferred to take a backseat.

This was a fortunate decision, as the final 186-page ATSB report into the
incident is an exceptionally complete and lucid document by any standards. For
the technically minded, it is like a detective story in that there is enough basic

information for the reader to draw his or her own conclusions regarding what might or might not have happened had this or that parameter been different.

How the Events Unfolded

Qantas 1, a Boeing 747 flight from Sydney to London with an intermediate stop at Bangkok, first nosed into the exceptionally heavy rain at a height of 200 feet, and with just 930 yards (850 meters) to go before crossing the threshold of the slightly shorter of Bangkok Airport's two parallel runways. The aircrew, consisting of the captain, first officer, who was the pilot actually piloting the aircraft (PF), and second officer, were aware they would encounter difficult conditions, since the storms over the airport had long been visible on their weather radar. Now, at late final, the usually crisp white runway lights were only visible to the PF for brief moments after each pass of the windscreen wiper blades.

Another Qantas aircraft, call sign "Qantas 15," had been about three minutes ahead of them in the landing sequence but had decided to go around because of poor visibility. The crew of Qantas 1 were still on the approach control frequency and were not aware of this. Therefore, when they moved to the tower frequency and were informed that a Thai Airbus had landed ahead of them and "braking was good," they thought the interval was the usual three minutes, when in fact it had become six. This is a long time in a tropical storm. Had the tower told them their colleagues just ahead had abandoned their landing, their mind-set might well have been different—there would have been less pressure to land.

As Qantas 1 descended in the downpour to 140 feet, the captain became concerned that while the aircraft had speeded up it had not descended sufficiently fast and said to the first officer:

You are getting high now!

Shortly afterward came the automatic voice warning that they were at a hundred feet.

The captain said:

You happy?

The first officer replied:

Ah, yes.

According to the ATSB report, the first officer later stated that he felt he was getting near his personal limits by this time but was happy to continue with the approach, as the captain appeared to be happy. He maintained that he had the feel of the aircraft, and it made more sense for him to continue rather than hand over control at that point. The second officer also reported that he was comfortable with continuing the approach at that stage. (The first officer had decided to carry out the approach manually rather than use the autopilot to "keep his hand in.")

On crossing the runway threshold, they were thirty-two feet *above* the ideal height, almost fifteen knots *above* the target speed, and nineteen knots above the reference speed, V_{REF}. These excesses were individually just within company limits. At the same time, the distorting effect of the rivulets of water on the windscreen would certainly have made it difficult for the first officer to judge distances correctly.

The fifty-foot-altitude warning sounded, and the nose went up slightly, resulting in the aircraft prematurely beginning its flare[1] and prompting the captain to say:

Get it down! Get it down! Come on, you're starting your flare.

Acknowledging this, the first officer began to retard the engine thrust levers in preparation for touchdown. The rate of descent, which had already dropped to approximately five feet a second, slowed even further due to the flare.

The first officer later reported that although the reduced visibility made it difficult to judge the landing flare, they were already in it and thought it best to pursue it and allow the aircraft to settle onto the runway. He believed that they had more than enough runway remaining for them to stop.

The thirty-foot warning marking the height they would normally have begun their flare sounded, with the longer-than-usual interval between the fifty-feet and thirty-feet calls indicating a slower-than-normal descent. The captain, no doubt getting concerned about the delayed touchdown, increased the autobrake setting to "4" without advising the other crew members, as it did not materially affect the touchdown.

With only ten feet left before the dangling main wheels would first touch the runway, the captain ordered a go-around. He felt the aircraft was "floating," and he could not see the far end of the runway. In addition, he was not happy with the speed, which again was within company limits, but at the upper limit of what he personally was prepared to accept.

Instead of using the takeoff/go-around (TOGA) switch, which would have reconfigured the aircraft automatically for takeoff, the first officer initiated the go-around by pushing the engine thrust levers forward. This is quite common practice, as the automatic TOGA go-around controlled by the aircraft's computers can be a little alarming to passengers, since it is abrupt and indelicate, on the assumption that it may be an emergency. Anyway, the first officer reacted very quickly, so it did not make much difference, other than that a manual go-around is easier to cancel.

With the engines needing some eight or so seconds to spool up from idle, the aircraft's main wheels would inevitably brush the runway before the aircraft could regain enough speed for positive lift. Just then, when it seemed they were quite rightly going to forgo the landing in the name of safety, a letup in the rain allowed the captain to see right to the far end of the runway. Reassured, he decided to cancel the go-around.

Instead of announcing his intention, he put his hand over the first officer's, left hand resting on the throttles, to push them back. For a moment, the first officer was unsure who was flying the aircraft. Worse, one lever slipped from his grasp and remained where it was.

The 747's systems interpreted the fact that one lever was forward as an intention to take off and disarmed both the automatic braking and automatic deployment of the spoilers. (Spoilers are flat panels hinged at the front set on top of the wings that flick up to "spoil" the flow of air over the wing. This not only produces an air braking effect but also pushes the aircraft down, improving the grip of the tires on the runway.)

Having to pull the recalcitrant lever back to join the others in the idle position possibly made the first officer forget to apply reverse thrust.

Braking manually as hard as they could, the two pilots at the controls were shocked to find the usually exceptionally effective carbon brakes were hardly slowing the aircraft at all. This was particularly troubling, as having landed so far down the runway in the first place, and furthermore, having for a moment one engine thrusting them forward with no spoilers and no wheel brakes, meant even more of the remaining runway had been used up. With the tires aquaplaning on the layer of water on what little runway remained, they ran onto the short overrun area and then onto the grass at an incredible 100 mph.

Fortunately, the heavy rain that had initially been their undoing ultimately proved their salvation, for the rain-sodden ground allowed the huge wheels to sink deeply into it, with the result that the aircraft finally came to rest some 240 yards (220 meters) farther on, without encountering any serious obstacle on the way.

The rain and wet ground may indeed have saved them a second time by quickly dousing any nascent fires. With the aircraft having come to a stop, the captain immediately reviewed the situation, made more difficult by the fact that wires for communicating with the cabin crew, as well as those for the PA (public address) system, had been severed, as they passed close to the crushed nose-wheel section. He could only get information piecemeal by messenger.

Only after waiting some twenty minutes and after the arrival of rescue vehicles did he order the evacuation.

The ATSB board of inquiry thought this delay in evacuation had been unwise, as the captain could not have been sure a fire would not break out. In addition, the batteries supplying power for the emergency lighting system were on the verge of giving out, as the designers did not anticipate emergency evacuations taking so long. If there had subsequently been a fire, an evacuation without even emergency lighting would have been a nightmare. In the event, waiting for transport avoided injuries and the danger that the captain later mentioned of passengers being struck by aircraft when tempted to "walk over the adjacent busy runway towards the brightly lit terminal."

On board, there had been 3 aircrew, 16 cabin crew, and 391 passengers. None sustained significant physical injury.

Despite the soft ground, the aircraft itself sustained a considerable amount of damage and stress, as evidenced by the $75 million initial estimate for the cost of repairs.

Verdict—Not What You Might Expect

From the above account, one might immediately assume that the flight crew was responsible for the near disaster, and even culpable. Instead, the ATSB, relying very much on Professor Reason's Swiss cheese accident model—according to which accidents happen where all the holes (mistakes and faults) in the cheese line up—concluded that Qantas had not properly prepared its Boeing 747-400 pilots for landing on "contaminated" (in other words, water covered or icy runways) and, partly to reduce costs, had introduced a new, "less conservative" (that is, more risky) standard landing procedure without proper consideration.

The ATSB report noted, "With the introduction of the more powerful carbon brakes, Qantas had changed their standard landing procedure to Flaps 25 idle reverse thrust rather than the previous more conservative Flaps 30 full reverse Thrust."

Flaps are extensions to the leading edge and trailing edge of the wings that configure the wing to give more lift, especially at low speed, when taking off or landing. "Reverse thrust" simply means that cowlings on the engines move so that the thrust from the engine pushes the plane backward rather than forward. Before the introduction of the better-performing carbon brakes, passengers would almost invariably hear this engine roar just after landing. "Idle reverse" does little to slow the aircraft but means the transition to full reverse can be accomplished quickly.

Qantas made this change in landing procedure for financial and, to some extent, noise abatement reasons. It could save money because carbon brakes wear less if applied continuously rather than on and off. In addition, with a flap angle of 25 degrees, there is less wear on the flap mechanism. It would also reduce maintenance costs for the thrust reverser mechanism. However, extra wear on the tires would negate some of these gains.

Comparison in the ATSB report with five other major airlines flying routes in Asia and worldwide—unnamed due to reasons of commercial confidentiality—showed the others to be more conservative:

1. Many stressed the need to be warier of storms and the need to use Flaps 30/full reverse thrust in heavy rain situations.

2. The other airlines had more experience than Qantas of flying into difficult airports.

Some Qantas pilots believed the more conservative Flaps 30 also made it easier to land precisely at the desired point—Qantas 1 landed well beyond the ideal touchdown point. The crew of Qantas 1 did not even discuss the Flaps 30/full reverse option, even though their weather radar had revealed the storms over the airport when they were far away, with plenty of time.

The ATSB report said this was probably because, with the introduction of the new standard, landings were hardly ever made with Flaps 30, and they were unlikely to try something with which they were unfamiliar.

According to reference data supplied by Boeing, the aircraft *could not have stopped in time* on the contaminated runway using the standard Qantas Flaps 25 idle-reverse landing procedure. This remained true even if it had landed at the ideal touchdown point 400 yards (366 meters) from the threshold. However, it *could* have stopped if they had used full reverse thrust in addition and everything else had been ideal—that is to say, no canceled go-around and so on.

The whole incident could have been avoided had the captain not dismissed the first officer's earlier suggestion that they hold off to the south until the weather improved, saying it was only a shower. Their aircraft was not the only one trying to come in, so it was just a matter of opinion, bearing in mind that conditions can suddenly change in that part of the world.

The ATSB report cited the following adverse factors in the reverse order of their occurrence. The author's comments are in parentheses.

1. Reverse thrust
   Had they not in the turmoil forgotten to engage reverse thrust, even in the idling mode, as specified in their standard landing instructions, there would have been a slight deceleration effect rather than the slight acceleration produced by idle forward. Many Qantas pilots said they occasionally forgot to engage reverse thrust when something distracted them. Normally, this would not have serious implications and therefore would not be uppermost in their minds. Of course, full reverse rather than idle reverse thrust would have been better.

2. Nonverbal cancellation of the go-around
   The captain, perhaps because it was a habit developed in the course of his frequent work training pilots, cancelled the go-around merely by putting his hand over that of the first officer to pull back the engine thrust levers. It was unlucky one lever remained forward with the consequences already mentioned, and as said, the first officer for a moment wondered who was flying the aircraft. Theoretically, this direct

action by the captain would have averted the delay between his issuing the command to cancel the go-around and its execution.

3. Cancellation of go-around

   Pilots generally do not consider canceling a go-around to be good practice, as it can lead to confusion and other problems, as was indeed the case here.

   However, put yourself in the captain's place. There he was—so he thought—finally safely on the ground after a difficult landing. In the atrocious weather conditions, a second landing might be even trickier, with not much reserve of fuel for a third attempt. Moreover, both he and the first officer were convinced there was more than enough runway to stop, which there would have been had it not been for the standing water on the runway. Ordering and subsequently canceling the go-around meant more runway was used up due to the engines beginning to power up, not to mention one engine continuing to pull because of the "lost" thrust lever, which in turn made the aircraft "think" it was taking off, which prevented immediate deployment of the spoilers and application of the brakes on touchdown.

4. Landing too far down the runway

   The first officer came in somewhat fast and too high and initiated the flare early, consequently touching down far along the runway. He attributed this partly to the heavy rain. (One point not stressed by the investigators was that the depth of the water at the beginning of the runway where aircraft normally land—and where the Airbus six minutes ahead found braking had been "good"—was doubtless less deep than at the other end, because landing aircraft would have splashed it away. If he had touched down earlier, braking might have been sufficient to reduce the speed enough to prevent aquaplaning on reaching that presumably deeper water.)

5. No prior decision to use safer Flaps 30 with full reverse thrust

   The crew's experience of trouble-free landings in heavy rain in places such as Bangkok and Singapore may have made it overlook the possible severity of some patches of rain in Thailand.

6. Qantas had abandoned use of windscreen water repellent

   Qantas had abandoned the use of water repellent on Boeing 747s, deactivating the systems several years before for financial reasons and ostensibly to protect the environment—water repellents consist of fluorocarbons that cause depletion of the ozone layer. (It is perhaps true, as some Qantas pilots maintain that water repellents do not in general make much difference.

[The author's personal experience of driving in Thailand showed that in torrential downpours they make an incredible difference to visibility through the windscreen, even well after the wiper blades have passed. Thus, merely the use of water repellent could have greatly changed the scenario: the first officer might not have landed so far down the runway, and the captain might have been able to see the far end of the runway early on and not have ordered the go-around in the first place.]

7. Second officer's wife

   The ATSB report discounts the presence of the second officer's wife in the cockpit as an adverse factor in the incident. (Could concern for his spouse have led to a moment's inattention, causing the second officer to miss noting the first officer had not applied idle reverse?)

8. Autopilot not used

   The first officer flew the aircraft manually to get more hands-on practice. (Furthermore, the diffractive effect of the layer of water on the windscreen would have made it difficult for the first officer to judge the distance correctly. The aircraft would surely have come in at the slower correct speed and touched down near the optimum point on the runway had the first officer flying the aircraft [PF] opted to use the automatic pilot. This would have given an extra 636 meters, not to mention the already stated fact that the beginning of the runway might well have had less water on it, partly because aircraft landing there would have dispersed it. Had it not been for the standing water, his landing would have been just about acceptable.)

Conclusion

Qantas, in the inquiry's view, seemed to have erred in not training pilots suitably for coping with contaminated runways. Though Boeing said the aircraft could not have stopped in time with the given configuration even had everything, apart from the aquaplaning, been perfect, in the author's opinion the overrun would only have been slight, and most airports have some spare space at the end of their runways for such eventualities. It was the combination of so many other negatives that made this a potential disaster, which would have been the case had the soggy ground not slowed the aircraft. Had there been solid obstacles earlier in the path of the aircraft as it hurtled off the runway, any one of these factors could have meant the difference between life and death for hundreds of people.

In response, Qantas introduced changes in its training and management to avoid such an incident happening again. As usual, other airlines learned from this mishap at no cost to themselves.

In May 2000, after being repaired in China, that same aircraft had to turn back on a flight out of Hong Kong because of generator problems. According to the *Sydney Morning Herald*, a Qantas engineer, who did not wish to be named, told it that he and his colleagues had predicted such electrical problems "because of the quality of workmanship in China." One should note, however, that *the Sydney Morning Herald* said Qantas's chief executive, James Strong, denied that Qantas had had the aircraft repaired rather than scrapped, just so Qantas could maintain its claim that it had never had a hull loss. According to Mr. Strong, the $100 million cost of the repairs only represented 40 percent of the cost of the aircraft. However, there is no mention in this of the scrap value, as parts could surely have been reused.

In addition, there were probably considerable additional costs in terms of the Qantas management time required to oversee the repair project in China. People who have flown in the repaired aircraft have noted how the front looks newer than the back.

Personnel Changes at Qantas

Very often airlines and organizations wait quite some time before officially announcing personnel changes after an accident. To bolster its position, the airline will initially express full confidence in those involved, only to let them go later, as was the case for the Singapore Airlines pilots involved in the takeoff from the disused runway at Taipei, where two of the three were later dismissed. Did the same apply at Qantas after the incident just described?

One wonders, because on May 30, 2003, the *Australian* featured the startling headline "Qantas Safety Tsar's Reign Ends." According to the article, Ken Lewis (the "tsar" in question) had "left the job after twenty-three years as head of safety and almost four decades at the airline." It added that he would continue to advise senior management until leaving the airline.

The paper emphasized that he was highly respected throughout the industry, "having qualified as a meteorologist, worked as a flight attendant, as a ground simulator instructor and navigation instructor, and is a qualified air safety instructor." He had held positions on international safety bodies and was at the time president of the Australian Society of Air Safety Investigators.

Though he was certainly a good man—incidentally, more problems seem to have occurred since his replacement—one wonders whether his qualifications would have given him the mind-set and weight to influence senior Qantas management (including the bean counters) in the manner the investigators thought necessary in this particular case.

The article in the *Australian* also mentioned that the airline was replacing its chief pilot and returning him to line flying. Possibly, the moves were prompted by the fact that the airline was under investigation regarding another incident, in which a 737 landing in rainy conditions veered off the runway for a while at Darwin after a heavy touchdown just beyond the normal runway threshold. Again, no one was injured, but there was some damage to the tires and flaps.

However, on January 13, 2008, the *Sydney Morning Herald*, in an article about the slide in Qantas's share value and Merrill Lynch's sell recommendation, said the carrier a week previously had suffered "arguably the biggest dent to its once-enviable safety reputation in decades" after one of its Boeing 747s lost electrical power on approach to Bangkok. It was referring to a Boeing 747-400 London-to-Bangkok flight that lost electrical power from all four engine-driven electrical generators fifteen minutes out of Bangkok and had to rely on backup power to land there.

Battery power would only have lasted for about an hour, and had the aircraft been a long way from an airport and in bad weather, the situation could have been precarious. Unlike many twin-engine aircraft, the four-engine 747 is not fitted with a RAT, a ram air turbine, to generate emergency electric power while gliding.

Investigators found the loss of electrical power had been caused by water entering the electrical equipment bay through cracks in the drip shield above it. Checks found similar cracks in the shields of other 747-400s in the Qantas fleet, and all were repaired. Whether there is any direct connection is not certain to the author, but a court case was to proceed regarding a Qantas engineer with allegedly fake qualifications who had worked on that model of aircraft.

Airlines do go through periods where one problem after another crops up and often end up better as a result. These incidents at Qantas are being cited merely to remind passengers that they should be wary of safest-airline claims based on raw accident statistics that do not take into account the nature of the routes flown. However, if they are flying those intrinsically safer routes, there will generally be less risk than on the average flight, whatever the carrier.

---

[1] The landing flare (usually performed at a height of about 30 feet) consists of raising the aircraft nose to produce (1) *extra lift* to break the descent and (2) *extra drag* to slow the aircraft so it subsequently sinks onto the runway. It is rather like a big bird landing, except that birds can lower their "flaps" and do a flare dramatically at the very last instant, which passenger aircraft, at least with present technology, cannot.

Air France A340 Overruns, Catches Fire in Gully (Toronto, 2005)

All 309 occupants barely manage to evacuate, many with their carry-on baggage

> Even though on reaching the end of the runway the Air France A340 at
> Toronto was traveling twenty knots slower than Qantas's 747 at
> Bangkok, there was little clear ground beyond, and that ground was not
> soft enough to have a significant braking effect.
>
> *Air France Flight 358, August 2, 2005*

Just as with the Qantas 747 at Bangkok, the Airbus A340 from Paris was being flown manually as it came in to land in a storm at Toronto's Lester B. Pearson International Airport. This meant that when the throttles were pushed forward to compensate for a sudden switch to a tail component to the wind, they stayed forward longer than would have been the case under autothrottle, which in turn contributed to the aircraft passing over the runway threshold at seventy to eighty feet, some forty meters higher than usual.

The rain and poor visibility then made it difficult to bring the aircraft down quickly, and as a result it touched down more than two-fifths of the way down the 9,000-foot (2,743-meter) runway at 143 knots (274 km/h) IAS (indicated airspeed), with only 5,250 feet (1,600 meters) of runway left.

The spoilers duly deployed after three seconds and maximum manual breaking was applied. Yet idle reverse thrust was only selected 12.8 seconds after touchdown (at IAS 118 km/h, with only 670 meters of runway remaining), and full reverse only after 16.4 seconds. This delay can be attributed to the pilot flying (PF) concentrating on keeping the aircraft on the runway in the relatively strong crosswind, and the fact that attention was not drawn to the failure to apply reverse thrust, as the pilot not flying (PNF) was not making the customary announcements confirming deployment of the spoilers and thrust reversers.

Unable to stop in time, the aircraft departed the runway at eighty-six knots corrected ground speed, passing over a grassy area and then a road before ending up in a minor ravine. Most of the damage to the aircraft occurred in the ravine.

With fire breaking out, an emergency evacuation was ordered. Ignoring instructions from the cabin crew not to do so, almost half the passengers retrieved their carry-on baggage. One man even blocked an aisle as he busied himself rearranging items in his case. Ignoring angry comments from passengers standing behind him and orders from the flight attendant to leave his baggage and go to the emergency exit, he persisted, obliging the attendant to redirect passengers through the middle bank of seats to the other side of the aircraft to gain access to the only available emergency exit in the aft cabin.

Moments after the last person exited the aircraft it was engulfed in flames, so it was a close-run thing, and the bringing of luggage almost resulted in fatalities.

Of the 309 people on board, 12 (2 crew and 10 passengers) suffered serious injuries, nine of which were incurred at the time of impact, and three during the evacuation. The two members of the cabin crew who were seriously injured were hurt at the time of impact but were still able to perform their duties. Passengers with serious impact injuries were nevertheless able to walk.

At the time, the airport was on red alert because of rain and lightning, and some have argued that the control tower should not have permitted the landing, with others saying the decision was up to the captain, who had sufficient fuel to divert.

Delay in Producing Final Report

Considering that everyone survived, and that investigators did not have to recover wreckage from inaccessible places, such as the bottom of an ocean, the inquiry took an inordinate length of time to publish its conclusions. Strong representations made by the parties and, notably, the airline were said to be the reason for this delay.

The freelance French aviation writer François Hénin has said the accident was essentially due to a failed approach—the aircraft came in too high and landed too far down the runway—and a lack of communication between the pilots.

Interestingly, Hénin points out how Air France PR flooded the French media with glowing accounts of the "truly remarkable job" the Air France cabin crew had done when their performance had merely been satisfactory, thus diverting attention from what could have been a tragic disaster with the airline apparently at fault.

Hénin cites this poor landing as an example of laxness at the airline. Some have suggested the delayed touchdown and initially too fast approach was largely due to a microburst producing a tailwind during the latter part of the approach. However, the failure of the PNF to make the customary callouts confirming deployment of the spoilers and thrust reversers certainly did constitute laxness.

Chapter 4
MIDAIR COLLISIONS
AND TCAS

Midair Collisions at Grand Canyon and New York City (1956, 1960)

At the time air traffic control in the US was very primitive, in places non-existent

> These midair collisions made the nation realize the air traffic control system was deficient and supervision of civil aviation in the US needed rethinking.
>
> *TWA Flight 2/UA Flight 718, June 30, 1956*
> *UA Flight 826/TWA Flight 266, December 16, 1960*

TWA Flight 2 Collides with UA Flight 718
at Grand Canyon

On June 30, 1956, two aircraft took off at around nine o'clock in the morning from adjacent runways at Los Angeles International Airport. The first was Transworld Airlines Lockheed Super Constellation Flight 2 bound for Kansas City, with sixty-four passengers and six crew. The second, only three minutes later, was United Airlines Douglas DC-7 Flight 718 bound for Chicago, with fifty-three passengers and five crew.

The DC-7 was the first airliner able to fly almost one hundred passengers across the US from coast to coast nonstop. US Airways was justifiably proud of it, though it was soon to be superseded by the Boeing 707 and other jet airliners. Its powerful engines were fault prone, and ultimately more DC-6s than DC-7s remained in long-term service.

Each aircraft climbed out of Los Angeles along a different controlled airway, with the one taken by the TWA Constellation taking it northeast to a waypoint at Daggett, roughly in line with its eventual route. That followed by the DC-7 took it southwest to one at Palm Springs, more off its eventual route. After they had adjusted course at these waypoints, it so happened that, due to the DC-7's 18 knot (20 mph) higher airspeed, the aircraft would cross paths simultaneously an hour after takeoff in an area just before the so-called Painted Desert line. This should not have presented a problem, with the Constellation flying at nineteen thousand feet and the DC-7 at twenty-one thousand feet.

The term "Painted Desert line" is somewhat confusing, as it suggests a line in the desert. In fact, it was the name given by air traffic control to a virtual line about two hundred miles long running north-northwest between the VOR radio beacons at Winslow, Arizona, and Bryce Canyon, Utah. Passing to the east of the Grand Canyon, it traversed an area of beautiful rock formations in striated colors that was called the "Painted Desert" (El Desierto Pintado) by an expedition under Francisco Vázquez de Coronado in 1540, hence the name.

In 1956, air traffic control in the US was very primitive and in many areas virtually nonexistent. After all, the country was enormous, and airliners were so few in number that apart from areas where they would be funneled (bunch up), there was plenty of room for them to avoid each other when flying under visual flight rules, following the principle of "see and be seen."

Furthermore, other than in the vicinity of large cities or the airport of departure or arrival, the controllers did not contact the aircraft directly but usually through the intermediary of the airline's dispatchers. In vast uncontrolled areas with no radar coverage, pilots would report to their companies when passing over the waypoints on their route.

At 9:21, the Constellation reported that it was approaching the Daggett waypoint and requested a change in cruising height from nineteen thousand to twenty-one thousand feet to try and avoid cumulonimbus clouds building up over the Grand Canyon area. The dispatcher contacted air traffic control at Los Angeles Center, who in turn contacted Salt Lake City control center, which covered the airspace TWA was about to enter. The Los Angeles controller, aware the United Airlines DC-7 would be crossing paths with the Constellation at precisely that altitude, doubted it would be possible, and the Salt Lake City controller confirmed.

The Los Angeles controller contacted the airline to say the request for twenty-one thousand feet could not be approved, to which the TWA dispatcher replied, "Just a minute. I think he wants a thousand on top; yes, a thousand on top until he can get it."

Here the captain of the Constellation was exploiting a regulation that said that if flying under visual flight rules, where the pilots are responsible for looking out for each other's craft, they can request to fly one thousand feet above the clouds or overcast. On being assured TWA 2 would be at least one thousand feet above the clouds, the Los Angeles controller duly gave permission, and as a result the Constellation ended up cruising at precisely the altitude that had just been refused. However, unlike the DC-7, the Constellation was warned of the presence of the other in the vicinity. The Constellation even confirmed to the dispatcher that the United Flight 718 was at twenty-one thousand feet. That was not much help, as the DC-7 would be overtaking from the side and behind.

Although "one thousand above" suggested the Constellation flying under visual flight rules would have been above all clouds, this was not true, for some clouds rose to twenty-five thousand feet, and the aircraft would have to skirt around them. The pilots of both aircraft would also have been positioning their aircraft to give their passengers some wonderful views of the scenery below, notably the Grand Canyon.

At approximately 9:58, United 718 made a position report to the CAA communications station at Needles stating that the flight was over Needles at twenty-one thousand feet and estimated that it would reach the Painted Desert line at 10:31. Surprisingly, only a minute later, at 9:59, TWA 2 reported to the TWA dispatcher at Las Vegas that it had passed Lake Mohave at the Arizona border at 9:55, was "a thousand feet on top," at twenty-one thousand feet, and would be reaching the Painted Desert line at 10:31.

No further communication was heard from the Constellation.

At 10:31 an unintelligible transmission was heard by aeronautical radio communicators at Salt Lake City and San Francisco, who had a contract with United to handle radio traffic. When it was played back later, they worked out that it must have been from the captain of the DC-7 saying they were about to crash: "Salt Lake, United 718 . . . uh . . . we're going in."

With no radar coverage, knowing where the aircraft were depended on receiving their reports over the radio as they passed waypoints, and sometimes pilots forgot to give them, which was the reason that the respective dispatchers were not overly concerned when there was no report at 10:31 of the aircraft passing over the Painted Desert line. However, when there were no reports at subsequent waypoints concern grew.

At 10:51, air traffic control at Salt Lake City, which covered the area where the two aircraft should be flying, received a call from United saying that the DC-7 was twenty minutes overdue at Painted Desert. Shortly afterwards they received a call from TWA saying the Constellation was also twenty minutes overdue at Painted Desert. Two aircraft "forgetting" to report was highly unlikely, and with everyone getting more worried, Salt Lake City tried to raise them every couple of minutes, while dispatchers tried various frequencies. After an hour of fruitless attempts to make contact, Salt Lake City telegraphed local authorities to see whether they had any information. There were no reports of anything untoward, and because the aircraft were not tracked by radar, looking for debris in that vast area would be like looking for a needle in a haystack.

Palen and Henry Hudgin, brothers who operated a small plane taking tourists on overflights of the Grand Canyon, heard about the search and remembered seeing smoke earlier in the day. They went back to look that evening and found the wreckage of an aircraft high up on an escarpment on the side of the canyon. The following morning, they saw the wreckage of another aircraft on the bed of the canyon, clearly identifiable as the Constellation due to its unique triple empennage.

Though marveled at by tourists—usually from above—the Grand Canyon is inhospitable, dangerous, and inaccessible. The first task was to get people in to see if there were any survivors, however unlikely that might be. Reaching the debris of the DC-7 high up on the side of the canyon proved especially difficult, and United Airlines brought in a team of Swiss mountaineers, who happened to be climbing not too far away. However, they found no survivors. In fact, there were hardly any identifiable bodies at either site, and much of the wreckage had fused to the rocks in the fires. The total death toll was 128.

Investigation

At the time the Civil Aeronautics Board (CAB) was responsible for air accident investigations, but it was in its infancy and lacked the tools, such as flight data and cockpit voice recorders, that it has today. There were no witnesses either.

It was immediately obvious from the fact that the empennage of the Constellation was some distance from the rest of the wreckage that it had come off in flight. Likewise, the left wing outside panel of the DC-7 had separated in flight. From red paint markings from the Constellation on the DC-7, and those from the DC-7 on the Constellation, it was possible to work out that the DC-7 had struck the Constellation coming downward and moving right to left, shearing off the tail empennage.

The details other than the direction and angle are not so important. They show that the faster DC-7 was coming from above and slightly behind, meaning the Constellation would not have been able to see it. The person in charge of the difficult investigation carried out experiments to determine how difficult it would be for the pilots of the DC-7 to see the Constellation and concluded that as it would be in the peripheral vision, it might not be obvious if not moving from side to side. There were, of course, other explanations: for instance, the two aircraft could have been on either side of a cloud and come upon each other with little time to react, or the pilots might have been distracted by having to monitor their engines, or looking down to ensure their passengers got the best view of the Grand Canyon.

Air traffic control seemed to be a system with rules that everyone had followed, or in the case of the captain of the Constellation, exploited, with no one responsible other than the pilots over vast areas. Even as the two aircraft climbed out of Los Angeles in a controlled airway under instrument flight rules, they were asked to switch to visual flight rules, making avoiding other aircraft their responsibility.

People wonder why the Constellation was warned of the presence of the DC-7 and not the latter of that of the Constellation. Apparently, this was because it was not controlled airspace and not the controllers' business, and other aircraft could have been present as well, anyway. When the controller was asked about this, he said his statement about the presence of the DC-7 had not in fact been an "advisory" but merely an explanation as to why he had not been able to grant the request for a change in altitude to twenty-one thousand feet!

As a result of this crash and numerous similar but less deadly ones that had occurred in previous years, the Federal Aviation Agency was created in 1958. Its name was changed to the Federal Aviation Administration (FAA) in 1967, and responsibility for investigating air crashes was transferred from the CAB to the newly established NTSB.

Air traffic control was subsequently greatly improved, and the whole country was eventually covered with sophisticated radar able to warn controllers of a risk of collision. Technology has made a great difference, notably the introduction of TACAS (traffic [alert and] collision avoidance system, now referred to as TCAS), which alerts pilots to the presence of other aircraft in their vicinity, and if a collision looks imminent even orders one to go up and the other down so they do not turn into each other, as people often do when coming suddenly toward each other in the street.

UA Flight 826 Collides with TWA Flight 266
at New York City

This midair collision at New York City was very different from the one at the Grand Canyon in that both aircraft were under direct air traffic control and flying under instrument flight rules. It was the first time information from an airliner's flight data recorder featured significantly in an air crash investigation.

The first aircraft involved was a United Airlines DC-8 four-engine jetliner, with seventy-seven passengers and a crew of seven, that had departed Chicago O'Hare Airport and was bound for New York's Idlewild Airport, now called JFK.

The second aircraft, which had thirty-nine passengers and five crew, was a slower Trans World Airlines piston-engine Lockheed Super Constellation that had come from Dayton, with a stopover at Port Columbus, and was bound for New York's LaGuardia Airport.

Though air traffic control was not informed, the pilot of the DC-8 had told United Airlines that one of its VOR receivers was not working.

The aircraft approached New York at the same time as the DC-8, at 10:25, allowed to take a shortcut to a point called Preston, where it was to hold, circling at five thousand feet, and at no more than 210 knots. Preston was not a beacon but the point where radials from beacons and an airway intersected.

For one reason or another—United Airlines later claimed one of those beacons was not working properly—the DC-8 overshot Preston, and at a higher speed than it should have been traveling, going eleven miles beyond it.

The TWA Constellation coming in to land at Idlewild was allowed to turn right slightly early to intersect the line leading to the runway, at which point it would turn left to line up with it. Though the TWA pilots had been warned of jet traffic to the right, no one imagined it would be where it was.

The engine of the DC-10 struck the fuselage of the Constellation, which broke up. Though parts were found far away, most fell onto the relatively open ground of Miller Army Airfield below. All forty-four people on board died, but no one on the ground was killed.

Having lost an engine and a great part of a wing, the DC-8 flew on, with some observers believing the pilots were trying to land, though the aircraft was doubtless uncontrollable. It crashed on the Park Slope section of Brooklyn, setting fire to houses and buildings, immediately killing all but one of the 128 occupants but, considering it is such a densely populated area, luckily only six people on the ground.

Pictures of the carnage in the streets were shown all around the world, with the story given extra legs by the fact that an eleven-year-old boy was miraculously thrown from the fuselage onto a bank of snow, which broke his fall. Residents rolled him in the snow to cool him and extinguish his smoldering clothes and found he could talk. He was a wonderful boy in all respects. A miracle amidst disaster. Sadly, he died of pneumonia the following day because his lungs had been seared by burning jet fuel.

Aftermath

A new regulation was brought in that required pilots operating under instrument flight rules to report all malfunctions of navigation or communication equipment. A 250-knot speed limit near airports was also implemented.

Furthermore, the accident led to the installation of DME (distance measuring equipment) in aircraft showing the distance from radio beacons, and the FAA was prompted to modernize the air traffic control system through a task force reporting to President Kennedy.

Collisions in General

As this and the next chapter show, collisions between aircraft are potentially very grave, because whether they are in the air or on the ground, twice the usual number of people may be involved.

PSA 182 727 Collides with Cessna (San Diego, 1978)
Pilot error but contributing factors

> The improvements in air traffic control following the two midair
> collisions described in the previous section meant there was never
> another collision between airliners in the US.
> However, about twenty years later a Boeing 727 trijet collided with a
> Cessna piloted by a young man wearing a hood for instrument flying
> training.
>
> *Pacific Southwest Airlines Flight 182/Cessna, September 25, 1978*

The Pacific Southwest Airlines 727 twinjet was approaching San Diego Airport, which unlike most major airports in the United States is nestled close to the city, making it more difficult to see small aircraft against a background of buildings. Being so confined, it is a challenging airport for pilots besides being the busiest single-runway airport in the States.

It was a bright, sunny day with visibility ten miles. On board the aircraft were 7 crew and 128 passengers, including 29 of the airline's employees. The pilot flying was First Officer Robert Fox, thirty-eight, and in the left-hand seat was Captain James McFeron, forty-two. With them was the obligatory flight engineer, and in the jump seat, and a source of distraction, was an off-duty PSA captain.

At 8:59, ATC alerted them to the presence nearby of a Cessna 172 Skyhawk, a very small GA aircraft that had taken off from Montgomery Field executive airport six miles from downtown San Diego and was legally operating under visual flight rules (VFR). This meant that they had not had to file a flight plan and only had to verbally signal their intentions to the controllers who could give them orders. However, the pilot flying the Cessna was being trained to fly under instrument flight rules (IFR) and to make it realistic was wearing a hood restricting his vision to his instruments and controls. An instructor able to see outside was with him.

Though being piloted under IFR simulation, the Cessna was to the outside world performing a missed approach under visual meteorological conditions from San Diego's runway and climbing away to the east, and in contact with San Diego Approach Control. The Cessna without informing the controllers made a change in heading that bought it ahead of the faster airliner but below. The PSA 727 was descending.

Meanwhile the atmosphere in the cockpit of the PSA 727 had been very relaxed, with laughter and the off-duty captain telling an anecdote. They confirmed to ATC that they had seen the Cessna but not that they had subsequently lost sight of it. Nowadays, the sterile cockpit rule applied under ten thousand feet would have precluded this type of banter.

Just before they came down onto the Cessna there was the following conversation:

09:01:11 First officer:

Are we clear of that Cessna?

09:01:13 Flight engineer:

Supposed to be.

09:01:14 Captain:

I guess.

[Laughter and unclear.]

09:01:20 Off-duty captain:

I hope.

09:01:21 Captain:

Oh yeah, before we turned downwind, I saw him at about one o'clock. Probably behind us now.

09:01:47

[Sound of impact.]

The impact with the Cessna severely damaged the 727's right wing, severed hydraulic lines, and made the aircraft uncontrollable. It pitched down to the right, with the fuel inside the wing catching fire. As a last gesture, one and half seconds before the three-hundred-miles-an-hour, nose-down, fifty-degree right bank impact with the ground, the captain said to the passengers, "Brace yourself!"

The 727 came down in a residential part of San Diego, killing three women and two boys. None of the occupants of the aircraft survived, making the total death toll 144. Five people on the ground were injured.

The investigators could not agree on the findings. It was generally agreed that the Cessna was difficult to see. Some faulted the Cessna for having changed course; others blamed the controllers, who were not told that the PSA pilots had lost sight of the Cessna—the approach controller was aware of an automated conflict alert nineteen seconds before the collision but took no action, because alerts were usually nothing of the sort, and he believed the PSA pilots had the Cessna in view.

Regulations were tightened for general aviation aircraft flying into and near major airports, with them obliged to have transponders so that collision avoidance systems work. The FAA installed ILS at Montgomery Field and two other small airports in the county, so San Diego would no longer be used for GA training.

Worst-ever Midair Collision (New Delhi, 1996)
Kazakhstan Ilyushin IL-76 Hits Saudia 747—total death toll 349

> Delhi had up-to-date radar but had not installed it. Consequently, instead of continuously seeing an aircraft's altitude on the blip on their screen, air traffic controllers had to rely on what pilots told them and could not see changes as they happened.
>
> *Kazakhstan Airlines Flight 1907/ Saudia Flight 763, November 12, 1996*

Besides having archaic radar, Delhi Airport operated under difficult conditions, because the Indian Air Force had appropriated much of the airspace, and both incoming and outgoing airliners were funneled along the same narrow path, thus greatly increasing the possibility of a midair collision. The midair collision over Lake Constance we describe later in this chapter was similarly in part due to funneling, because the Swiss Air Force had restricted the airspace for civilian flights.

The Saudia (Saudi Arabian Airlines) flight, a Boeing 747-100 with 289 passengers and 23 crew, had taken off at 18:32 local time from Delhi, bound for Dhahran, Saudi Arabia. Coming the other way on its approach to Delhi Airport was a Kazakhstan Airlines charter flight, an Ilyushin IL-76TD, with 27 passengers and 10 crew. Both aircraft were being handled by the same approach controller.

The first-generation Saudia 747 had 3 aircrew, 2 pilots and a flight engineer, with the pilots handling communications with air traffic control. The Kazakhstan Ilyushin had 2 pilots, a captain and first officer, but since it was originally a military aircraft it had a dedicated radio operator sitting at post behind the pilots without any flying instruments. Handling communications with air traffic control, the operator had to lean toward the pilots' shoulders to check the heading and altitude. Complicating matters further, controllers at Delhi found working with aircrew used to flying only in the old Soviet Union difficult, as they were used to dealing in meters rather than feet, and their understanding of English left much to be desired. Though no one has suggested it, the fluent Indian English used by the controllers may not have helped.

Both aircraft were in the same corridor, coming in opposite directions. The Ilyushin wanted to descend to land at Delhi, while the Saudia 747 wanted to climb to its cruising height. To be safe, the approach controller decided to let them pass each other before proceeding. He therefore allowed the Kazakhstan Ilyushin to descend to 15,000 feet and maintain that altitude, and allowed the Saudia 747 to climb to 14,000 feet, a 1,000-foot separation being recognized there as an adequate safety margin. The Kazakhstan radio operator reported they were at 15,000 feet when in fact they were at 14,500 and still descending.

Even though not in theory necessary, the approach controller warned the Kazakhstan aircraft of the presence of the 747:

> *Identified traffic twelve o'clock, reciprocal Saudia Boeing 747, ten nautical miles. Report in sight.*

He got no reply.

For some reason that has not been established—possibly the pilots misunderstood, bearing in mind they were not talking directly to the controller—the Kazakhstan Ilyushin continued to descend and found itself unwittingly below the 747, in which case, although not intended, it should have passed safely underneath.

By a stroke of very bad luck, it was then that the radio operator noticed their altitude was wrong. The captain applied full power to climb, and as they rose their tail clipped the left wing of the 747, shearing both off. The 747 spiraled downward, the remains of the wing on fire, breaking up before hitting the ground at over 700 mph.

The Ilyushin did not break up but was uncontrollable, hitting the ground with such force that no one survived, though four people with fatal injuries were found alive. At that stage in the flight the passengers had probably undone their seat belts. Two passengers strapped in their seats were found alive on the 747 but later succumbed to their internal injuries.

The total death toll was 349.

Aftermath

The shock brought about action that should have been taken long before.

1. The Indian military ceded airspace so that Delhi Airport could have more corridors and never again have incoming and departing aircraft using the same one.

2. Though it took two more years, the languishing radar equipment showing the altitude and identify of aircraft was finally installed.

3. It became mandatory for airliners using the airport to be equipped with TCAS, traffic (alert and) collision avoidance systems.

Any one of these three actions would have prevented the deadliest-ever midair collision.

Two Japan Airlines Jumbos in Near Miss (Japan, 2001)
Police treat one pilot as a criminal for not obeying the TCAS advisory

> TCAS (traffic collision avoidance systems) are now fitted to all airliners
> to alert pilots of the presence of other aircraft in their vicinity, warn
> them of any danger of collision, and, as a last resort, instruct them on
> what they must to do to avoid a collision.
>
> *Japan Airlines Flight 907/ Japan Airlines Flight 958, January 31, 2001*

In terms of the number of lives lost, Japan Airlines holds the record for the worst-ever crash involving a single aircraft. The airline also almost claimed the record for the worst-ever midair collision, when two of the airline's aircraft were involved in a near miss. Had they collided, the death toll could have been as high as 677.

Not so long ago, the airline went through a patch where it was subject to warnings from the Japanese aviation authorities regarding safety. The near miss in question involved the following aircraft:

1. **Flight number JL907**, a Japan Airlines Boeing 747 with 427 people on board that had just climbed out of Tokyo's domestic Haneda Airport and was bound for Okinawa to the south and seeking permission to finalize its climb to thirty-nine thousand feet. On board the flight attendants were just starting to serve drinks.

2. **Flight number JL958**, a Japan Airlines DC-10 with 250 people on board that was coming in from Pusan, South Korea, to land at Tokyo's Narita International Airport, northeast of Haneda Airport.

At 15:46, a trainee controller (under supervision) duly authorized JL907 to climb to thirty-nine thousand feet (Flight Level 390). Two minutes later the nearby DC-10 crew reported they were at thirty-seven thousand feet (Flight Level 370), meaning their paths would cross in the vertical plane, which of course would not matter if they maintained separation in the horizontal plane.

However, six minutes later the trainee controller discovered they were on courses that could result in a collision. In probably a panicky reaction to the aural warning of the potential conflict, the controller, who had intended to tell JL958 (the DC-10) to descend, mistakenly told JL907 (the Boeing 747) to do so. The TCAS in the descending Boeing 747 gave an aural conflict resolution advisory to climb. Though TCAS resolution advisories are mandatory, the captain of the 747 continued his descent in accordance with the erroneous ATC instruction.

Meanwhile the DC-10 that had continued when the controller attached the wrong flight number to the order to descend initiated a descent in accordance with instructions from its onboard TCAS. (When the trainee controller noticed the DC-10 was continuing to fly level, he ordered it to turn right, but apparently, this message did not get through.) The supervisor had tried to order the 747 to climb, but in vain, as he said, "JAL 957," which applied to neither—he meant JL907.

Seeing the two aircraft were about to collide head-on, the JAL 747 captain forced his aircraft into a steep dive and succeeded in missing the DC-10 by 345 to 550 feet (105 to 165 meters) laterally and between 65 and 200 feet (20 meters and 60 meters) vertically. Drinks trolleys on his aircraft hit the ceiling, and one boy was thrown four seat rows. On the JAL 747, five passengers and two crew members sustained serious injuries, while about a hundred crew and passengers sustained minor ones, mostly limited to bruising. No one was injured on the DC-10, which continued on to Narita as scheduled, while JL907 returned to Haneda Airport.

To the consternation of many in aviation circles, the Japanese police treated the Boeing cockpit like a crime scene. The captain, who probably thought he had done well to save the aircraft by putting his aircraft into a steeper dive, must have been surprised to find he was going to be put on trial, with prosecutors demanding a custodial sentence, though they subsequently relented. Prosecutors pursued the air traffic controllers, with at one time a hundred civil service demonstrators protesting outside a court in their favor. Various trials and appeals concerning the air traffic controllers have continued year after year, with one even in 2008.

TCAS has been very successful in preventing midair collisions, and as usual it is surprising the pilots, who originally thought it would be a nuisance, delayed its introduction for so long. It also alerts pilots to the presence of other aircraft before specific mandatory action to avoid it is ordered, but "A" for "alert" is omitted.

There are now improved versions of TCAS that are able to handle separation in the horizontal plane as well as the vertical and tell aircraft to go left or right as well as up or down. Ideally, it should include a program to prevent an aircraft being ordered to fly into the ground should an incident happen low down. However, some projects designed to incorporate this have been canceled because of cost.

Distraught Father Assassinates Controller (Lake Constance, 2002)
One of the pilots obeyed the controller instead of the mandatory TCAS

> The controller issued last-minute instructions to ensure the two aircraft did not collide, but with collision possible, TCAS in the aircraft gave its own perhaps more appropriate orders. One pilot obeyed, and the other followed the controller with disastrous results.
>
> *Bashkirian Airlines Flight 2937/DHL Flight 611, July 1, 2002*

One evening a middle-aged stranger came to the suburban residence near Switzerland's Zurich Airport where that air traffic controller, Danish-born Peter Nielsen, lived. (He had recently returned from medical leave to assume other duties at Skyguide, the Swiss air traffic control company.) After a brief exchange of words at the front door, the unknown visitor proceeded to stab Nielsen to death in full view of his wife.

A senior police officer soon dismissed the notion that it had been a hit man, saying, "Hit men don't get emotional and they don't use a knife." Soon it was realized that the middle-aged man must have been the father of one or more of the many children killed 623 days earlier in a midair collision over Lake Constance, lying between Switzerland and Germany and touched at its foot by Austria. Many could sympathize but not condone.

The terrible collision had occurred at 35,400 feet in a virtually empty sky at 23:35 local time on July 1, 2002 and had been between a Tupolev-154 airliner with sixty-nine people, including fifty-two children, on board and a DHL cargo plane with just two pilots. Everyone on the two aircraft lost their lives. Among the dead were the wife, son, and daughter of forty-eight-year old Viktor Kaloyev. Those who knew Kaloyev said he had been implacably distraught since losing everything he had to live for, and, indeed, it was he who murdered Nielsen.

A court subsequently sent Kaloyev to a psychiatric hospital. However, was his act of vengeance misplaced? Was he right to focus on the air traffic controller? As is usual in an air accident, a whole series of unfortunate events and failures, on their own of little consequence, led to the midair collision for which Viktor Kaloyev held Nielsen responsible.

Five and half minutes before the collision, Nielsen had authorized the DHL cargo aircraft, flying north, to climb to thirty-six thousand feet to save fuel. Meanwhile the TU-154 was flying in a westerly direction at the same altitude, meaning they were at right angles.

Their speeds and relative positions were such that they might collide or, rather, risked a lack of separation. Air traffic controller Nielsen would usually have realized this quite soon.

With his companion taking a rest, as allowed by company regulations, outside the control room, Nielsen was handling all the air traffic in the Zurich area and had to watch over two screens and slide his chair between them. This would not have been too difficult if traffic had been virtually nil, which was usual at that time of night, or if he had been able to devote all his time to controlling the air traffic rather than placing calls through public phone lines to which he was not accustomed and did not have the special features mentioned.

Also, in the roughly five minutes prior to the collision, he was responsible for four aircraft on one frequency and a fifth on another. Then, moments before the collision, a sixth aircraft called in.

The air traffic control system was in fallback mode for servicing, which meant aircraft needed to be farther apart than usual. In addition, the short-term conflict alert system, which would have warned Nielsen of an impending collision, was not working. It would have given an audible warning and shown the echoes of the aircraft in red. It seems Nielsen was not aware that this was not functioning.

Furthermore, management had allowed engineers to switch off the special phone system, so they could perform overnight maintenance. Though based on quite an old analogue system, it used dedicated lines to link the neighboring control centers, enabling the automatic rerouting of calls if one line failed, and included a priority ringing system so a controller could tell if a call was especially urgent, as in the case in question.

Using unfamiliar public phone lines, Nielsen had wasted considerable time just before the collision contacting a nearby German airport to hand over a delayed incoming Airbus. Worse still, the phone outage meant a controller one hundred miles away in the German Karlsruhe Center who was aware of the possibility of a conflict could not get through to warn him of the danger, despite eleven desperate attempts. Nielsen finally also noted the potential danger of collision, and about forty-four seconds before impact he told the TU-154 to descend immediately, as there was crossing traffic.

The DHL 757 continued to implement the TCAS instructions to descend, but the TU-154, after seeming to have hesitated, obeyed Nielsen instead and also descended. Twenty-two seconds before impact, the TCAS in the DHL 757, sensing the increasing danger, ordered the DHL to increase its rate of descent.

Eight seconds before impact, the TCAS ordered the TU-154 to further increase its rate of climb, though it was descending.

For the occupants of the TU-154, including the many children, death was surely relatively quick. That was not true for the DHL pilots, as their aircraft subsequently flew on relatively intact for some distance and crashed eight kilometers away from the location of the debris from the TU-154, which had broken up in the air. The DHL cockpit voice recorder features the voices of the pilots even after the impact between the two aircraft.

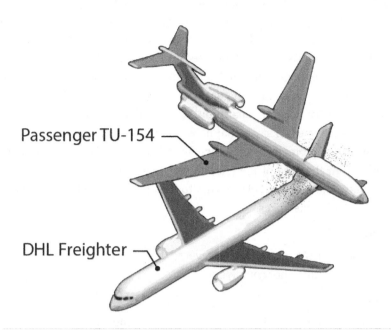

Passenger TU-154

DHL Freighter

Figurative portrayal based on diagram provided by Germany's BFU.

In view of the above, it is difficult to blame Nielsen, who nevertheless found the shock and sense of responsibility difficult to bear. Interviewed by a German magazine two weeks after the accident, he expressed his sorrow but said he was part of a system and networks with many interrelated features even though, as an air traffic controller, he was responsible for ensuring accidents didn't happen.

The presence of an extra captain of very great seniority in the jump seat of the TU-154 would seem to have been the tipping factor as regards the failure to obey the TCAS instructions. Apparently, he stopped the first officer doing so, which partly explains the hesitation just mentioned. The senior captain in the jump seat may have acted thus because he was from the old school and had considerable experience of flying in the Soviet Union, where TCAS is little used and obeying the air traffic controller would be the norm and built into his psyche. Be that as it may, psychologically speaking, after starting to take evasive action by going one way, it is not easy to immediately readjust and switch to doing the opposite.

That said, one must be very unlucky to collide with another aircraft at 11:35 p.m., when there should be a lot of empty space in the sky. One must be even unluckier to collide at right angles, where the horizontal separating effect of any variation in relative speeds is maximal. Some people partly blame the Swiss Air Force for appropriating so much space for itself that commercial airliners are funneled through the rather limited space over Lake Constance.

Though there is a lot of empty sky at night, cargo aircraft, by the very nature of their tasks, often do fly at night. They are especially vulnerable to collisions, because radar systems are taken out of service then for maintenance and everyone tends to be less alert, not only because it is nighttime but also because there is usually not enough action to keep people on their toes.

Sadly, had Nielsen been incompetent or careless and had not finally noticed the potential conflict (lack of separation), the accident would probably never have happened. All one can say is that he was largely a victim of circumstances, since so much equipment was either in fallback mode for servicing or unavailable, as in the case of the telephones. Regulations and manning levels should not have made it possible for him to be left to cope with everything on his own in such circumstances.

Safer Today

Newer versions of TCAS automatically reverse their commands should they find that one of the aircraft is not complying.

Rate of Descent

It is likely the tail-engine TU-154 trijet, like the Boeing 727 and de Havilland Trident, could achieve exceptionally rapid rates of descent and the TU-154 was dropping must faster than the Boeing 757.

Chapter 5
GROUND COLLISIONS

Worst-ever Multi-Aircraft Disaster (Tenerife, 1977)
The KLM captain's impatience was a key factor

> A whole series of events led towards this terrible disaster. Take any one
> away and it would not have happened.
>
> *KLM Flight 4805/Pan Am Flight 1736, March 27, 1977*

One Sunday afternoon in March 1977, a terrorist bomb and the possibility of another had made the authorities temporarily close Las Palmas Airport in the Canary Islands. Most of the incoming flights were diverted to Los Rodeos, a less important airport on nearby Tenerife, turning that relative backwater into a hive of activity. Aircraft languishing there waiting for Las Palmas to reopen were blocking key taxiways, including the normal route for taxiing to the far end of the runway for takeoff.

Though it had a good, long runway, the airport's ground handling facilities were not designed for aircraft as large as the Boeing 747. As a result, a Dutch KLM 747 and a Pan American 747 parked on the apron were taking up so much of the available space that the Pan Am 747 would not be able to squeeze past the KLM to get out. They were both waiting to resume their journey to Las Palmas.

The KLM 747 had just come from Amsterdam, a four-hour journey, with a group consisting mostly of young Dutch tourists. The 248 people on board included 48 children, 3 babies, 2 pilots, a flight engineer, and 11 cabin crew.

The Pan Am 747 had come from Los Angeles, with a stop in New York for refueling and a change of crew before the eight-hour transatlantic flight to what should have been Las Palmas, where the mostly elderly passengers were to join a cruise liner. The 396 people on board included the 2 pilots, a flight engineer, and 2 company employees in the cockpit jump seats. The 747, *Clipper Victor*, had a dent in its nose made by a champagne bottle striking it to celebrate the inaugural commercial Boeing 747 flight, from New York to London on January 21, 1970. It was one of the first jumbos.

The KLM 747 also supposedly had some fame associated with it, in that a photo of its Dutch captain, van Zanten, was being used in KLM's advertising material, including that in the in-flight magazine the passengers must have been perusing during the long delay. Much has been made of this publicity photo, with suggestions that van Zanten was a self-important stuck-up prig—a captain-of-the-*Titanic*-like figure—as maintained by the Spanish side. The author, as surely were many others, was seduced by this simplistic portrayal until he read *Disasters in the Air*, by Jan Bartelski, a pilot with KLM, who held important posts with the International Federation of Airline Pilots Associations (IFALPA).

Bartelski's role at IFALPA, where admittedly defending the interests of pilots would be paramount, is reflected in his very pro-pilot approach to accidents, but many of the points he makes cannot be lightly dismissed. His thirty years at KLM did also give him some inside knowledge.

According to Bartelski, van Zanten was rather a serious and introverted man, and the only reason for the publicity department using his photo was likely to have been his availability for photo sessions. As a training captain, van Zanten was usually freely available at the home base; other captains would be either away flying aircraft or resting at home. Indeed, the photo of van Zanten included in Bartelski's book gives the impression of a rather accommodating person.

Again, he says van Zanten was not as senior as suggested. It is well known that, because of the influence of powerful unions, much at the major airlines depends on seniority (rather than ability). KLM had promoted van Zanten out of turn to the rank of captain when the captain in charge of 747 training retired, and consequently the Dutch pilots' union insisted he only fly routes when no other captain was available. This limited his amount of line experience even more than normal for a training captain.

First Officer Meurs, also a captain, seconded Captain van Zanten. Though very experienced on other aircraft, Meurs had only flown ninety-five hours on the 747, having shortly before converted to the aircraft under the instruction and authority of Captain van Zanten himself. While this supposedly made Meurs particularly deferential toward him, Meurs was an outspoken and extrovert type, and van Zanten followed his advice at several points while proceeding to the end of the runway for takeoff. Assisting them was Flight Engineer Schreuder.

In charge of the Pan Am 747 was Captain Victor Grubbs, a fifty-seven-year-old with over twenty-one thousand hours of piloting experience. First Officer Bragg and Flight Engineer Warns were with him on the flight deck, together with two company employees in the jump seats.

The waiting dragged on all through the afternoon, with the KLM crew becoming increasingly worried that their permitted flying time would expire. If it did, they would have to stay overnight at Los Rodeos or Las Palmas and fly back to Amsterdam the following morning, upsetting any plans they might have for that day. In the peak season it would be virtually impossible to find overnight accommodation for the passengers.

A number of potential accidents due to fatigue had made the Dutch authorities establish strict legal limits for hours of duty, removing the discretion the captains previously had had in this regard. The regulations were so complex that captains often had to ask the airline's operations center for a ruling to cover themselves. Contacting their HQ by high-frequency radio, the KLM crew was relieved to have a ruling that it would be okay provided it got away before a certain time later that evening.

Finally, with news that Las Palmas Airport had reopened, other aircraft began to take off for that airport, which made Captain van Zanten realize that refueling delays at Las Palmas might jeopardize their chances of getting off from there before their permitted flying hours were up. He therefore opted to refuel at Los Rodeos instead, with enough fuel to continue on to Amsterdam from Las Palmas.

The Pan Am crew found it could not taxi to the runway with the huge KLM jet blocking its path. After the long wait for flights to Las Palmas to be authorized, it was not at all happy to be told by the KLM crew—with no hint of apology—that refueling would take some thirty-five minutes. On hearing that, the Pan Am first officer and engineer got out and paced the ground to see whether there just might be room for them to squeeze by. Visibility was falling, and when refueling had finally finished and the two aircraft could leave the apron for takeoff, it was down to three hundred meters.

Los Rodeos Airport, at an altitude of some two thousand feet, is subject to cloud rather than typical lowland fog. However, it can be quite worrying, as visibility can be very poor at moments and improve or worsen suddenly.

Since the taxiways parallel to the runway leading to the holding point for takeoff were blocked by parked aircraft, the air traffic controller told the two aircraft to enter the runway at the other end and back-taxi up it. The KLM 747 in the lead was to go right to the far end, perform a U-turn, and wait for permission to take off.

The Pan Am 747 was told to follow the KLM 747 but turn off at the third taxiway on the left, to be out of the way and give the KLM a clear runway to take off. As Los Rodeos was merely a diversion airport for emergencies, the Pan Am crew only had a small-scale plan showing the layout. What is more, the taxiways did not have signs identifying them. For the pilots looking at their plan, the obvious exit to take seemed to be the fourth taxiway, since it turned off at a comfortable 45 degrees, with another easy turn onto the main taxiway after that. The third exit, on the other hand, would require two difficult 135-degree turns and initially take them back toward the terminal.

Unsure whether the controller had said the first or third exit, the Pan Am pilots had asked him to repeat his instructions. The controller did this, saying, "Third taxiway. One, two, three, third."

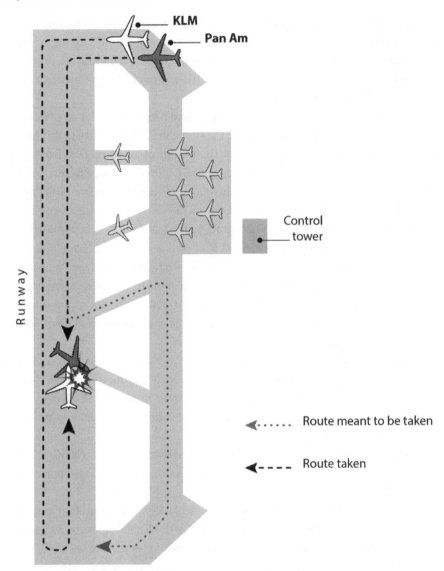

Tenerife's Los Rodeos Airport, now called Tenerife North Airport

On reaching the third taxiway, the Pan Am crew was loath to take it in view of those awkward 135-degree turns, not to mention the additional distance involved, and proceeded slowly onward to the fourth turnoff. Anyway, the controller had told them to report when they left the runway, and it would be inconceivable that he would give the KLM 747 permission to take off before confirming it was clear.

Some commentators have suggested the controller asked the Pan Am to make the awkward 135-degree maneuver because he had little experience of handling aircraft as large as the 747. However, as the Dutch investigators subsequently showed, this 135-degree turn was by no means impossible. Oddly, no one seems to have suggested that consciously or out of habit, the controller might have been trying to keep the two aircraft well separated in the fog.

Had the Pan Am 747 been instructed to take the fourth and last turnoff and missed it in the poor visibility, it would have come face-to-face with the KLM 747, with hardly any room to maneuver at best. As aircraft cannot go backward on their own, a tractor might have been required to separate them, and this would have meant closing the runway and further delays for all the other waiting aircraft.

On reaching the end of the runway, van Zanten duly made the tight 180-degree U-turn to face back down the runway ready for takeoff, directly facing the Pan Am 747 trundling toward him. The poor visibility meant the aircraft were out of sight of each other and out of sight of the control tower.

Van Zanten spooled up the engines a little and allowed his giant aircraft to inch forward. According to the Dutch comments made later, this application of a small degree of power was a normal maneuver to check the functioning of the engines and not an indication he was about to initiate takeoff. Whatever its significance, First Officer Meurs said:

> Wait a minute; we don't have an ATC clearance.

Van Zanten replied:

> No, I know that. Go ahead and ask.

He then applied the brakes fully.

Meurs then called the control tower, saying:

> KLM4805 is now ready for takeoff—we're waiting for our ATC clearance.

The tower replied:

> KL4805. You are cleared to the Papa beacon. Climb to and maintain Flight Level 90. Right turn after takeoff. Proceed with heading 040 until intercepting the 325 radial from Las Palmas VOR.

Van Zanten apparently assumed this meant they were also cleared for takeoff, and as Meurs began confirming the ATC clearance by reading it back to the controller, he released the brakes and this time advanced the throttles to give high, but not maximum, takeoff power.

With Meurs occupied with confirming the ATC clearance, van Zanten then said to Flight Engineer Schreuder:

> Let's go! ... Check the thrust.

The first officer, though preoccupied with reading back the ATC clearance to the tower, added a phrase that could have saved them. However, it failed to do so, because he used a Dutch idiom that in English had a different sense:

> Roger, sir, we are cleared to the Papa beacon, Flight Level 90, until intercepting the 325. We're now at takeoff.

Meurs, using the Dutch idiom, meant they were actually taking off, and, indeed, when the two words "at takeoff" left his mouth, the aircraft had already been gaining speed for six seconds. The controller, like anyone used to speaking English, naturally assumed "at takeoff" meant the KLM 747 was at the takeoff position at the end of the runway, awaiting clearance to proceed.

In a run-of-the-mill fashion, the controller then said:

Okay. Stand by for takeoff. I will call you.

However, the crew of the Pan Am 747, feeling vulnerable and fearful that the KLM 747 might take off at any moment, interrupted the tower's reply to the KLM with the frantic words:

We are still taxiing down the runway!

The simultaneous radio transmissions produced a radio squeal called a heterodyne, with the result that the KLM 747 only heard the single word "okay" from the controller, and not the rest of the controller's sentence or the words from the Pan Am 747 regarding the fact they were still taxiing down the runway. However, the tower did hear the Pan Am transmission and, thinking the KLM 747 was still holding, replied:

Roger, Pan Am 1736, report runway clear.

Pan Am:

Okay, will report when clear.

Controller:

Thanks!

This exchange, which took place when the KLM jet had already been accelerating for twenty seconds, was being followed by KLM Flight Engineer Schreuder, but not by the two pilots, who were concentrating on the tricky takeoff in the fog.

Schreuder expressed his concern to his colleagues somewhat hesitantly in Dutch, with words to the effect of:

Did he not clear the runway then?

Not understanding what he meant, van Zanten said:

What did you say?

The flight engineer clarified his question:

Did he not clear the runway—that Pan American?

Van Zanten and Meurs very affirmatively replied:

Yes, he did.

Having expressed his doubts deferentially as a question rather than as a strong positive statement that the Pan Am 747 was possibly still on the runway, the flight engineer was wrong-footed and merely repeated the questioning statement.

Captain Grubbs, in the Pan Am 747, not knowing that Dutch-speaker Meurs had meant they were actually taking off when he had said "We are at takeoff" nevertheless sensed van Zanten was extremely anxious to get away and might take off at any moment.

Addressing his crew, he exclaimed:

Let's get the f**k out of here!

The others agreed, commenting sarcastically about van Zanten keeping them waiting for so long (while refueling) and then developing a sudden urge to get away.

Seconds later the crew of the Pan Am 747 sighted the first glimmer of the KLM 747's landing lights piercing the fog some 580 meters away. Those lights grew brighter and brighter as Captain Grubbs pushed his throttles fully open to try to pull out of the way. He did not stand a chance of succeeding in the eight seconds that remained before the inevitable impact, since his engines needed five of those seconds just to spool up enough to begin accelerating the massive jumbo, with its considerable inertia.

Grubbs and his colleagues could only pray that van Zanten would manage to get off the ground in time to pass over them. One of them even muttered:

Get off! Get off!

It was only as his KLM 747 attained the takeoff decision speed, V_1, of about 140 knots that Captain van Zanten in turn saw the Pan Am 747, which by then was less than five hundred meters ahead of him. Even though he had not quite reached rotation speed, V_R, he yanked the control column so far back that the rear underside of his aircraft made a long gouge in the concrete runway.

Lifting off at that slow speed with the added burden of the 55,500 liters of fuel he had cannily just loaded was never going to be easy. Even so, his aircraft heaved itself up sufficiently for the nose wheel to pass over the Pan Am craft. All to no avail, as the number four engine pod suspended below the right wing struck the Pan Am 747's humped back just behind the flight deck, crushing the first-class upper passenger cabin behind the pilots and ripping off the flight deck roof. When the first officer almost instinctively reached up for the engine fuel shut-off levers, he found nothing there!

Though van Zanten had placed his aircraft in a sharp nose-up attitude, its trajectory at that slow speed and great weight was virtually horizontal. In consequence, the massive main landing gear wheel bogies dangling well below the nose wheel were bound to slam into the roof of the Pan Am's main cabin and crumple it as they rolled over it. Again, that sharp nose-up attitude meant the tail section of the KLM was lower still, so that it in turn sliced neatly between the double tracks of carnage created by the landing gear bogies.

With fire breaking out in both aircraft, the KLM 747 continued its trajectory for a second or two before crashing down onto the runway, with engines and other pieces falling off. After the KLM 747 had slithered to a halt some 450 meters from the point of collision, its fuel tanks exploded, engulfing the fuselage in fire. None of the 248 people inside even managed to open a door or emergency exit, let alone jump out.

As already mentioned, passengers usually have a ninety second time frame to escape when an aircraft on the ground is seriously on fire. In the case of the Pan Am 747, there was only a minute. In the main cabin, where many of the passengers were quite elderly, those still alive on the side opposite the one that received the full impact of the KLM's main wheels had their escape hampered by debris, and few managed to get to the exits.

Helped by the collapse of the floor of the upper-deck first-class cabin behind the flight deck, the three aircrew and the two company staff with them in the jump seats were able to escape, along with those in the lower-deck first-class section below. Those in the upper-deck cabin that had received the direct impact of the KLM's right engine pod did not stand a chance.

A number of the Pan Am passengers had to jump to the ground from a considerable height, sustaining further injury. In fact, nine out of the seventy people who did get out alive subsequently died from their injuries, making the final death toll for that aircraft alone 335.

Adding 335 to the 248 dead in the KLM 747, the total death toll was 583, making it the worst-ever aircraft accident if one excludes 9/11, where, in terms of occupants of the four aircraft, less than half that number were killed—and that was not an accident.

Conclusion

Had the transmission of the Pan Am saying they were still taxiing down the runway not overlapped that of the tower to the KLM 747 saying, "Okay. Stand by for takeoff. I will call you," so that the KLM only heard "Okay," the collision might never have happened. Also, had the first officer not used the confusing phrase "We are at takeoff," instead of "We are taking off," it might have been averted.

While it has been generally assumed that using the phrase "We are at takeoff" to mean "We are taking off" is quite usual for a Dutchman speaking English, it is possible that the first officer, who was quite used to speaking English, had reverted to using Dutch idiom under the stress engendered by having to read back the airways clearance while the impatient captain was initiating a tense takeoff in fog. One of the US investigators said the voice of the first officer had changed around that time, as if he were stressed or worried. Having stopped van Zanten taking off once because they had not received their airways clearance, perhaps he hesitated to do so a second time and thought he would cover his back by announcing they were taking off—unfortunately, with the wrong choice of words—in which case the controller could tell them to abort if dangerous.

The fact that the KLM aircraft declined, on the grounds of being too busy, to receive that airways clearance earlier meant that Captain van Zanten was preoccupied with handling the aircraft, and the first officer was preoccupied with reading back the clearance.

Thus, van Zanten was effectively carrying out a tricky takeoff in fog with help from the flight engineer rather than the first officer.

Flight Engineer Meurs had said, "Did he not clear the runway—that Pan American?" but with the KLM 747 gaining speed he would have had to be much more forceful to get both pilots to back down. Both pilots had unhesitatingly said the Pan Am had.

Underlying the disaster was van Zanten's impatience, arising from his fear that they would exceed their duty time and have to stay the night in Las Palmas.

The Pan Am crew might have made more of the fact in their exchanges with the controller that they were wondering whether they were taking the appropriate exit. However, they did twice indicate they were still on the runway, with their warning on the second occasion doing more harm than good. Perhaps they began to think better of what they had done, and that is why the captain said, "Let's get the f**k out of here!"

The Dutch investigators said the air traffic controller could have performed better and that the sound of a football match being broadcast in the background suggested he might have been distracted from his tasks. Had there not been a football match playing on a radio in the control tower, and had it not been a critical stage in the match, it is possible the control tower would have made an intermediate check of the Pan Am 747's progress.

Another factor in the disaster may have been that the controllers had been having a grueling day dealing with the many diverted flights, and things were finally easing up, which is just the time accidents tend to happen, as people are a little less sharp once the pressure is off. According to Bartelski, the Spanish authorities very likely sought to protect their controllers by only providing investigators with a poor copy of the control tower recording tape, which could not be synchronized with the cockpit voice recorder (CVR) tapes in the two aircraft.

One point made by Bartelski and not generally known is that following the Tenerife accident, all KLM pilots were made to undergo strict practical hearing tests in addition to the usual audiograph tests, and as a result two older captains had their licenses withdrawn. Because of medical confidentiality, Bartelski says, there is no direct evidence that a hearing deficit should be considered as a likely contributory factor.

However, he points out that if van Zanten had such a hearing deficit, he would have been more likely to miss the word "report" in that possibly lifesaving transmission from the tower to the Pan Am 747:

Roger, Pan Am 1736, report runway clear.

747 Takes Off on Blocked Runway in Heavy Rain (Taipei, 2000)
Wish fulfillment—you see what you expect to see

> Singapore Airlines (SIA) has one of the best reputations in the industry, with a young fleet and superb flight attendants, who see their acceptance by the airline as a qualification for a good marriage. The news of an absurd disaster at Taipei, Taiwan, in which one of its aircraft took off in atrocious weather on a disused runway with construction equipment parked in the middle came as a great shock.
>
> *Singapore Airlines Flight 006, October 31, 2000*

An English edition of the Japanese *Yomiuri* newspaper wrote, "In what is probably a world first, the operators of Tokyo's Narita Airport have decided to use paint and nets to camouflage part of the new 2,180-meter runway to keep aircraft from accidentally landing on an unused section. The camouflage is designed to make the superfluous section of the runway look like grass."

Why should Tokyo's International Airport go to so much trouble and expense when all they had to do was to use the traditional X (no entry) signs to denote that section of runway was out of use? According to officials at the airport, one reason was an incident in 2000, when a Japan Air System jet accidentally landed at Haneda domestic airport on a new but unused runway despite it having those X's. Another reason may have been the death trap at Taiwan's Taipei Airport into which a Singapore Airlines Boeing 747 was lured on the stormy night of October 31, 2000.

That night one of the worst typhoons that Taiwan was to experience in recent years—dozens of people were killed—was on its way and already bringing torrential rain and gusty winds. If SIA transpacific flight 006, bound for Los Angeles, did not get off quickly, company regulations might forbid it from doing so on account of the increasing crosswind component. Canceling the departure would mean waiting until the crew had its mandatory rest and leaving the next day.

That said, the captain did not seem to be overly hurrying things along to get away before conditions deteriorated further. He apparently told the catering people, who were having trouble loading the victuals due to the terrible weather, to "slow down and take their time." He did, however, take extra precautions, one of which—or as some might say, both of which—were to prove fatal. Firstly, he decided to handle the takeoff himself instead of letting the first officer do it, as had been planned. This meant he would be preoccupied with concentrating on the physical handling of the aircraft rather than on the overall situation.

Secondly, and this was certainly the most fatal decision, instead of choosing Runway 06, on the southern side of the terminal building, which SIA invariably used in view of its proximity to its boarding gate, he opted to use the "safer" Runway 05L to the north. This slightly longer runway should have given a greater margin of safety in the wet and slippery conditions and had the advantage of being a Class II runway, allowing operations in poor weather.

While the captain and first officer had quite often flown out of Taipei, they had not used Runway 05L, on the northern side, for two or three years. The layout they were used to using to the south was much simpler, consisting of just a taxiway and a runway parallel to it, instead of a taxiway, a minor runway, and the major runway, as on the north.

To reach the main 12,008-foot Runway 05L, they would have to taxi past the end of the shorter 05R, which had just been taken out of service (with that end used as a supplemental taxiway), and the middle-to-far end, used to park construction equipment, including concrete blocks.

The trap into which they were to fall was a perfect line of closely spaced green taxi lights leading in a neat curve to disused Runway 05R. In the heavy rain these were very inviting, especially to pilots who were used to Singapore, which has a system whereby controllers turn taxiing lights on and off to show the pilot where to go, so all he or she has to do is to *follow the green*. On the other hand, at Taipei the green taxiway lights leading straight ahead that they should have followed to take them to the operating runway were few and far between, with one not working at all and another quite dim.

Despite the rain and slippery conditions, the SIA 747 captain negotiated the taxiways to the other side of the airport and made the ninety-degree right turn to enter taxiway N1, which would first take them past the end of the narrow disused Runway 05R and then on to the wide, active Runway 05L, which was quite a bit farther on.

Of course, had visibility been better, the crew of SQ006 would have been able to pick out the bright lights of the operating runway straight ahead. Indeed, the pilot of a cargo plane belonging to another company that took off eight days after the SQ disaster, but in not quite such poor visibility, said he too had been tempted to follow the bright bright-green lights leading to the disused runway. He had not done so because in the better conditions the bright lights of the operating runway ahead were visible straight ahead.

As every motorist knows, distances seem much greater when moving very slowly in poor visibility, and the crew perhaps thought they had come a fair distance when they mistakenly followed those beckoning green lights and lined up for takeoff from the disused runway. All three pilots assumed it was the runway.

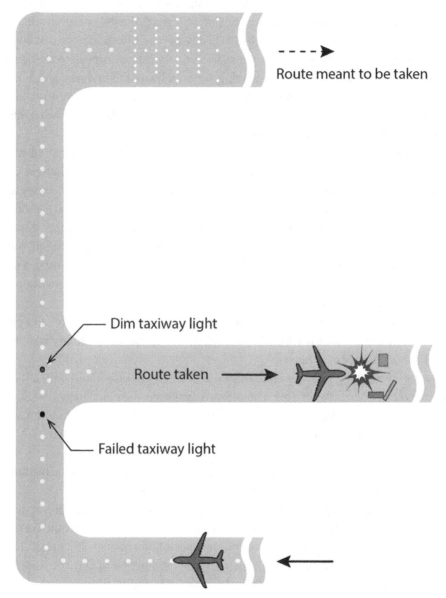

Route meant to be taken

Dim taxiway light

Route taken

Failed taxiway light

SQ006 had taxied from southern side of the airport from where it usually took off.

There was a delay of about a minute before they started their takeoff roll, during which it seems they were conscientiously calculating the crosswind component of the headwind to make sure they were not breaking any company rules. Had they not been so preoccupied with the letter of the law, they might have thought it odd the center lights were green and not white, as they should be on an operating runway. It seems—though it is not 100 percent provable—

that there were no runway edge lights. The captain later said he was "80 percent sure" he saw the runway edge lights; the first officer said he did not see any lights other than the center lights. The disused runway was fifteen meters narrower than it should have been, but perhaps the very absence of runway edge lights made this less obvious. Lights shining on raindrops might have given the impression they were difficult-to-see edge lights. On an operating runway there would normally, but not necessarily, also have been the bright touchdown lights, indicating where aircraft should land.

Something the crew did notice and that could have saved the situation was the failure of their para-visual display (PVD) to activate (to unshutter) on turning onto the "runway." The equipment sounds complicated but is merely an instrument with a horizontal rod somewhat like the sign outside a barbershop. A computer senses deviation from the runway centerline and indicates this to the pilots by varying the speed and direction of rotation of the striped rod, producing the optical illusion that the lines are moving to the left or right. This instrument can be helpful for keeping on the runway centerline when visibility is extremely low, even down to fifty meters.

Though this helpful piece of equipment was not working, the pilots did not worry any more about it when they found, on lining up, that visibility was not bad enough for them to need it.

After rechecking that the crosswind did not exceed authorized limits, they commenced their takeoff roll. The aircraft gathered pace normally, but four seconds after reaching the takeoff decision speed, V_1, of 142 knots, the captain uttered an expletive and the two words

Something there!

He tried to lift off, but since the aircraft was full of fuel it was too late. Instants later the aircraft's nose wheel hit the first concrete block.

Then, as the huge fuel-laden aircraft was trying to rise, contractors' equipment, including two excavators and a bulldozer, ripped into its hull like the iceberg into the *Titanic*, by which time the speed had reached 158 knots. The 747 fell back, crashed into more construction equipment and concrete blocks, and broke its back. With fire spreading from the left wing, it exploded and ended up, perhaps fortunately, in three pieces.

Some of the passengers and crew were able to scramble out of the breaks in the fuselage and certain exits. Some exits were unusable due to raging fires outside, while the powerful wind made it difficult to open others—some have suggested the Asian flight attendants were not strong enough. In cases where exits could be opened, the escape chutes often did not deploy properly, with one blowing back into the aircraft and pinning down a flight attendant, others flapping around, and some catching fire.

Nevertheless, the survival rate was 54 percent. Of the 179 (159 passengers and 20 crew members) on board, 96 survived and 83 died. Most of the survivors were injured in some way, but as in many terrible disasters, some—16 in all—escaped unscathed. The pilots were among the survivors, and someone saw one of them assisting a flight attendant who had lost a leg.

This was a big shock for SIA, an airline highly respected throughout the industry for the expertise of its management and its efficiency. Except for a crash involving its subsidiary, SilkAir, where there were suspicions that the pilot crashed the aircraft on purpose in a kind of suicide, this was its first fatal crash in its twenty-eight-year history. SIA was also famous for having one of the youngest fleets, at a time when British Airways was still flying some twenty-seven-year-old Boeing 747s.

Many motorists have taken a wrong turning on badly signposted roads in heavy rain, but how could highly trained pilots do such a thing? The captain had 11,235 hours of flight time.

After some bad PR due to their Los Angeles office prematurely reporting no casualties to news organizations phoning from around the world, SIA won plaudits, at least locally, by admitting responsibility, with its chief executive, Cheong Choong Kong, saying at a press conference in Singapore after returning from Taipei, "They are our pilots. It was our aircraft. The aircraft should not be on that runway. We accept full responsibility."

In addition, the airline offered to pay $400,000 to the families of those killed and to compensate generously the survivors for their expenses and so on.

Points Related to this This Disaster

A feature of the Boeing 747 that many must have noticed is that the pilots are very high up off the ground, rather like the drivers of some SUVs looking down on others from their lofty perch. The reason for this was that the design for the 747 was originally based on one for a military freighter (the C5A) that could be loaded from both front and rear with large items, such as vehicles.

This height off the ground may have made the task of those SIA pilots seeing the markings on the ground even more difficult. Interestingly, the Airbus 380 design for a true double-deck superjumbo has the cockpit set lower down.

It was quite clear that the taxiing lights at Taipei's international airport left much to be desired. The airport did not have ground radar, which would enable controllers to see the location of aircraft in bad visibility, but Taipei airport is not alone in this regard, and its absence was not contrary to international regulations.

When the Taiwanese authorities finally produced their 508-page report into the disaster, it raised so many points that it was difficult to see the wood for the trees. In addition, many of these points seemed to lay the blame on the SIA pilots rather than the airport. The Taiwanese allowed the Singapore side to participate in the investigative phase, but not in the analysis phase. The Singapore side was unhappy about that and produced a commentary rebutting many of the Taiwanese side's conclusions.

The Singapore view was that it was a systems failure, with many factors involved. The Taiwanese view was that while the airport may have had its shortcomings, the pilots should have, and would have, realized they were on the wrong runway if they had taken more care and paid more attention. Also, according to them, SIA had not trained the captain in poor visibility taxiing.

In some respects, it is paradoxical that the crew may have made the mistake because they were being extra cautious, checking and rechecking the crosswind component to make sure they were allowed to take off under company regulations. Unfortunately, their checklist did not include checking repeatedly, and by all possible means, that they were on the correct runway. One could say that the crew should have been trained to consider whether failure of the para-visual display (PVD) to activate (to unshutter) on turning onto the "runway" should be seen as a reason to think they might not be on the active runway, but probably most airlines would have failed to think of that.

One point that might seem strange to some was the Taiwanese Chinese investigators asking the pilots whether they had any cultural problems regarding their relationship with other crew members. This really alluded to the crew being partly Malaysian and partly Chinese.

SIA later dismissed the captain and first officer, after having overtly given them full support during the inquiry. It did not dismiss the third officer even though his facile explanation as to why the para-visual display was not working made them miss an opportunity to realize they were not on an active runway.

Some say this accident was a classic case of scenario fulfillment in that people see what they expect to see.

UK Skyvan Pilots Unable to Follow French ATC (CDG, 2000)

Freighter entered middle section of the runway and was swiped by French airliner needing its full length

> Shortly before the accident described below, Air France's safety department had recommended its pilots consistently use English in the interest of safety. Not only did some pilots oppose this, but politicians and the press said it threatened the role of the French language in the world.
>
> *Air Liberté Flight 8807/ Skyvan freighter, May 25, 2000*

It was 02:50 and, with very little traffic, the controllers at Paris's Charles de Gaulle Airport were not using the southern control tower. In addition to the usual cargo flights, there was the unusual problem of coping with some twenty thousand Spanish football fans returning to Spain after watching a match between Valencia and Real Madrid in Paris.

Speaking in French, a controller cleared an Air Liberté MD-80 charter flight carrying 150 of those fans for takeoff from the far end of Runway 27. Speaking in English, another controller told the English crew of a Shorts 330 Skyvan cargo plane to hold before entering the second section (3,300 feet farther down) of that same runway, since the small twin-turboprop cargo plane did not need as much runway as the airliner to take off.

The controller had told the crew of the Skyvan that they would be number two. The cockpit voice recorder (CVR) of the cargo plane showed that its English captain was wondering who number one was and whether it was the Boeing 737 that had just landed. No doubt thinking that was the case, he moved onto the runway just as the MD-83 (on Section 1 of the runway) was gathering speed on its takeoff run.

In thirty-three seconds, the MD-83 attained V_1 and was just about to reach rotation speed, V_R, when its captain saw the cargo plane nosing onto the runway. With only five seconds remaining before certain impact, he proceeded to abort his takeoff—something virtually never done after attaining takeoff decision speed, V_1. However, it was fortunate that he did, as a third of his left wing was shorn off as it struck the nose of the Skyvan, killing its first officer[1] and seriously injuring its captain. Fortunately, with plenty of runway left at such a large airport, the MD-83's crew were able to bring their damaged aircraft to a halt without any loss of life.

The subsequent inquiry mentioned poor coordination between the controllers and that the rainy conditions, light pollution (caused by floodlights and ten vehicles with rotating lights involved in construction work near the threshold) made observation more difficult for the controllers. In addition, the MD-83 had delayed its takeoff because of an autothrottle problem, and this made the situation evolve rather differently from what the controllers had anticipated. This was perhaps one of those dangerous halfway situations where conditions were not good enough to see perfectly but not bad enough to use all means, including ground radar, to check what was happening.

In addition, it would have been a quiet time, with everyone rather relaxed. Interestingly, psychologists note an inverted U-curve for the effect of stress:

1. No stress; people are not switched on.

2. Medium amount of stress; they are at their best.

3. Too much stress; their interactive skills diminish.

Even so, the key finding of the inquiry seemed to be that if communication between the French aircraft and the control tower had been in English, the crew of the cargo plane would have known the MD-83 was about to take off and would not have entered the runway.

Readers will note that the accidents in this chapter have a number of common threads.

---

[1] Traditionally, the captain sits on the left and the first officer (copilot) on the right. This is because holding patterns at airports usually involved left-hand turns and this permitted the captain to see where they were going by looking out of his side window.

Cessna Crosses Runway in Path of SAS MD-11 (Milan, 2001)
Air traffic controllers sent to prison

> Use of Italian to communicate with another general aviation aircraft
> meant that the pilots of the Cessna were less aware of what was going on
> than they otherwise might have been.
>
> *SAS Flight 686/Cessna Citation CJ2, October 8, 2001*

Twenty-seven minutes elapsed before the air traffic controllers at Milan's second airport, Linate, realized the two aircraft they had lost had collided at the airport itself.

Milan's Linate Airport is prone to early-morning fog, and at the time of the accident there was no operating ground radar system. The old system had broken down, and the new and much better system bought from Norway some four years earlier had not been set up, due to administrative delays.

After the accident, the so-called administrative and technical problems quickly evaporated, and the system was set up in a couple of months. Part of the delay had been due to an inability to decide whether the airport or air traffic control should be responsible for setting it up.

The brand-new—only twenty-eight airframe hours and twenty cycles—twinjet Cessna Citation, with its two German pilots and two passengers (a Cessna sales manager and a prospective customer) had landed early that morning in fog, with its pilots not qualified to land in such poor visibility. The Cessna landing in a northerly direction had come to a stop just beyond the taxiway (R6) leading off left to the general aviation apron to the west. However, as there was hardly any traffic at the time, the air traffic controllers allowed it to do a U-turn on the runway and back-taxi to and take taxiway R6, this time to the right. This enabled them to avoid taking a circuitous route via the commercial airliner apron, situated to the northeast.

When they were ready to set off for their next destination, Le Bourget (Paris), visibility was still too bad for the pilots to take off with their pilot's license ratings. However, it was not against the rules for them to taxi in the hope that visibility would improve by the time they were to take off.

Contrary to international convention, the numbering of the taxiways at Linate was not consistent. They should have been numbered in a clockwise direction. R6 should have been R5, and R5, farther around the clock to the north, should have been R6. The Cessna pilots had duly parked on the West Apron after arriving directly from the runway.

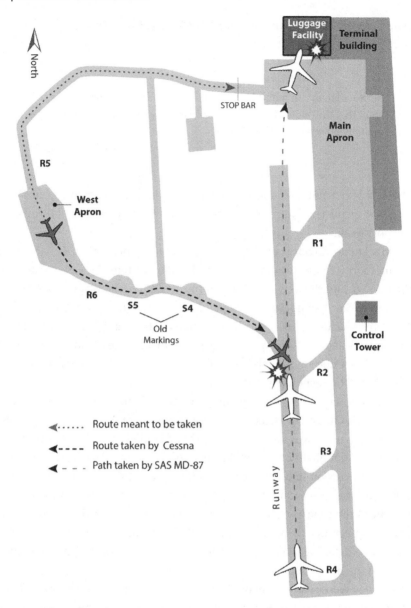

Cessna taxied south from West Apron instead of north as instructed by controller.

However, when it was time for them to leave the West Apron for departure, the airport was busy, so the ground controller instructed them to taxi north via taxiway R5 and call him back when they reached the stop bar to the "main runway extension." Extension was the point on the taxiway in line with the end of the runway that aircraft taking off would overfly.

Ground control:

DVX taxi north via Romeo five. QNH1013 . . . Call me back at the stop bar at the main runway extension.

DVX:

Roger, via Romeo five and 1013 and call you back before reaching main runway.

The controller did not pick up on the fact that the callback omitted the word "north," but possibly the word was only to make his instruction clearer—for him the pilot had correctly repeated the key information, R5. Some investigators have suggested the German pilot missed the word "north" because it is very short when spoken in English by an Italian.

Seven seconds later the ground controller gave a similar authorization to taxi via R5 to another light aircraft, parked on the West Apron north of where the Cessna had been. However, the controller gave his instructions in Italian and told them to wait until the German Cessna had passed by before moving off. The German pilots in the Cessna could not follow this exchange in Italian.

After a while the controller checked with that second light aircraft, asking them whether they were duly proceeding along taxiway R5. They replied they were still waiting for the Cessna to pass. Their pilot then asked in Italian whether the controller knew where the German was.

The controller replied that the Cessna was on the main apron and added:

I should say you can go.

Frustrated by the unnecessary wait, the Italian pilot replied:

I should say so! We move.

If this exchange had been in English, it might have given the Cessna pilots another hint they were on the wrong track.

The Cessna, erroneously proceeding southeast along taxiway R6 instead of north via R5, came to a spot a little before the runway marked S4 and informed the controller they were "approaching the runway . . . S4." The controller might have misheard, but the German pilot did use the term "Sierra 4" so it could not have been so unclear. Anyway, S4 did not have any significance. The marking "S4" had been painted on that taxiway many years before when it was thought extra parking spaces might be needed for airliners. However, the project was terminated when the flights were transferred to Milan's other airport. Serving no purpose, it was not mentioned on airport plans used by the controllers.

The controller told them:

Continue your taxi on main apron. Follow the A line.

The Cessna read back the instruction correctly:

Roger, continue taxi in apron, A line . . .

The controller replied:

This is correct. Please call me back entering main taxiway . . .

About that time the controller told an aircraft actually on the main apron to call back on entering the main taxiway leading down to the takeoff point.

The Cessna pilot could have been in no doubt that he was going to cross a (the) runway, but he had the green center lights continuing past the stop bar lights and could have been under the impression that he should cross the runway as quickly as possible. Having taken the same route in the opposite direction when landing, he should have known he was not at the northern runway extension but possibly in the path of an aircraft taking off or landing.

The Cessna moved forward toward the runway and began to cross, unaware that a SAS MD-87 airliner waiting to start its takeoff roll had received its takeoff clearance and was already moving forward. They were unaware of this fact, as the takeoff clearance had been, as is normal, on the tower frequency, which the taxiing Cessna pilots would not be using.

The point where the Cessna was traversing the runway proper was one and half kilometers from the point where the SAS aircraft had started its takeoff roll just thirty-seven seconds earlier. Consequently, the SAS MD-87 was traveling at about 146 knots and well into its rotation when it encountered the Cessna directly in its path.

The nose landing gear lifted off two seconds before the impact, and though the main landing gear was still touching the ground, the weight was coming off the main wheels, with their shock absorbers extending. (The liftoff of the nose wheel had automatically sent a signal to the SAS home base in Copenhagen indicating that the flight had duly taken off!)

A fraction of a second before impact, there was a distinct nose-up elevator input and crew exclamation on the MD-87, so they could only have seen the tiny Cessna at the very last moment. Following the collision, the crew of the MD-87 advanced the throttles farther, but the extra power from the engines mounted at the rear was not forthcoming, as the right-hand number two engine had fallen off and the only remaining engine was failing, no doubt due to ingestion of debris.

Nevertheless, the MD-87 continued its climb to a height of ten and half meters (thirty-five feet) before plunging down onto the runway eight and half seconds after separating from the Cessna. Not only did the MD-87 not have enough thrust from the remaining engine to climb away, the loss of the heavy, rear-mounted number two engine meant the center of gravity had moved too far forward for the aircraft to fly. Realizing the hopelessness of the situation, the pilots pulled back the throttle levers.

Having vainly accelerated to a speed of 166 knots, the MD-87 hit the runway, with the right-hand main wheels missing and failing hydraulic pressure. This meant that the aircraft tipped to the right and the right wingtip struck the ground, causing the aircraft to slew around. The crew adroitly applied reverse thrust to the remaining left-hand engine, but with the tail-mounted engine so close inboard, unlike those mounted engines mounted far out on the wings, the directional correction it provided was limited.

Added to that, lack of hydraulic pressure meant the brakes were ineffective. Veering slightly to the right, the SAS MD-87 was still traveling at 139 knots when it slammed into the luggage-handling building located 460 meters beyond the end of the runway and not far to the right of the runway axis. Just six seconds had elapsed from the moment the MD-87 fell back on the runway and its collision with the building. Four baggage handlers in the building, and all 104 passengers and 6 crew on the airliner, were killed.

The situation for the two pilots and two passengers in the frail Cessna was hopeless from the start, and all were either immediately killed by the impact or within a minute or so by the fire and smoke. All told, the death toll was 118. Had it not been for the fine airmanship of the MD-87 crew in trying to straighten out the trajectory, it could have been even worse, with the aircraft veering farther to the right and striking the passenger terminal.

Analysis

As mentioned at the beginning of this section, the failures of the controllers and those running the airport were so many and so glaring that they were held criminally responsible. However, did the investigators fully consider the possible fatigue of the Cessna pilots?

The official report only said they presumed the Cessna pilots' duty time had started at 04:30 local time. One supposes they worked this out from the time it would have taken the pilots to fly to Linate from Cologne.

One of the German pilots was sixty-four and the other thirty-six. There is no knowing what they were doing before they picked up the aircraft for the middle-of-the-night departure from Cologne, but it is quite likely they would have been pretty tired by the time they asked for clearance to taxi at 06:05 at Linate. That was the moment they made the fatal mistake of not noticing the air traffic controller had said they should go *north* and call him back when they reached the stop bar at the main runway extension.

What the controller quite meant by "*runway extension*," may not have been clear. Since the Cessna pilots did not confirm the word "*north*," they possibly had not noted it. It is quite possible that after their brief break, one or both German pilots were not fully alert. Without evidence from a cockpit voice recorder, it is difficult to be sure of the exact situation; they could have been distracted by the passengers' presence.

Nevertheless, how the Cessna managed to take taxiway R6 instead of R5 is something of a mystery, unless one assumes that the pilot looking at the plan of the airport was not concentrating, because of tiredness or some other reason. It was confusing, if not looking at the plan, to have taxiway R5 after R6 in the clockwise direction. In addition, as they had followed that route on coming off the runway on their arrival, that route might have seemed more logical, as it was only about a third of the distance to the point where they would start their takeoff. There is a slight parallel with the crew of the Pan Am 747 at Tenerife in that their taking the easy taxiway at 45 degrees, rather than the difficult one at 135 degrees leading back toward the terminal, would seem more logical and be much shorter.

The Cessna pilots only had one chance to determine what number taxiway they were on, and that was where the yellow lines on the apron forked, with the yellow line for taxiway R5 forking to the left (and north), and the yellow line for taxiway R6 forking to the right. The nearest taxiway light in the direction of R6 was only 80 meters away, while that for R5 that they should have taken was 350 meters away, thus perhaps luring them in the wrong direction, just as with the Singapore Airlines pilots at Taipei. However, if the Cessna pilots and the Singapore Airlines pilots had been looking at the plan of the airport, they surely would have seen their error.

Once the Cessna pilots made the mistake—for whatever reason—there was no further indication along the whole length of the taxiway R6 giving its number. Whatever the truth, the controller pigeonholed them in his mind as being on R5, and nothing they said, apart from the meaningless (to him) statement that they were at S4, suggested otherwise. Even if they missed the stop bar on taxiway R5 at the runway extension, the controller would assume they could not come to much harm, as any aircraft taking off should be well into the air by then.

Chapter 6
CONTROL SYSTEMS
COMPROMISED

DC-10 Cargo Door Opens Inflight (Windsor-Detroit, 1972)
Would have been fatal had the cabin floor not been reinforced for a piano

> Systems operating essential airfoils (ailerons, elevators, rudder, etc.) are usually in triplicate, and the chance of them failing of their own accord is minimal. The hydraulic power they usually require is provided in most cases by two, or more often, three circuits.
> However, if these hydraulic lines pass close together, some catastrophic event can, in the absence of systems to isolate the damaged section, take away all hydraulic power and make the aircraft impossible to control.
> *American Airlines Flight 96, June 12, 1972*

On June 12, 1972, an American Airlines DC-10 bearing flight number 96 took off from Detroit for Buffalo with only fifty-six passengers. Normally, there would have been no need to open up the rear cargo hold with so little baggage to load, especially as they would not want to shift the center of gravity too far back. However, the manifest included a coffin, and perhaps out of consideration for the baggage handlers at Buffalo the decision was taken to load it into the rear hold, situated under the floor of the aft passenger cabin.

Though the point of departure and the destination were both in the United States, the route was almost entirely over Canada, as the frontier came southward to follow the center of Lake Erie. With the first officer at the controls and flying on autopilot over the Canadian side of the lake, they were near Windsor, Ontario[1], and coming up to twelve thousand feet on their routine climb to twenty-three thousand feet when the pilots heard a bang from behind and felt the impact of a rush of air on their faces. The flight deck door sprang open, and a lot of dust started flying around. An explosive decompression had obviously occurred.

Back in the passenger cabins, the rush of air was propelling papers and other loose items rearward, with the situation being worst right at the back, where two flight attendants in seats by the rear doors were for a moment sucked downward toward the floor, buckling into the cargo hold below. The flight attendant on the left could actually peer into the cargo hold and see the gaping black hole where the door had once been.

Unaware of this, the pilots' immediate thought was that something had struck them, but whatever the cause of the decompression, they first had to determine what controllability they had:

1. The rudder pedals were jammed against the stops at the full-left-turn position.

2. All three throttles had sprung back to their idle stops.

3. Engines one and three, mounted on the wings, seemed to be okay and, despite having begun to spool down due to the movement of the throttles, responded when these were pushed forward. Engine number two in the tail was no longer functioning and its throttle was immovable.

4. The vital (apart from the rudder) control surfaces in the tail, namely the elevators, were barely working and felt very heavy.

5. Despite the rudder pedals being jammed at full left, the rudder itself angled to the right was making the aircraft yaw to the right. However, by increasing the power on the right engine, reducing that on the left engine, and applying a fair amount of left aileron to make the aircraft bank to the left, it was possible to get it to fly relatively straight.

Subsequently, by adjusting engine power and varying the amount of aileron, the pilots found they were able to steer the aircraft very roughly where they wanted to go, with an inevitable time lag between the input and the response due to the engines needing time to spool up and down.

The DC-10 is notable for having the engines mounted under the wings and very low slung, with this compensated for by having the engine in the tail very high up and at a slight angle. This leverage allowed the pilots to raise or lower the nose by juggling engine power. However, without the compensatory engine high up in the tail, there was a danger of pushing the nose up too far and not being able to push it down. The pilots had to be very wary and proactive—if they got into too sharp a dive, they might not be able to pull out of it, or in trying to do so might not be able to push the nose down afterward and end up stalling.

By making a long, shallow approach, the captain managed to bring the aircraft down onto Detroit's runway, which was 10,500 feet (3,200 meters) long and 200 feet (60 meters) wide, at just thirty knots over the normal approach speed and even managed a flare by having both pilots heaving back on the leaden elevator controls at the last moment.

The touchdown was a good one in the circumstances, but once they were on the runway, application of left aileron could not prevent their trajectory veering to the right, because of the rudder sticking out to the right. As they left the runway, the first officer, realizing they would be in trouble before the reverse thrust and brakes could slow them enough, increased reverse thrust on the left engine beyond the normal limit and at the same time disengaged reverse thrust on the right-hand engine by pushing the throttle lever forward. This succeeded in countering their deviation, and it was not long before their airspeed had dropped enough for the protruding rudder to have little effect, so that the nose-wheel steering, always wanting to veer left under the command of the jammed rudder pedals, came into play, bringing the aircraft back to the runway. They ended up half on and half off the runway, and some five hundred meters from the end.

The captain had not jettisoned fuel, because of uncertainty regarding the extent of the damage to the rear of the aircraft. Even had he opted to do so, the fuel dumping limitation (already mentioned) on the DC-10 would not have allowed him to drain the tanks enough to prevent fire in a crash landing.

The few occupants were able to evacuate in less than a minute, making it a lucky outcome for all concerned. Proficient handling of the stricken aircraft by the crew had saved the day, but why had the incident happened in the first place, and why did the apparent failure of a cargo hold door have such consequences?

The simple answers are:

1. The door-locking mechanism was badly designed.
2. The floor did not have vents in it to allow air to pass through it without it buckling first.
3. The control lines for the rudder, elevators, and rear engine were attached to the underside of that floor.

It so happened that the cabin floor had been specially reinforced to support a piano for some event. Otherwise, it would have no doubt crumpled further, as was to occur a couple of years later to a fully laden DC-10 that had taken off from Paris (See next account).

The Door-Latching Mechanism
Recovery of the cargo hold door from a field under the flight path meant the investigators had all the elements required to determine the immediate cause of the near disaster. The hinges and interfaces between the door and the fuselage were all in good condition, and the investigators quickly concluded the C-latches must have opened for some reason.

As their name implies, the C-latches affixed to the bottom of the door are shaped like the letter C, and in theory they swivel around when the door is locked so as to encircle a very solid crossbar (spool) on the doorjamb, so that the greater the force trying to prize the door open, the firmer it should hold. This supposes proper engagement of the C-latches in the first place.

Unfortunately, the give in the various levers and components of the locking mechanism, and poor design, meant those doors on DC-10s could appear to be locked when in fact the C-latches were not properly home. This was partly because the original design had envisaged using hydraulic power to operate the primary (nonmanual) locking system, and this was subsequently changed to electric on the demand of the airlines on grounds of economy (cost and weight). Hydraulic operation is generally much more positive.

McDonnell Douglas

McDonnell Douglas, the DC-10's manufacturer, had had numerous complaints from airlines about the door in question. The problem with the aft cargo hold door and some other early problems with the DC-10 were said to be due to the frenetic pace at which the aircraft was brought into production, with Lockheed's very similar TriStar in direct competition, and the Boeing 747 lurking in the background.

Although the competing Lockheed TriStar L1011 was an excellent aircraft, McDonnell Douglas had the advantage that US airlines were so used to buying its aircraft that they were likely to choose them out of habit and inertia. This did not prevent the airlines playing off the two companies against each other, with the result that prices became so low that neither aircraft would ever be a big money-spinner, unlike the Boeing 747, which was in a category above.

In addition, delays in developing the power plant for the TriStar had almost led to the demise of the UK engine manufacturer Rolls-Royce, which had to be bailed out by the British government, and in turn meant that Lockheed got off to such a delayed start that sales never took off. Despite offering inducements (bribes) to sell the aircraft abroad, total sales only came to some 250 units, compared with the 500 needed for the project to break even. This ultimately led to the great Lockheed, famous for airliners as the beautiful Constellation, withdrawing completely from the civilian airliner market.

Inducements paid in Japan in the expectation that All Nippon Airways (ANA) would purchase the TriStar led to the resignation and subsequent trial of the Japanese prime minister, Tanaka Kakuei. In parallel, a very senior manager at Hong Kong's Cathay Pacific Airways had to resign in disgrace for accepting inducements, albeit on a much smaller scale.

Ironically, the two did their respective airlines a favor by stopping them from jumping on the dangerous DC-10 bandwagon before its problems were resolved. The DC-10 that crashed outside Paris, as described in the next narrative, with the loss of all on board had originally been destined for Japan's ANA!

Like the fatal *"put on your manager's hat"* decision to launch the Challenger Space Shuttle, many engineering ethics courses cite as classic examples for study the decisions subcontractors made concerning the DC-10 cargo hold door, when engineers were fully aware of the problems.

The Subcontractors

The contractor for McDonnell Douglas was Convair, and six weeks after the Detroit/Windsor incident in 1972, Convair's director of product engineering, Daniel Applegate, submitted a memo to his superiors officially delineating problems with the cargo hold door, which had become apparent as early as August 1969. These had been confirmed in a ground test the following year. According to some reports, those tests also revealed what could happen to the passenger cabin floor.

Following the discovery of the problems by its engineers, Convair's management was halfhearted in pursuing the matter with McDonnell Douglas for fear that under the terms of its contract it might have to pay the cost of the modifications. McDonnell Douglas's management was in turn fearful about the effects any delay might have on its sales. Undoubtedly, the best option would have been to redesign the whole thing, but as such systems have to go through a lengthy approval process with the FAA, it was too late to start over, and it therefore fiddled with an "approved" design, as modifications would only be minor and quickly certified. (Another instance where regulations meant to ensure safety have the opposite effect.)

The opening of a cargo hold door at altitude is always a serious matter, as the slipstream usually wrenches it off. If it is a front cargo hold door, it could damage the control surfaces on the wings or disable the engines under the wings or, if at the rear, damage the horizontal stabilizer and elevator on the tail, as happened in the Detroit/Windsor case.

What manufacturers did not properly consider at the time was the disastrous knock-on effects of such a decompression. Not only would the passengers risk being sucked out or struck by flying objects, they would also be subjected to the physiological effects of explosive decompression. The downward buckling of the passenger cabin floor under the pressure differential could also damage the vital control links and hydraulic lines leading to the tail, generally attached to its underside. This in turn could make the aircraft uncontrollable.

Within three weeks of the Detroit/Windsor scare, the NTSB made two urgent recommendations:

1. Modification of the DC-10 door-locking mechanism so that it is physically impossible to bring the vent-flap-locking handle to its stowed position without the C-latch locking pins being fully engaged.

2. Vents (holes) should be incorporated in cabin floors to considerably relieve sudden pressure differentials, such as those caused by the opening of a cargo hold door in flight.

The Gentleman's Agreement

The NTSB could only advise. It was up to the FAA to make these two modifications mandatory.

Just when the FAA was about to issue an airworthiness directive (AD) making interim and long-term solutions mandatory for all US operators of the DC-10 (which foreign operators would have followed), discussions between the FAA administrator and the president of the Douglas division of McDonnell Douglas led to the senior FAA technical staff being overruled. Douglas and the FAA were no doubt being subjected to pleading from US airlines, who would not want to take their aircraft out of service in the peak summer season. So the FAA did not issue that airworthiness directive. Instead, McDonnell Douglas almost immediately issued recommendations, in particular the installation of a "lock mechanism viewing window." This gentleman's agreement between the FAA administrator and McDonnell Douglas's Douglas division president sufficed to prevent a repeat accident *in the United States*, but not overseas.

A fascinating and troubling book, *The DC-10 Case: A Study in Applied Ethics, Technology, and Society*, describes the scandal and human factors that had resulted in an inherently dangerous aircraft and allowed it to continue as such. Here we use background information from that book, without the space to do it justice.

DC-10 Cargo Door Opens Fatally Inflight (Paris, 1974)
All hydraulic lines attached to cabin floor damaged as it buckled

> This disastrous DC-10 crash outside Paris was a rerun of the
> Detroit/Windsor incident, but without the passenger cabin floor having
> been reinforced for a piano.
> *Turkish Airlines Flight 98, March 3, 1974*

Demand that day for seats between Paris and London was particularly high, as one of the two main airlines serving the route, British European Airways (BEA) crew members were on strike. This was part of a negotiation tactic to get a better deal for its staff in its merger with British Overseas Airways (BOAC), to form what is now British Airways (BA). There was some friction, as BOAC people appeared to see themselves above those at BEA, who were apparently paid less to fly less exotic routes. Fans returning to England from a rugby match in Paris the previous day were putting added pressure on the few available seats.

Hence when a Turkish Airlines DC-10 arrived en route for London, stranded travelers clamored for seats, and as a result it departed for London with most seats filled, which was unusual. Taking off in an easterly direction from Orly Airport, situated to the south of Paris, it skirted the city and continued some twenty-five miles before turning north toward England, with everything seeming routine.

The Detroit/Windsor incident repeated itself just as the almost-full-to-capacity wide-bodied DC-10, likewise climbing to cruising height, came abreast of the northern limits of Paris's suburbs at a height of some nine thousand feet and a speed of three hundred knots. However, this time the floor had not been reinforced and so opened up much more. Six hapless—or perhaps fortunate—passengers were sucked downward and disappeared through the chasm in the floor and out of the failed door to subsequently hit the ground still strapped in in the two rows of their three-abreast seats.

Coinciding with an emergency transmission from the aircraft, French air traffic controllers saw flight number 981 transmitted by the DC-10's transponder disappear from the secondary radar. However, the primary radar, which works by bouncing a radio wave off an aircraft, showed the echo separating into two blips: a large one and a much smaller one. The large one veered northwestward before disappearing after a minute, while the weaker one (the six passengers in their seats no doubt held together by the broken-off rail used to attach the seats to the floor) remained stationary before disappearing after a couple of minutes or so.

The large blip (the aircraft itself) had come down in Ermonville Forest, some *thirteen miles* northwest of the point where the decompression had occurred.

The Crash Site

The high sink rate and forward speed meant the trees could not cushion the terrible impact. Air crash sites are usually horrible, but this one was particularly so. The absence of a conflagration meant the scattered distribution of body parts was immediately obvious.

In *Air Disaster: Vol. 1*, Macarthur Job says how Police Captain Lanier, who was one of the first to arrive, described the terrible scene and saw two severed hands, a man's and a woman's, clasped together. In such circumstances, there was no hope of any survivors.

The Pilots

What had been the situation on the flight deck in those final minutes? Could the pilots have saved the aircraft? Indeed, it seems they had finally found a way to at least bring the aircraft out of its fatal plunge but did not have enough height left. The buckling of the floor had, as in the Detroit incident but to a greater degree, damaged or severed the triplicate hydraulic lines serving the tail and the control cables both for the tail flight-control surfaces and rear engine.

From the cockpit voice recorder (CVR) and the flight data recorder (FDR), investigators ascertained that pressurization-warning klaxons sounded in the cockpit when the cargo door blew out and that the crew thought there had been a failure of the fuselage. Not that it made any difference in this case. Apparently, the situation developed as follows.

As the DC-10 went into a twenty-degree dive, its airspeed increased alarmingly, and despite the engines being throttled back, the overspeed alarm sounded. With not quite a minute having elapsed since the nightmare started, the captain astutely decided their only hope was to ignore the aural overspeed warnings and go for even more speed. Pushing the throttles forward, he said, "*Speed!*"

Whether it was the speed itself or the nose-raising torque of the two viable engines slung so low below the wings, or a combination of both, the aircraft gradually began to level out, with the rate of descent decreasing so much that the g-forces would have forced the pilots and passengers hard down into their seats. They were almost level when they struck the trees of Ermonville Forest at 430 knots. Had they had a greater height margin, the story could have been different, with the captain's call for speed allowing them to fight on.

All 339 people remaining on board died in the subsequent crash. Perhaps because the captain had cut the engines just before impact there was no major outbreak of fire.

As the high-speed impact had so traumatized the bodies, it was evident that this absence of major fire would not have made much difference to the occupants. However, the relative lack of smoke and fire did make it surprisingly difficult to locate the exact site of the crash in the relatively vast forest. The total death toll, including the six passengers ejected earlier, came to 345.

Comparison

With so little means of control—just the engines—the situation facing the Turkish pilots was more akin to that faced by those of the Sioux City DC-10 than that at Detroit. However, it differed in that the Sioux City aircraft became uncontrollable at the great height of thirty-seven thousand feet and not at nine thousand feet. Besides, it did not start to dive on its own. This gave the Sioux City pilots more than ample time to decide what action to take, whereas in the Paris case, only seventy-two seconds elapsed between the decompression and impact with the ground, and only ten seconds between the initiation of the almost successful addition of power to bring the aircraft out of its dive and that impact. Had the Paris pilots been able to recover in time, would they have been able, or lucky enough, to replicate the Sioux City feat?

At first, with the aircraft having disintegrated into tiny pieces, as had the passengers, it looked as if the investigation would be difficult. Postmortems of the relatively intact passengers ejected right at the beginning in their seats provided valuable evidence, and, notably, indicated that the crash was not due to a bomb, as had first been feared. Furthermore, recovery of the cargo door with its mechanism virtually intact, as in the Detroit/Windsor case, was a great help.

The investigators found:

1. The stiffening of the linkage from the flap handle had not been done as prescribed in Service Bulletin 52-37. Some modifications had been made, but an item not up to aeronautic quality had been used, and in fact this had made the door even more vulnerable to improper locking than the one on the Detroit DC-10.

2. The microswitch system for indicating (in the cockpit) whether the door was locked was not adjusted optimally.

3. The door did have the prescribed window for checking whether the latches had gone home. However, the cargo handler responsible for closing the door was not qualified or told to perform that check, which was normally done by a Paris-based Turkish Airlines engineer, who was on holiday, or in his absence by the flight engineer, who failed to do so.

The gentleman's agreement mentioned above between the head of the FAA and the president of the Douglas division of McDonnell Douglas had meant the radical measures recommended by the NTSB after Detroit were not made mandatory. However, the scandalous aspect was that even the half measures proposed by McDonnell Douglas's Service Bulletin to airlines had not been carried out by them on an aircraft ordered and delivered long afterward. (Eventually, it transpired that three engineers at Douglas or its subcontractor had signed off the work as having been completed.)

The airline and the manufacturer paid out some of the highest compensation ever for an air disaster, without admitting culpability. Had the disaster occurred in the United States, the damages would have been far higher.

Though it is clear the main problem was one of engineering ethics, there were as usual multiple factors, not least the fact that the flight engineer did not do a visual check through the observation window prior to departure from Paris. The FAA hurriedly made the measures recommended by the NTSB after the Detroit incident mandatory. These included having vents in cabin floors to relieve extreme pressure differentials. Passengers are safer now thanks to that.

DC-10's Textbook Speed Seals Fate of AA191 (Chicago, 1979)
Would not have flipped if flying slightly faster

> The photo of the DC-10 banking at ninety degrees with one engine
> missing told all. An engine had fallen off the wing, but that should not
> have prevented it from flying.
>
> *American Airlines Flight 191, May 25, 1979*

Just as the American Airlines AA191 afternoon flight taking off from Chicago for Los Angeles was rotating at a V_R of 145 knots, the *rear* support of the pylon holding the number one (left) engine failed. The powerful takeoff thrust then pushed the engine assembly forward and upward as it pivoted on the *front* support, which in turn fractured. Finally, the engine, with the pylon still attached, passed over the top of the wing and, because of its weight and the low airspeed, tumbled to the ground without damaging the DC-10's high-up tail assembly. The wing minus its engine dipped for a moment but then came up again as the pilot used the ailerons to climb out at 159 knots—6 knots above the safe climb-out speed (V_2), which was the minimum allowed.

Despite the separation of the engine, there was every indication they would be able to return safely to the airport, and the controller observing the takeoff from the tower asked the pilots if they wanted to come back, and if so on which runway. According to the flight data recorder (FDR), the aircraft then accelerated, reaching a maximum speed of 172 knots nine seconds after becoming airborne. At 140 feet agl (above ground level) and still climbing, the pilot apparently let the airspeed fall back, in accordance with the airline's engine failure procedure.

On reaching a height of about 325 feet, with the airspeed dropping back to the recommended 159 knots, the left wing from which the engine had fallen inexplicably began to drop. The aircraft yawed to the left and was soon sideslipping in a ninety-degree bank, with the wings perpendicular to the ground, as shown in the photo on the next page.

The final reading, obtainable from the data recorder three seconds prior to the left wingtip hitting the ground, showed 112 degrees' bank and a nose-down attitude of 21 degrees. The aircraft then flipped over and exploded, and a large fireball engulfed the debris.

Would-be rescuers said the crash scene was a horrible sight, with no bodies intact and even the remains charred beyond recognition. All 271 people on board were killed, together with two bystanders, making it the worst air disaster up to then in US history.

(Courtesy NTSB)

Left wing stalls as slats on that wing retracted

It could have been worse, as the aircraft came down just short of a large caravan park, and there were fuel storage tanks farther on.

For the hundred or so investigators sent to the scene, there were two main questions:

1. Why had the pilots lost control when the aircraft should have been able to fly in that configuration at that airspeed?
2. Why had the engine and pylon fallen off in the first place?

Change-of-Configuration Stall

The outer leading-edge slats on the left wing had for some reason retracted, and this had led to a *change-of-configuration stall*—that is, a stall where an alteration to the position of the flaps and/or leading-edge slats means the aircraft can no longer fly at the given airspeed.

Paradoxically, had the captain not followed the rule book to the letter and reduced speed on the loss of the engine, he would have been above the stalling speed for that configuration, and the ailerons would have been sufficient to compensate for the discrepancy in lift due to the retracted outer slats on the left wing. He could have flown all the way to Los Angeles.

The rule (later modified) had been made to protect the other engines, which automatically spool up to maximum or above maximum rated power when an engine fails on climbing out on takeoff to compensate for the loss of power. The pilots are told they should level out and decrease the power on the remaining engines, which the pilots of the DC-10 duly did.

However, as in so many crashes, there were other unfortunate factors. Firstly, on twisting upwards and skidding over the wing the engine assembly had severed a cable passing just inside the damaged leading edge of the wing that should have transmitted the information that the slats had retracted to the pilots.

Secondly, the stick shaker (stall warning system) did not activate, and the instruments were not working properly because the electric circuit had been locked out. Also, with the inner slats still deployed there would have been virtually no buffeting that the pilots could sense. That electrical system could have been unlocked by the pilots, but they were too busy, and for the engineer to do so, he would have had to turn his seat around and unbuckle himself, which was hardly possible in the time available.

Not knowing what was happening, and with no time, the pilots could not have been expected to realize where the route to salvation lay. Also, the dissymmetry in lift and resultant banking meant that the minimum control speed (V_{MC}), as well as the stalling speed, would rise so quickly that even getting the wings level would quickly become hopeless. Pilots given the full background were able to save the aircraft in the simulator, but those not so briefed reacted as the actual pilots had and crashed.

DC-10 Design

The DC-10, unlike many other aircraft, did not have a mechanical locking device to keep the slats from retracting under external force in the event of a failure. However, in most cases of failure the hydraulic fluid trapped by a valve would prevent that happening.

Anyway, even if the slats did retract, the ample power available from the engines would immediately give them sufficient speed to both avoid a stall and have controllability via the ailerons. The possibility of an engine failure, coupled with an undesired slat retraction, was considered improbable.

However, in the case in question the separation of the engine and pylon damaged the hydraulic lines and caused the slats to retract. Some maintain McDonnell Douglas was taking risks by having so many vulnerable lines just behind the leading edge of the wings.

DC-10 Maintenance and Forewarnings

Reviews of eyewitness reports and examination of the wreckage indicated that the pylon's rear bulkhead flange had been cracked beforehand. In view of the considerable safety margin built in, this could hardly have happened during normal operations and must have happened during maintenance. Indeed, the pylon (together with the engine) had been removed and replaced two months earlier to substitute the bearings. Investigators found deviations in the maintenance work at a number of airlines. For instance, maintenance workers, on finding certain bolts difficult to remove, would do so in a different order from that specified, thereby placing excess stress on others.

The FAA grounded all United States–registered DC-10s subject to a check of the rear flanges in question. Four American Airlines aircraft and two Continental

Airlines aircraft had cracks there. Continental had found their cracks in the rear flanges and replaced them. Other airlines had not experienced that problem.

Forklifts

The reason soon became obvious. The two airlines had devised what they thought was a superefficient way to remove the engine and pylon simultaneously, using a forklift, disregarding the McDonnell Douglas maintenance manual, which said they should be removed individually.

The DC-10's engines are cantilevered far in front of the wing, which means a considerable mechanical moment (levering) is imposed if the engine is lifted more than a certain extent, and that can put tremendous stress on the rear flange. In addition, vibrations from the forklift can be transmitted to it. The accident was partly attributed to poor communication between the parties, namely Continental, the FAA, the manufacturer, and the airlines, regarding the earlier problems at Continental.

DC-10 Grounded for Thirty-Eight Days

This dramatic disaster, which unlike the previous one at Paris was not so much the fault of the manufacturer, led to three airworthiness directives and an almost unheard-of thirty-eight-day grounding of the fleet. The damage to McDonnell Douglas's reputation in the eyes of the public was so great that many orders for the DC-10 were canceled. The company then dropped behind the competition after failing to develop any more radically new airliners, and was finally taken over by Boeing.

Finally, Extremely Safe

The modifications and changes in maintenance procedures introduced following this disaster, coupled with those made in the light of earlier ones related to the cargo door blowing out, meant the DC-10 became statistically one of the safest aircraft, and pilots who flew it seemed to appreciate its qualities.

Worst-ever Single-aircraft Disaster (Japan, 1985)
Aircraft "Uncontrollable" after Rear Bulkhead Failure

> This disaster in Japan, where a Boeing 747 staggered drunkenly around the sky for half an hour, with passengers writing last wishes on their boarding passes, must have been a terrifying experience and is the worst single-aircraft crash ever in terms of souls lost.
>
> *Japan Airlines Flight 123, August 12, 1985*

**Off-duty flight attendant Yumi Ochiai
felt her hair lift off her neck
and a momentary sense of weightlessness
as the staggering 747,
with its 524 terrified occupants,
began its final downward plunge.
Shaving the trees on one mountain ridge
it ended up on another.
Her pelvis broken, and trapped between seats
in the remains of the broken-off tail section,
she was fitfully aware of the sound of young children,
their cries fading as the injuries, shock,
and cold of the night took their toll.
To be brave, a young boy,
who was not to survive,
cried out, "I'm a man!"**

The Boeing 747 SR (short range), a jumbo specially adapted and reinforced to carry as many as 550 passengers on domestic short-haul routes in Japan, was the early-evening shuttle from Tokyo's domestic Haneda airport to Osaka, Japan's second largest city, four hundred kilometers away to the south. The flight was being flown by First Officer Yutaka Sasaki, thirty-nine, an experienced pilot training for promotion to captain. In the right-hand seat, acting as copilot, was Captain Masami Takahama, forty-nine, a JAL instructor with more than 12,400 hours' experience. Hiroshi Fukuda was the flight engineer.

Unusually, the occupants were mostly women and children, for it was the Obon August holiday period, when the Japanese go back to their hometowns to visit family graves and see relations. The aircraft was virtually full, and most of the passengers were clad in only the lightest of clothes in anticipation of a clammy midsummer evening. As in a number of flights that have ended in disaster, the flight number, JL123, is easy to remember.

Haneda, situated at the edge of Tokyo Bay and very close to the center of the city, had served as Tokyo's international airport until the construction of Narita Airport in the face of fierce opposition some fifty miles (eighty kilometers) away. With little need for noise abatement over the sea, JL123's climb out of Haneda was a simple affair. Ten or so minutes after takeoff, the busy pilots were able to relax. Everything seemed normal.

The first indication the Tokyo controller had that everything was not so came without any forewarning at 7:25 p.m., thirteen minutes after takeoff, and just as JL123 was leveling out at its cruising altitude, when the echo for the aircraft on his radar screen switched to 7700, the emergency code.

Shortly afterward came the following disjointed call from the aircraft:

> Tokyo. Japan Air 123. Request immediate . . . ah . . . trouble. Request return back to Haneda . . . Descend and maintain Flight Level 220.

JL123 then asked for the vector (course) back to Oshima Island, the waypoint on the easiest route over the sea back to Haneda. The controller gave the crew permission to descend and indicated the appropriate course, but instead of turning 177 degrees to go back on its tracks, JL123 merely made a very slow right turn of some 40 degrees. Surprised by this noncompliance, the Tokyo controller repeated his instructions.

JL123 still did not comply, and with the aircraft heading for dangerous mountains, Japan Airlines' operations center called them on the company frequency, without extracting more information, except that the crew thought a rear cabin door (R5) was broken and that they were going to descend. Watched by the air traffic controllers in Tokyo, the 747 meandered in an area of treacherous mountains not far from Japan's famous Mount Fuji. The pilots kept repeating they were *out of control* while at the same time requesting directions back to Haneda. Periodically losing considerable height and partially regaining it, the aircraft at one point made a tight 360-degree turn.

Finally, after some thirty minutes the echo on the screens showed JL123 rapidly losing altitude, before sinking out of radar view. The controllers vainly hoped the 747 had merely gone into a valley, but getting no reply on the radio, they finally accepted that it must have crashed.

A US transport aircraft taking off from the US Air Force Yokota airbase thirty-five miles away reported seeing fire in the midst of the mountains "that was very likely to be the crashed aircraft."

Who saw what and when is uncertain. The author read a report that five minutes after the crash the Japanese Air Self-Defense Force had scrambled two F4 Phantom fighters to investigate, and that twenty minutes later they were able to report the presence of fires and the general location as "*299 degrees and 35.5 miles from Yokota US* airbase's TACAN[1]."

Certainly, Yokota airbase, in the outer west-northwest suburbs of Tokyo, was relatively near, and it has been said, perhaps in connection with the excellent TV series *Air Crash Investigation*, that a helicopter from Yokota reached the site twenty minutes after the crash but was called back so that the Japanese Self-Defense Forces could take over. (The US military has a privileged but extremely sensitive presence in Japan and scrupulously avoids making any comments that could in any way cause friction, and so it is not easy to check these things out.)

Macarthur Job's *Air Disaster*: Vol. 2 also mentions a Japanese helicopter reaching the site and reporting the wreckage was scattered on a forty-five-degree slope, with fires and virtually no likelihood of any survivors, and that in view of the darkness, rain, and steepness of the terrain, landing there would be impossible. Subsequently, other helicopter crews, working from maps, gave many erroneous locations. In such terrain, without GPS and with one peak looking very much like another, one had to know the area and have reasonable visibility to get the location right. It was becoming pitch dark. Officials, by their very nature, were wondering whether the crash had occurred in Gunma Prefecture or Nagano Prefecture. Which police force would be responsible?

At the time Japan did not have a civil emergency unit to deal with such events. Also, lingering antagonism between the police and fire departments (which at one time had been a single organization) meant, according to some, that the police did not immediately call in the more suitable fire department helicopters. Furthermore, use was not made of US forces, with their considerable experience of plucking downed pilots at night from the mountains of Vietnam, perhaps because of face or the fact that no one wanted to take the responsibility if things went wrong. While in hospital Yumi Ochiai also mentioned hearing the sound of helicopter rotors and seeing lights, which could have been those of the US helicopter ordered back or those of the first Japanese helicopter.

In addition, one has to remember no one was expecting there to be survivors. Perhaps we are wiser nowadays, with experience showing that at least some of those on board can survive the most horrendous crashes. Adding to the confusion was the absence of high officials, away on the Obon national holiday, the holiday those passengers had been looking forward to enjoying.

To complicate matters, the crash site was in a particularly inaccessible place. No roads came anywhere near, and it was dark. Finally, at 05:37 the *following* morning a Nagano Prefecture police helicopter reported the fuselage was scattered in an area some seven hundred meters east of the prefectural frontier and was all in Gunma Prefecture. With the site pinpointed, the main rescue teams could set off on what they expected would be a body recovery rather than rescue mission. The earlier reports regarding the fires and the sighting of wreckage scattered over a wide area implied there was little chance anyone could have survived.

Because of the steep tree-covered slopes at the crash site, the authorities had determined that helicopters could not land close to the site even in daylight. To make matters worse, the weather then closed in.

At 08:49, the process of lowering seventy-three paratroopers by rope from giant helicopters hovering over the site began. They were to be the first people on the scene, nearly *fourteen hours* after the crash. Just as the first paratroopers to descend were confirming by radio at 09:25 that there were no signs of survivors, two members of the Nagano Prefecture rescue team descended at a point one and half miles (2.3 kilometers) away. Subsequently, at 10:15 the neighboring Gunma Prefecture police team arrived on foot, having scaled the mountains.

Survivors
Twelve-Year-Old Schoolgirl Symbolizes Disaster

At 10:54, and almost sixteen hours after the crash, a firefighter from Ueno Village and a member of the Nagano Prefecture police rescue team unexpectedly found a survivor trapped in the broken-off part of the rear cabin that had slid some way down the ridge from the main crash area. She was the off-duty JAL flight attendant who heard the cries of the dying children. Next, they found a mother and her eight-year-old daughter under nearby wreckage. All had fractures.

Most surprising of all was the discovery of twelve-year-old Keiko Kawakami in the wreckage of the tail section, with hardly an injury other than a slightly injured arm. Reports that rescuers found her wedged in a tree may be mistaken, as there is a photo of that wreckage with arrows showing the spots where she and the other three of the four survivors were found. She said she found herself in the wreckage with her mother dead and her father saying he could not help, as he could not move. When he died, she consoled her sister, who was to die too, saying their grandmother would look after them.

A photo taken from the ground showing Keiko being clasped around the waist by a paratrooper in combat gear as the two of them are being winched up into a giant helicopter came to symbolize the disaster. For years, weekly magazines used this appealing image on commuter train posters to advertise any issue with purported new facts about the crash, or failing that, with anything showing Japan Airlines in a bad light. Hopes of further survivors engendered by this good news soon evaporated. Others had survived the impact, and some of them almost surely could have been saved had help come earlier.

What had really happened to the passengers in those thirty minutes leading to the final plunge onto the mountain ridge? What was behind what to this day remains the worst-ever crash involving a single aircraft?

The Passengers

The flight data recorder and descriptions by the rescued off-duty flight attendant, Yumi Ochiai, reveal how terrifying the thirty-minute roller-coaster ordeal preceding the crash must have been. The aircraft would yaw, pitch, and roll, with each cycle taking about a minute and a half. In addition, there would be the vacillating scream of the engines as the pilots tried to steady the aircraft vertically and horizontally and attempted to nudge it in the direction they wanted. A number of the adults scribbled last words to loved ones on the back of their boarding passes or other scraps of paper.

The Flight Crew

Thanks to the cockpit voice recorder (CVR), the situation in the cockpit is well known, as there are both the crew's conversations between themselves and their exchanges with the air traffic controller—a couple of times the captain tells his colleagues to ignore the air traffic controller and concentrate on keeping the aircraft in the air rather than on ATC.

Although the pilots are speaking in Japanese, there are easily identifiable words from the captain, such as "*Flap up! . . . Power! . . . Power!*" Finally, at the end of the half hour, one clearly hears the verbal warnings from the aircraft's ground proximity warning system (GPWS) making a whooping sound, immediately followed by the words "*Pull up!*" repeated five times over a period of nine seconds. Perhaps the most distressing part is that there follows the sound of one impact, followed a second later by another, as the aircraft bounced from one ridge to another. Then silence.

For the crew it had been a desperate struggle, as the only way they could maneuver the aircraft appeared to be by varying the relative thrust of the engines. At one point they made, as mentioned, a tight 360-degree turn.

Tokyo ATC was calling them, and the nearer American Yokota Air Base was doing likewise. Though the Tokyo controller had suggested Nagoya Airport to the south, they seemed intent on going back to Haneda. Perhaps they were more familiar with it, perhaps because it was "home," and perhaps because it was at the sea's edge and they could ditch the aircraft in the water if their approach failed and they did not come a cropper on the unforgiving sea wall on its perimeter.

The crew knew they had no hydraulic pressure to control the aircraft's control surfaces but did not know why. This in itself was surprising to them, as aircraft manufacturers design these key hydraulic systems with built-in redundancy on the belt-and-braces principle, in that if one fails, one of the others will take over. The systems were in triplicate. In addition, the crew did not know why the aircraft was so unstable. It was yawing (veering violently from side to side), rolling (one wing tipping down one moment, the other wing the next), and pitching (nose down one moment, nose up the next). Indeed, it was misbehaving in all three axes with a phugoid motion. This porpoise-like pitching, and to some extent the rolling, could be explained by the lack of hydraulics to control the ailerons and elevators, but why was the yawing so considerable? Little did the crew realize that most of the tail plane had been lost.

However, in trying to go back to Haneda instead of going to Nagoya, where they could at least have ditched in the sea en route should they not be able to reach it, they ended up going over the mountainous terrain near Mount Fuji.

When, over the mountains, they repeatedly tried to turn toward Haneda, the aircraft seemed to want to go the other way. Finally, when the speed dropped sufficiently, the first engineer (perhaps on his own initiative) lowered the landing gear. This had the effect of stabilizing the aircraft somewhat but had the disadvantage of making the aircraft lose height and at the same time made it more difficult to steer away from the mountains. The pilots tried lowering the flaps, but this made things worse, so they retracted them just before the crash. (They were able to lower the landing gear and maneuver the flaps by using a backup electrical system.)

They probably were doomed from the moment the incident started thirty minutes earlier. Knowing what we now know, one can say that it was a great feat of airmanship even to stay aloft for so long. In addition, Japan is not a good country in which to be in such a predicament, for it consists mostly of steep volcanic mountains. What little flat ground exists is put to use. Had the aircraft managed to escape from the high peaks near Mount Fuji to make its way directly to Haneda, it could have crashed in built-up areas, with even greater loss of life. Housing in Japan is often frail and closely packed and the aircraft could have cut a long swath through suburban housing.

The Investigation

No one could understand what had happened, and rumors were rife, and some continue to be so. Had it been hit by a missile? On the other hand, was it a bomb? Was it a meteorite?

Then some solid information began to come in. People on the ground had seen the 747 staggering through the evening sky, and an amateur photographer even had a blurry photo of it *without its vertical tail fin*. A Japanese naval vessel then recovered a large portion of the tail from the sea in the area where the pilot had first declared the emergency some thirty minutes before the crash. Subsequently, searchers recovered other pieces in that area.

Never had there been a structural failure like this on a Boeing 747. The aircraft was relatively old, but there were 747s twice as old still flying safely. However, as Japan Airlines employed the aircraft on short domestic routes, with many takeoffs and landings, it did have a large number of flight cycles, and it was thought likely that the accident might be linked to that.

A flight cycle is one takeoff and landing and is significant because each time an aircraft climbs into the sky, the pressurized air in the cabin makes the fuselage stretch a little, so everything moves and warps slightly. When the aircraft descends, the opposite happens. Metals do not like constant flexing beyond a certain elastic limit. A phenomenon called "fatigue failure" occurs, where the metal becomes brittle and fails. Anyone who has tried to break a thick copper wire has experienced the phenomenon: one bends it one way and then the opposite several times and it suddenly becomes brittle and breaks.

After the world's first passenger jet (Comet) disasters due to metal fatigue in the late 1950s, manufacturers were very careful to avoid areas of stress concentration. Thus, it was unlikely Boeing had made such a design mistake. Nevertheless, it was very worrying for the company, as the safety of the entire fleet of 747s might be questioned, leading to enormous financial loss. A number of countries immediately ordered their airlines to carry out checks on the tail section.

Boeing sent telexes to all users with information about the maximum number of cycles other 747s had flown without structural trouble. The maximum number of flight cycles for any 747 was 22,970, which was somewhat more than the crashed aircraft's 18,830. Thus, though JL123's number of cycles was relatively high, it was not deemed excessive.

Investigators began to direct their attention toward the rear pressure bulkhead, which though made of quite thin material is a kind of plug keeping the pressurized air in the cabins from escaping through the rear of the aircraft. However, Boeing had designed it to last at least twenty years and the crashed aircraft was only eleven years old. The bulkhead could have failed because of corrosion—indeed, another type of aircraft flying out of London had crashed due to corrosion of the rear pressure bulkhead fourteen years previously—but there was no sign of corrosion on the JAL bulkhead.

Then one of Boeing's investigators discovered a small piece of the rear pressure bulkhead with repairs made using a splice and rivets, with merely a single row of rivets. How could there only be a single row of rivets when Boeing, for the very fatigue failure reason mentioned above, always mandated that there be a double row of rivets to prevent concentrated stress leading to fatigue cracking?

Records showed the aircraft had suffered a tail strike seven years earlier on landing at Osaka Airport with its nose too high in the air and had been out of service for three months for repairs. Boeing immediately thought they were off the hook, as it was assumed Japan Airlines had been responsible for the incorrect repairs. Then to their consternation they discovered the repairs had been carried out under the supervision of, and according to instructions from, their own engineers sent over from the States!

There followed a period of buck-passing. Japan Airlines said Boeing was responsible for the crash due to the faulty repairs; Boeing said it was Japan Airlines' fault for not having detected the crack. One of Boeing's arguments was that as the Japanese are chronic smokers, there must have been visible signs of tar on the edges of the developing crack on the far side (from the cabin). Japan Airlines countered by saying it was in a visually inaccessible place, and anyway, no one would expect to see anything wrong at such a spot in the absence of signs of corrosion, say from water from the toilets or galley. These fatigue cracks can take years to develop and might have been virtually invisible early on.

Allegedly, Japan Airlines accepted 20 percent of the blame on the grounds that for an extended period up to the disaster, whistling had been heard at the back of the aircraft in question and no one had checked it out. (A number of websites say this, but as they use identical wording, they must be copying one another, making it difficult to find the original source.)

Boeing

Finally, both companies realized their slanging match was mutually damaging and decided to compromise. After all, Japan Airlines was a major Boeing customer, and falling out with it could in the long term cost much. Another point was that the accident happened in Japan, where there is not the lawyer compensation culture found in the United States, and the payouts were not as great as they might have been. After arranging for some free modifications to existing 747s worldwide to ensure that failure, however unlikely, of another rear bulkhead would not result in catastrophic consequential damage, Boeing finally came out of the affair relatively unscathed.

Japan Airlines

The same was not true for Japan Airlines. No one could prove it was entirely its fault, but the Japanese press turned on it like a lynch mob. This was partly due to the haughty attitude the featherbedded airline had shown over the years. The media hounded it relentlessly, picking up on every possible fault. It took years for it to recover, and perhaps domestically it never quite has.

In Japan it is always felt that someone should take the blame and demonstrate his or her contrition. In the case of the 747 disaster, the police treated the crash site as a crime scene and only let the investigators borrow the debris. The main purpose of air accident investigations in the West is to determine the cause so a similar thing does not occur again, but even in the United States there is an increasing tendency to obfuscate, since the potential legal liability can be so costly. However, where the impartiality of the accident investigators might later be questioned, as in the case of the Airbus crash at a tiny air show in France, there is something to be said for judicial authorities ensuring flight recorders are not tampered with, assuming the judicial authorities can be trusted.

In the following months, someone at JAL did commit suicide, leading the press to conclude he was the man responsible for the bad repairs or for not finding the crack in the pressure bulkhead. In fact, the suicide had nothing to do with the crash. Yet if someone at JAL *had* felt personally responsible, that individual might well have committed suicide. For instance, around about that time there was a sad case where passengers who had had omelet for breakfast on a JAL flight staggered off the aircraft at Tokyo with serious food poisoning. A young JAL cook at the Anchorage stopover with an infected hand had handled the mixture of blended eggs, which had presumably been left standing uncooked. No one died, but he committed suicide nevertheless. Most safety experts would now say it was a systems or management failure, for the airline should have ensured the cook had gloves or had been kept off work—he had been too conscientious, returning to work.

Conclusion

The conclusions were that to prevent such accidents one should:

1. Find ways to avoid and detect such maintenance or repair mistakes through better management (oversight) and improved checks.
2. Ensure one failure does not result in collateral damage that could endanger the aircraft.

The author has heard it suggested that had Japan Airlines carried out maintenance less assiduously and left holes and weak parts in the fuselage behind the bulkhead, the air would have been able to escape through those without blowing off the tail fin!

Was Depressurization Really So Rapid?

A retired JAL pilot, Hideo Fujita, has written a book entitled *Kakusareta Shougen* (Hidden Testimony), in which he doubts that sudden depressurization ever occurred and says that the cause of the accident must lie elsewhere. His reasons for doubting that the decompression was as sudden as claimed are:

1. The pilots were able to continue physically demanding movements without donning their oxygen masks for some time, even though the aircraft was at twenty-one thousand feet, where the air is thin.
2. The rescued off-duty flight attendant, Yumi Ochiai, had heard a bang above her, not behind her, and had seen no papers flying around in any direction, which one would have expected had the air been rushing out of a big hole. (In the author's opinion, the panels of the tail fin could have been blown off without the immediate exodus of a large volume of air.)
3. Shortly after the crash, Japan Airlines said the rescued off-duty flight attendant had said she had seen air coming out of the floor vents, when

in fact she could not have seen them from her seat and had never heard of the term.

Whatever the truth, it is certain that the official cause of the crash being faulty repairs by individuals was very convenient for Boeing and even JAL. The Japanese police were not able to get their hands on the US individuals, and prosecutors were not intent on pursuing the matter. Nothing was ever made of the fact that more people could certainly have been saved had help come earlier. The report from a helicopter arriving early above the crash scene that there were surely no survivors was a factor. The Americans, who were probably the best placed to help, were kept away. The US military presence in Japan is such a delicate issue that no one on the US side would want to comment. Judging by the appreciation shown for the help given by the US forces in connection with the 2011 earthquake and tsunami, attitudes may have changed radically.

One must remember the slopes were too steep and tree-covered for helicopters to land even in daylight and that most of the potential survivors would have been in the tail section that had broken off and slipped down the steep slope to end up quite some distance from the main site. They would have been difficult to find in the dark, notwithstanding the cries and moans mentioned by off-duty flight attendant Yumi Ochiai.

Another footnote to the JL123 disaster is that for the Japanese families, the proper handling and fullest possible recovery of the bodies and all their parts was of the utmost importance. Senior people from the airline attended every funeral, and staff continue to visit the crash site on every anniversary. The impact on the airline's daily operations must have been considerable.

---

[1] TACAN a more accurate military version of the VOR/DME beacon.

Boeing 737 Rudder Quirk Took Years to Demonstrate (1991/1994)
Two disasters and almost a third

> The story of the exhausting investigation as much as of the incidents. Cockpit data recorders at the time did not reveal what the pilots actually did with the controls, so they were easily suspect.
>
> United Airlines Flight 585, March 3, 1991
> USAir Flight 427, September 8, 1994
> Eastwind Airlines Flight 517, June 9, 1996

United Airlines Flight 585
The First Incident at Colorado Springs

With no warning, the United Airlines Boeing 737 from Denver rolled sharply to the right as it approached Colorado Springs Airport. Then the nose pitched down. Abandoning the landing, the pilots increased thrust and reset the flaps to fifteen degrees to recover. But on down they went, in a tight spiral, sustaining accelerations over 4 g before crashing in a park less than four miles from the runway, killing all the occupants.

Mercifully, there had been only twenty-five people on board, and this is why the accident did not receive the attention it might have.

With their usual due diligence, the NTSB investigated the crash but, unusually, were unable to determine the cause, partly because at the time and for some time afterward cockpit data recorders were very primitive—in this Colorado crash only five parameters were recorded, namely heading, altitude, airspeed, g loads (vertical acceleration), and microphone keying. All parameters were sampled and recorded once per second, except vertical acceleration, which was sampled eight times per second.

In fact, it was not the recorder that was primitive but the aircraft, in that unlike today's airliners it was controlled by wire cables that physically pulled actuators, with no need for sensors that could provide data to the recorder capable of handling many more parameters.

The fragmentation of the aircraft and damage by fire made finding informative physical evidence difficult. The rudder power control unit (PCU), which was a key item, was too damaged for valid testing. Nevertheless, there were the radar tracks and reports from hundreds of witnesses.

What made this investigation, and the one for a similar crash three and a half years later, so inconclusive was there was no means to tell exactly what the pilots did, such as did they crazily stomp on the rudder pedal, and if they did, which one?

There were two main theories as to its likely cause:

1. The power control unit (PCU) caused a rudder "hardover," making the aircraft veer to the right and tip downward at the low speed it was flying at in preparation for landing.

2. Strong vortices of swirling air (mountain rotors), known to have been present in that general area but not necessarily at that location, had upset the aircraft.

Boeing considered the latter more likely, with the pilots very likely having stupidly stomped on one of the rudder pedals. However, a study of the fifty-two-year-old captain's and forty-two-year-old first officer's backgrounds suggested that neither he nor she would ever have done such a thing.

One should explain that rudders, unlike those on ships, are not much used on airliners. To turn, the pilot, human or automatic, uses the ailerons to bank the aircraft to the side toward which they want to go, after which the aircraft turns naturally. Otherwise, the rudder is mainly only used when landing in a crosswind.

The reason why airliners, such as the 737, with the engines attached way out to the wings rather than close in to the fuselage have such enormous rudders is to correct the tremendous torque produced when an engine fails and they should be able to fly (safely) on one engine pulling just on one side.

For simplicity, slim design, and cheap maintenance, the 737 was virtually unique in having what, on the one hand, was a very simple rudder system but, on the other, depended on a too-clever-by-half, two-in-one rudder power control unit. This control unit was designed by Boeing's engineers and manufactured for them by the Parker Hannifin Corporation per their specifications.

Other airliners had three PCUs and even split rudders so that if one half did an unwarranted hardover, the other half could easily compensate, with the effect anyway half what it otherwise would be. The 737 had a standby PCU that could be activated by the pilots if necessary, and if they had time.

Though Boeing considered a rudder hardover most unlikely on the 737, they claimed it could be handled, provided the pilots dealt with it correctly. However, this depended on their knowing what to do and the airspeed being high enough for the ailerons to be able to compensate.

USAir Flight 427
The Second Incident, at Hopewell

Tom Haueter, a senior investigator at the NTSB, was on call, having offered to stand in as a favor for a colleague wanting to go on holiday. Little did he know that would mean being bestowed with an investigation that was to dominate part of his life and even strain his marriage due to the unrelenting calls on his time, day and night, for months and months.

That day he was head of the go-team, consisting of experts in various domains always ready to go to any air crash at a moment's notice. They were in Washington, DC, and the crash had been at Hopewell Township, about nine miles from Pittsburgh International Airport, and almost an hour's flying time from where they were.

The first problem Haueter faced was mundane but not simple: finding a hotel they could use as a base for the investigations and a place for the team to stay and liaise. USAir had booked them all, and it was only after much effort that Haueter managed to persuade the airline to give one up—the Holiday Inn near Pittsburgh Airport. It was late in the evening, and the pilots of the FAA aircraft they would be using had run out of hours. They would have to fly down to Pittsburgh the following morning.

When Haueter did finally reach the USAir Flight 427 crash site, he found body parts everywhere—even hanging from trees. He declared the site a biohazard, meaning his staff would have to wear full-body biohazard suits for their protection as they fumbled through the wreckage looking for evidence, which had to be disinfected before being taken away for further study.

All 132 occupants—127 passengers and 5 crew—on board the Boeing 737-300 had died on impact.

For simplicity, we shall refer to this second 737 crash, which had much in common with the Colorado one, as the Hopewell crash.

USAir Flight 427, a Boeing 737-300, had departed from Chicago for Palm Beach, Florida, with a stopover at Pittsburgh.

As it approached Pittsburgh International Airport, it encountered the wake of Delta Flight 1083, a Boeing 727-200, about to land before them. The 727 was four miles ahead, and vortices produced would not in theory have been sufficient to engender a sustained changing of heading without input from the pilots.

The aircraft yawed slightly to the left, banked to the left, and rolled over twice leftward before hitting the ground upside down in a wooded area with a dirt road.

At first it seemed to Haueter and his colleagues that the investigation might be over more quickly than usual. The cockpit data recorder showed the aircraft had yawed and rolled to the left; this and other evidence clearly pointed to a rudder malfunction and, unlike in the first incident, the suspect power control unit (PCU) controlling the rudder had been recovered fully intact.

In fact, it was to take more than four and a half years before the NTSB would be able to draw the investigation to a satisfactory conclusion, and this process was greatly helped and given traction by a further incident, which fortunately resulted in no fatalities.

Eastwind Airlines Flight 517
The Third Incident

An Eastwind Airlines Boeing 737-201 reported a loss of rudder control while on approach to Richmond, Virginia. The airplane was on a regularly scheduled passenger flight from Trenton, NJ to Richmond, with forty-eight passengers, two pilots, and three flight attendants on board. Apart from a minor injury to a flight attendant, there were no injuries or damage to the airplane as a result of the incident. At the time of the event, the aircraft's airspeed was about 250 knots, and it was at four thousand feet MSL.

The captain reported that he was hand flying the airplane and he felt a slight rudder bump to the right. As he was asking the first officer whether he had felt the bump, the aircraft suddenly rolled to the right. He applied opposite rudder but said he felt the rudder resist. He said he then applied opposite aileron and used asymmetric power to keep the airplane upright. He stated that after he declared an emergency to the approach controller, he and the first officer performed the emergency checklist. The captain reported that as part of the checklist they turned off the yaw damper. He reported that the airplane became controllable but was not certain as to whether the problem went away at the same time as the yaw damper was turned off. The aircraft rolled over again but that time to the right. After another thirty seconds, the plane snapped back to level flight.

It is reported that the airplane has previously had problems with uncommanded rudder deflections. Previous reports had been of "rudder bumps" during departure and that the airplane would not trim properly. The FDR was removed from the airplane for examination.

The next day, people from the NTSB bombarded the pilots with probing questions. Here, at last, they had the aircraft and the pilots intact.

The Investigation(s)

The Colorado investigation had got nowhere, and after three and a half years the NTSB had officially admitted they could not determine probable cause—only one of four instances where that had been the case.

Still present were some of the factors that had made that investigation so difficult in the Hopewell disaster, one of which was that the cockpit data recorder, although providing slightly more information, did not record what the pilots did with the rudder pedals. This gave Boeing much wiggle room, with them able to claim (and to some extent believe) the pilots again stomped on the pedals mistakenly on being startled by the wake turbulence from the 727 ahead.

The NTSB investigators had faced a conundrum in that the Boeing 737 was too safe. Three and a half years had elapsed between the Colorado and Hopewell disasters, with the first having few fatalities, and so many 737s were flying safely worldwide. Even recommending a different rudder system would be very expensive, and anyway, the lengthy approval system meant that it would take time to come into effect, especially if it were ordered without convincing proof of its necessity.

To be fair, Boeing had not stinted on getting hundreds of engineers and experts to consider the matter and genuinely thought reasons other than the PCU were probably responsible. If they had definite proof, they would have acted, notably at the instigation of their lawyers, as they did later.

With even some at the NTSB having doubts after years of theorizing, tests, and meetings, the Eastwind incident was a shot in the arm for Haueter and the NTSB investigators, not only because it provided more material, but also because it gave credence to their belief that it was not the pilots but the PCU that must have been at fault.

> *Two similar cases might be coincidence, but three would be beyond coincidence.*

This knowledge helped sustain Haueter in his endeavor to find proof, and that proof came from dogged determination to pursue the search, though almost by chance. The revelation came in two parts.

Firstly, with the NTSB having been clutching at straws for so long, Jim Hall, chairman of the NTSB, who was a politician rather than a man with technical expertise himself, decided that a panel of the best experts in the domain of hydraulics and associated valves should review what had been done. Collaborating with the panel was a team under the NTSB's Greg Phillips, responsible for systems evaluation.

When the panel met, one of its members, sixty-seven-year-old Ralph Vick, mentioned to Phillips how thirty years before, while working at Bendix on a valve in the hope of a contract from Boeing for the 747, they had done a thermal shock test and found the valve jammed; however, when redesigned with different tolerances, it didn't. Could they try that?

Out of ideas, they did, albeit under somewhat primitive conditions, and found the new valve from the factory never jammed but the one from the crash did. What is more, when taken apart and examined by Hannifin Parker, the precision parts showed no marks indicating any jamming.

This was not quite the Eureka moment it was reported to have been in some accounts, for while showing it could jam without leaving any evidence, it did not explain the hardover. Furthermore, as Boeing insisted, the extreme conditions under which the thermal shock test had been carried out were impossible in the conditions under which the 737 flew.

Secondly, six weeks later, Phillips' team went to a Boeing laboratory in Seattle to run the test under more sophisticated conditions with better monitoring. Again, the factory unit passed, while the one from the crash failed the most extreme thermal shock test. The technician moving the valve back and forth felt it slow down; he didn't notice it bind, but a computer showed it had jammed momentarily. He repeated the action and felt the lever kick back when he tried to move it to the right. When he tried again, he felt it stick to the left and then jam. In his excellent book *The Mystery of Flight 427: Inside a Crash Investigation*, which took six years to research and write, Bill Adair explains what happened next.

"Despite their skepticism, Boeing engineers said they would examine the charts from the tests for anything unusual. A few days later, in a building overlooking Paine Field in Everett, a young Boeing engineer named Ed Kikta sat at his desk, reviewing the charts. He could see the test data on his computer screen, but he liked to print the results so he could study them more closely. The charts showed the flow of hydraulic fluid during each test: higher when it was pushing the rudder and down to zero when it was not. Kikta expected that when the outer valve jammed during the thermal shock, the inner valve would compensate and send an equal amount of fluid in the opposite direction, which would keep the rudder at neutral. That was the great safety feature of the 737 valve. It could compensate for a jam.

"But as Kikta studied the squiggly lines for the return flow, he saw dips that were not supposed to be there. When he matched them to another graph showing the force on the levers inside the PCU, he made an alarming discovery. When the outer valve had jammed, the inner valve had moved too far to compensate. That meant the rudder would not have returned to neutral, the way it was supposed to.

"The rudder would have reversed. That could be catastrophic. A pilot would push on the left pedal, expecting the rudder to go left, but it would go right."

That was the real Eureka moment.

Kikta's colleagues and boss confirmed the significance of his findings, and Boeing ordered the PCU manufacturer, Parker Hannifin, to check out the results. Their engineers confirmed them, adding that the levers in the PCU could flex, allowing the inner valve to line up with the wrong holes!

Boeing wasted no time. Taking a new 737 straight off the production line, they installed a PCU specially made by Parker Hannifin to simulate the jam. With Kikta on a trestle watching the rudder and PCU, they had a Boeing test pilot play around with the rudder pedals. At first nothing out of the ordinary was found, but when the test pilot pressed on the left pedal and then pressed down as hard as he could on the right one, it kicked back violently. Further tests showed that how the inner slide behaved depended on where the outer one jammed.

Aware how serious the situation was, Boeing went into overdrive, gathering their engineers to seek a solution. The next day they informed the FAA but asked for twenty-four hours to work out what should be done.

The fact that the NTSB was the last to be informed might seem to have been something of a slight. In fact, as the NTSB can only recommend, and it is the FAA that regulates at its discretion, according or not to whatever the NTSB says, it made sense for Boeing to do all they could to keep the FAA on side. Though Haueter had been in a meeting all day in the presence of someone from Boeing, that person never mentioned the latest findings—he later denied he knew. Haueter only found out by text message from his boss the following morning, who himself had just got the news from the FAA.

Boeing sent telexes to all airlines using the 737, warning them that under certain conditions some jams of the secondary slide could result in anomalous rudder motion.

Haueter carried out further tests and concluded that a rudder-induced disaster required an extremely rare combination of circumstances. It concerned just 1 percent of the PCUs with certain inherent characteristics where the aircraft encountered turbulence to which the pilot reacted in a certain way, and where the airspeed on coming in to land was too low for the ailerons to be able to override the effect of the rudder. (The airspeed above which the ailerons could override the ailerons being called the crossover speed.) By naturally pulling back on the control column in an effort to save themselves, the pilots were only making their predicament worse.

Haueter, with his colleagues' and bosses' support, finally got the NTSB board to make a final determination, thus ending an investigation that had taken five years, or almost ten if one includes the first incident.

However, publicity-wise, Boeing rather took the wind out of the NTSB sails by preempting them, saying they were taking measures to make the 737 *even* safer.

Oddly, unlike the McDonnell Douglas DC-10, where the second crash—largely due to the way the aircraft had been maintained by raising an engine with a forklift, putting too much strain on a component—resulted in the public and airlines losing faith in the aircraft, they never feared to fly on the 737, though some pilots were somewhat apprehensive.

We have only been able to touch on aspects of the longest air crash investigation ever. It finally ensured the most widely used airliner would be "even" safer.

Bill Adair's book explains it all in detail, and notably the party system, by which the relatively minuscule NTSB, with only limited resources, allows all parties to be part of its investigations. It thereby gets results it otherwise might not, despite each party fighting its corner until the truth becomes undeniable. No one knows an aircraft or component better than its maker, and nowadays that increasingly includes the software. However, it does have the drawback that the maker of a component that could have caused a crash for which it could be liable for enormous damages is sent that component for initial evaluation, as in the 737 SilkAir "suicide" disaster described elsewhere in this book.

There would seem to be a case for the NTSB to be better funded and able to fund more independent work.

Boeing redesigned the rudder mechanism, dispensing with the clever two-in-one PCU and replacing it with a dual input PCU and an automatically activated standby PCU. They also incorporated additional safety features, including some recommended by the NTSB. Work on existing 737s was done largely at Boeing's great expense.

There was one further scare, in rather different circumstances, implying the failure mode was even more complicated than that determined, but none after the modifications, apart from issues related to quality control in the manufacture of the control rods.

Northwest 747 Lower Rudder Hardover (Anchorage, 2002)
The aircraft was barely controllable at cruise. Would it be when slowed to land?

> The popular press tends to say a pilot is "fighting at the controls" when, in fact, by not diverting he or she caused the crash. In this case, the pilots were truly, and physically, fighting at the controls, miles from anywhere.
> *Northwest Airlines Flight 85, October 9, 2002*

Being the launch customer for an aircraft has its advantages, including publicity and very keen prices. However, a number of bugs might still have to be worked out, causing expensive delays and canceled departures. Though launch issues cannot be directly linked to what occurred in this instance, the aircraft happened to be the first production Boeing 747-400, delivered to Northwest Airlines on December 8, 1989.

The flight had taken off from Detroit, with 386 passengers and 18 crew, bound for Tokyo's Narita Airport, a flight of thirteen hours or so. Much of the early part had been over the US and Canada and the US again (Alaska).

Unlike the early 747s, which had ninety or more dials and switches to watch over, these later 747s had computers and no need for a flight engineer. However, some of the control surfaces had semi-direct manual control, unlike the latest fly-by-wire airliners, where a signal is sent electronically or otherwise to actuators that do all the work, though some sensation might be simulated to give the pilots the feel they are actually flying.

Because of the flight's thirteen-hour duration, there were two flight crews, and as the aircraft left Alaska and headed out over the Bering Sea, Senior Captain John Hanson and First Officer David Smith were settling in to relax and rest while Captain Frank Geib and First Officer Mike Fagan did the actual flying up front in the cockpit.

The aircraft was almost at the halfway point and above the Bering Sea when Captain Hanson felt it perform an odd maneuver. He and First Officer Smith immediately began putting on their uniforms, and no sooner had the aircraft recovered than the special chime sounded to indicate their presence was needed in the cockpit right away.

On arriving there, they found Relief Captain Geib literally fighting at the controls. The control wheel was about halfway over—something that would never happen in cruise—and Geib's right leg was straining on the right rudder pedal. Yet the instrument panel showed that the lower rudder was at the left (opposite) stop and, surprisingly, at a deflection of seventeen degrees, when at altitude the aircraft would never allow it to exceed about six degrees. They were at thirty-five thousand feet.

Captain Geib told them that the airplane, on autopilot, had suddenly begun an uncommanded roll to the left that reached almost forty-five degrees before he realized that the autopilot could not cope and disengaged it. By pressing as hard as he could on the right rudder pedal to move the upper rudder to the right and using the ailerons, he just managed to correct the bank. Had he not reacted quickly, the aircraft would have tipped right over and plunged earthward. With Tokyo more than six hours' flying time away and Anchorage only two, he had made a gentle left turn—the only direction in which he could—back to the latter. Too far away, they had to contact another Northwest airliner to ask them to relay their messages to the controllers.

Their manuals had solutions for almost every conceivable problem but said nothing about the one they had. Later they were able to contact Northwest by flaky HF radio, but they had no suggestions either. Unable to see the tail, they were in constant fear that the situation might deteriorate or that it might even break off, leaving them helpless.

As senior captain, Captain Hanson had taken over from Captain Geib, even though he had done a sterling job. Because of the physical effort needed to keep pressing down on the right rudder pedal, Hanson and copilot First Officer Fagan took turns.

Not knowing how the aircraft would behave when slowed for landing, they decided to approach Anchorage at about fourteen thousand feet, which, according to Hanson, "is a nice intermediate altitude. It's low enough that the air is nice and thick, and it's high enough that if you do lose control you can make one good honest attempt at recovery before the water."

They managed to land safely, even keeping the aircraft straight on the runway, so the feared emergency evacuation proved unnecessary.

Investigation

The fault causing the problem was found to be a most unlikely one, namely fatigue fracture of the lower rudder power control module manifold—a part that should have lasted forever. The module was redesigned with internal stops so that in the unlikely event of a reoccurrence, the deflection of the rudder would be limited and easily handled.

DHL Airbus Maneuvered by Engine Power Alone (Baghdad, 2003)
A remarkable feat that did not receive publicity it merited

Outside Homeland Security circles, not much attention has been paid to the feat whereby the three-man crew of a European Air Transport/DHL Airbus A300B4 freighter brought their aircraft safely back to Baghdad Airport by juggling engine power alone after it had been struck by a surface-to-air missile (SAM).

DHL Cargo Flight, Baghdad, November 2, 2003

Some superior surface-to-air missiles possessed by insurgents in Iraq had a maximum range of four and half kilometers at heights up to about ten thousand feet. Aircraft coming into Baghdad would keep well above that ceiling and descend spirally at the last minute with degrees of bank impossible to practice in simulators set with the conventional parameters. However, departure from Baghdad was riskier for workhorse aircraft unable to climb quickly (as opposed to fighters).

As a result, the DHL Airbus freighter that took off from Baghdad at midmorning on November 22, 2003 was only climbing through eight thousand feet at 9:15 a.m. when one of the two Russian-made SAM-14s fired from the ground struck it near the left wingtip. The crew, consisting of two Belgians, Captain Eric Gennotte and First Officer Steeve Michielsen, and Scottish flight engineer Mario Rofail, felt a judder, followed shortly by an alarm indicating trouble with the hydraulics. The flight engineer saw that two of the three independent systems showed zero pressure and a complete loss of fluid. Before the crew could go through the steps required for flying with just one hydraulic system operable, the pressure in the third and only remaining system dropped to zero. The instruments also seemed to indicate the fuel in the outboard left-wing fuel tank had disappeared.

Like in the cases of the "uncontrollable" JAL 747 and the Sioux City DC-10, their only hope lay in varying the engine thrust to direct the aircraft, except that compared to the JAL 747, they still had the complete vertical stabilizer (tail fin) to steady them—but not steer them—horizontally. Even so, the aircraft yawed to the left and the left wing began to sink, with the nose dropping. However, by adjusting the relative thrusts of the engines—more power on the left, less on the right—it was possible to correct this before the craft entered into a sideslip and spin, from which it would have been impossible to recover without the help of the rudder and ailerons driven by hydraulic power.

On limping back toward the airport, they were erroneously informed by the tower that the left engine was on fire. From their position on the flight deck, and with no cabin crew (and no windows, since it was a freighter), the crew were unable to see the extensive damage to the left wing. They had some measure of control, but, concerned about the alleged engine fire, they did not dare spend time experimenting, like the pilots of the Sioux City DC-10 had been able to do. They were intent on landing but found they were too high and going too fast to risk it without the use of air brakes and abort it.

This meant they had to make a time-consuming circuit getting far enough out to be able to make a shallow approach that they could manage. The pilot of a US Apache helicopter had informed them that it was not the engine that was on fire but the wing extremity. This was somewhat reassuring, but with fuel leaking from the wing there was the danger that fuel for that engine would run out. With no power pushing it up, that wing would tip downward, and they would plunge to the ground.

The only plus was that, having lowered the undercarriage for the landing, the aircraft after a few scary moments had become slightly easier to control.

On their second attempt, some sixteen minutes after the missile strike, they were aiming for Runway 33R but fortunately had veered to the left one and, caught up by turbulence at about 400 feet (120 meters), had to increase thrust to raise the right wing, touching down just off the centerline of 33L, the runway parallel to it on the left. "Touched" down is something of a misnomer, as they came down hard with a sink rate of some two thousand feet a minute instead of the usual maximum of three hundred feet a minute, and an airspeed of roughly 215 knots instead of the usual 140 knots or so.

Rofail immediately deployed full reverse thrust, but the Airbus veered off the paved runway to the left—had they been on the right-hand runway, they would have hit the fire station. The aircraft ran through rough soft ground, throwing up a plume of sand and dragging a razor-wire barrier, coming to a halt in about 3,300 feet (1,000 meters). With no hydraulics to operate the steering, they were unable to correct their course.

With no brakes, no spoilers, and no flaps, they only had reverse thrust to slow the fast-moving aircraft. Once off the runway, a 600-meter run through the sand with sparse grass helped bring the hurtling aircraft to a halt at a razor-wire barrier. The crew were able to evacuate safely down a second chute after finding the first damaged by the razor wire.

With emergency vehicles rushing to the scene at the airport perimeter, the three men evacuated via an emergency slide, stunned they had survived. Standing on the hot sand, they were about to make their way to the emergency vehicles, which, surprisingly, had held back.

"Don't move!" someone shouted. "You are in a dangerous zone with unexploded ordinance."

One of the drivers said he would back up to them—if the rear wheels did set anything off it would be far away from the cab—and drive out with them following in his tire tracks, which they duly did.

The nightmare was over.

Backstory

On that day, Claudine Vernier, a journalist for *Paris Match* magazine, who had been interviewing insurgents, was invited out with her cameraman to see them in action. Taken to an area near the airport, she was surprised to see masked men carrying missile launchers. The leader explained the difference between the SAM-14 and SAM-7, after which they made ready to fire a SAM-14 at an aircraft that had just taken off, in fact the DHL A300. Only then did the two journalists realize the insurgents were not bluffing.

Seeing they had hit the aircraft but that it was still flying, the insurgents fired a second missile, a less effective SAM-7, which missed.

Afterwards, the journalists were criticized for not having interceded, but Vernier said that would not have made any difference, except perhaps getting them shot in the head. They realized they had been set up. Their videos can be found on the Internet and were used in the *MAYDAY/Air Crash Investigations* portrayal.

Technical Points—Damage to Aircraft

The missile struck the rear tip of the left wing, setting the fuel tank on fire, but did not cause it to explode as, being full of fuel just after takeoff, there was no vapor. The slipstream meant the leaking fuel was burning along the trailing edge of the wing, with it being gradually eaten away, reaching the main spar, which was on the point of failing when they touched down.

A large section of the left wing was missing when they landed, making one wonder how it provided the lift that it did.

Aftermath—Airliners at Risk Almost Everywhere

The three-man aircrew were officially honored for their great achievement, which, with no passengers involved and not in full view in a major city, did not receive the publicity it deserved.

The crew, consisting of two Belgians—thirty-eight-year-old Captain Éric Gennotte and twenty-nine-year-old First Officer Steeve Michielsen—and a Scot, fifty-four-year-old Flight Engineer Mario Rofail, claimed they owed their success largely to how well they worked together as a team.

This highlights the fact that the "extra pair of hands"—in this case the presence of Flight Engineer Mario Rofail, and in the Sioux City miracle landing (described in Chapter 10) the presence of off-duty company DC-10 check and training captain Dennis Fitch, who, too, helped manipulate the controls—made all the difference.

Rofail said in an interview with the UK's *Flight International*, "Situations like this are unique every time. You cannot train for them."

However, some think you can and looked into ways pilots might be trained to cope. Not only that, it has been suggested that aircraft computers could be programmed so that pilots can switch to "thrust control," letting the autopilot fly the aircraft according to what the pilots input into the flight management system without them touching the controls.

The trouble is that total failure of the hydraulic systems is so rare it is not deemed worth the expenditure. Interestingly, when an engine of a Qantas A380 superjumbo disintegrated after the aircraft took off from Singapore, damaging many of the hydraulic systems, the fact that a number of functions were powered electrically rather than hydraulically helped make the incident survivable.

Another thing not deemed worth the expenditure for civilian operations, except for special cases, such as El Al and other Israeli airlines, the US president's Air Force Ones, and aircraft used by other notables, is having antimissile defenses incorporated in the aircraft.

These normally depend on detecting the missile and deploying chaff and flares to confuse it, or firing lasers to interfere with it. However, that is often not enough, for the pilots have to be trained to take avoiding action as well.

Israeli airlines found some countries unhappy with the fact that their flares might accidentally start fires.

While such measures can defeat heat-seeking MANPADS, such as those used against the DHL A300 at Baghdad, they would not be effective against radar-locking missiles, such as the BUK surface-to-air missile that shot down Malaysian Airlines Flight MH17 over Ukraine.

At the moment, the only option is to avoid conflict zones and rely on intelligence to foil plots elsewhere, which could be anywhere.

The crew were showered with praise and received a number of awards in recognition of their great feat of airmanship. Flight Engineer Mario Rofail retired, but the other two continued flying. DHL resumed service to Baghdad a week later, carrying letters and other items for the troops.

Chapter 7
FIRE AND SMOKE

Fire in Varig 707 Toilet on Approach to Orly (Paris, 1973)
Dense smoke in cockpit forced blind pilots to land short.

> With the high cost of aircraft, and passengers' and crews' lives so
> valuable, it is surprising that simple safety features, such as smoke
> detectors in the toilets, were not in use at the time of this accident and
> were only made mandatory in the United States after an in-flight fire in a
> washroom ten years later.
>
> *Varig Flight 820, July 11, 1973*

Friends and relatives of passengers on the Brazilian Varig flight from Rio de Janeiro about to land at Paris's Orly Airport were impatiently watching the arrivals board. They could not understand the delay, as it was well past the expected time still being indicated.

These were old-style mechanical boards, where the letters and numbers flip over until they arrive at the right one. When those for their flight finally stopped, those expecting to see "Landed" saw just "Contact Company."

This was somewhat different from what happened when an Air France A330 plunged into the sea in 2009 on exactly the same route. Even though the aircraft had been lost many hours before its scheduled arrival in Paris, the signboards used the much kinder term "Delayed." In 1973, surprise changed to alarm as the assembled greeters began to grasp that something terribly wrong must have happened. Only later did they learn that a fire had broken out in one of the aircraft's aft toilets not far from them on the final approach to the airport.

Informed by the pilots of the onboard fire, air traffic control (ATC) duly authorized the captain of Varig Flight 820, Gilberto Araujo da Silva, to make a quick straight-in landing. Despite being able to breathe thanks to their full-face oxygen masks, the pilots opted to crash-land five kilometers short of the runway, as the smoke was so dense they could not even see their instruments and had to stick their heads out of the cockpit windows to see where they were going.

Ten occupants—all crew members—managed to escape. Fire crews, arriving some six to seven minutes later, were only able to rescue four unconscious people, just one of whom subsequently survived. In all, 7 crew and 116 passengers died. Most of the deaths were by inhalation of carbon monoxide and other toxic products resulting from the fire. For some passengers, it would have been like a prison gas chamber, since hydrogen cyanide was one of the gases given off by the smoldering plastics.

This accident led to the overdue installation of smoke detectors in aircraft toilets, a review of air supply to flight decks, and a review of the plastics used in cabin interiors.

Surprisingly, it was not this event, outside the United States, that made the FAA mandate that aircraft lavatories be equipped with smoke detectors and automatic fire extinguishers but the 1983 Cincinnati Air Canada Flight 797 in-flight fire in the United States ten years later. The FAA then also mandated that within five years jetliners be retrofitted with fire-blocking layers on seat cushions and floor lighting to lead passengers to exits in dense smoke.

In the case of Air Canada 797, the DC-9 was flying at thirty-three thousand feet from Dallas to Toronto when a fire developed in the area of a rear toilet. Before long the cabin began to fill with thick black smoke as the aircraft made an emergency descent. As in the Varig case just described, the smoke was so dense the pilots had difficulty seeing their instruments. In their case they did manage to effect a landing at Cincinnati Airport, but shortly after the doors and emergency exits had been opened, a flash fire swept through the cabin before everyone had had time to escape, with the result that half of the forty-six people on board died.

Possibly the Varig incident awakened the authorities to the obvious need for fire detectors in aircraft toilets, while the Cincinnati incident confirmed it. Sometimes it needs an accident in the States to make Congress press for action, though it is not certain whether that was true in this case.

Did Captain Fatally Delay Evacuation for King? (Riyadh, 1980)
All die when home and dry

> This example highlights the danger of not evacuating a burning aircraft
> as soon as possible. In this case, not only was cockpit resource
> management (CRM) poor or rather, nonexistent, but the second officer
> was dyslexic and had been thumbing through the flight manual,
> repeating, "No problem, no problem."
>
> *Saudia Flight 163, August 19, 1980*

The Saudia L1011 TriStar had come in to Riyadh from Karachi, Pakistan. At about 10 p.m. it lifted off for Jeddah, in Saudi Arabia. On board were 287 passengers, 11 cabin crew, and 3 aircrew. For such a short domestic hop there would be relatively little fuel on board and no need to jettison it if an emergency landing became necessary.

It had a Saudi captain, a Saudi first officer with very limited experience of the L1011, and an American second officer, who reputedly had been a captain at the airline, but on being found to be dyslexic had been allowed to stay on as a flight engineer. One would have thought being dyslexic would be a greater handicap for an engineer than a pilot.

While climbing through fifteen thousand feet to their cruising height, the flight crew heard an alarm. A warning light indicated the presence of smoke in the rear cargo hold, C-3. This first indication of possible trouble occurred just seven minutes after takeoff.

The crew then spent some four minutes verifying the alert, with the second officer thumbing through the flight manual to find the procedure to follow. Since they would have to double back this delay added eight minutes to the time it would take to get back to Riyadh. They had reached twenty-two thousand feet when the captain decided to return to do just that. Two minutes later the wisdom of his decision was confirmed by the presence of smoke to the rear of the passenger cabin.

Perhaps because the first officer had little experience of the L1011, the captain did not delegate tasks. He not only flew the aircraft himself but also performed the other tasks, such as communication with the cabin crew and the airport. He was undoubtedly overstretched but managed to touch down normally on the runway, despite having shut down the rear engine due to a jammed throttle lever—the fire had burned through the cables.

Unbelievably, once on the ground the captain did not try to stop the aircraft as quickly as possible but let it trundle down the runway for two minutes forty seconds. Even more unbelievably, the crew left the engines running for a further three minutes fifteen seconds after that, thus holding off the rescuers, who then found their unfamiliarity with the exits meant they could not gain access for a further twenty-three minutes.

The sequence of events was as follows. Slight discrepancies in the timings are due to them being derived from different sources.

In the air

Lapsed time	Local time	
	22:08:00	Takeoff from Riyadh
00:00:00	22:14:54	C-3 cargo hold smoke alarm
00:05:06	22:20:00	Return initiated
00:07:06	22:22:00	Smoke at rear of cabin, passengers panicking
00:10:32	22:25:26	Number two throttle jams. Fire already in cabin
00:12:46	22:27:40	Captain tells all to stay seated. Passengers fighting in aisles
	Final approach	Captain tells crew *not* to Evacuate
00:21:30	22:36:24	Touchdown

On the ground

Lapsed time	Local Time	
00:00:00	22:36:24	Touchdown
00:02:40	22:39:04	Continues down runway for two minutes forty seconds
00:05:55	22:42:18	Engines cut
00:28:38	23:05:00	Rescuers finally succeed in opening door 2R
00:31:30	23:08:00	Fire engulfs interior, no doubt due to ingress of oxygen

By the time the rescuers gained access, the 301 occupants were long dead. The influx of oxygen caused the fire to burn even more intensely, with the result the aircraft ended up with its upper half mostly burned away.

The captain had specifically instructed his colleagues on the flight deck not to evacuate. So confident was he that evacuation would be unnecessary, he did not even tell the cabin crew to get ready just in case. There was a final transmission after the aircraft had stopped that an evacuation was about to take place, but no one opened, or managed to open, a single exit—some said the doors could not be opened because the air pressure inside the cabin was higher than outside.[1]

Whatever the judgmental mistakes and failures of coordination on the part of the aircrew, a better form of heat and acoustic insulation above the cargo hold might have prevented the fire propagating so rapidly. Indeed, Lockheed subsequently replaced it with high-strength glass laminate.

Then there is the question of how the fire started. Interviews recorded for a 1999 BBC Panorama TV program called *Die by Wire* suggest the fire might have been caused by the Kapton insulation used for the electric wiring.

Comments by Investigators and Others

Review of the CVR showed a serious breakdown in crew coordination almost from the first sign of trouble.

1. The captain failed to delegate responsibility to the other crew members, deciding to fly the aircraft and try to assess and remedy the problem as well.
2. The first officer had very limited experience on the L1011 and did not assist the captain in flying the aircraft or monitoring communications or systems.
3. The second officer, who was thought to be dyslexic, spent nearly all his time searching through the aircraft's operations manual, the whole time repeating to himself "No problem."

The airline modified the procedures for coping with such emergencies and stepped up its training for evacuations. In addition, it made sure the C-3 baggage areas were sealed off.

It seems incredible that with conditions bad enough for passengers to be panicking and fighting in the aisles, the captain did not bring the aircraft to the most immediate stop possible and evacuate. The pilots were farthest from the fire. One possibility is that with the captain doing everything himself, he did not realize the seriousness of the fire, especially as the fire warnings ceased because the intense fire had destroyed the sensors.

Was this an instance where using video cameras to give pilots an indication of the situation in the passenger cabins could have been helpful?

Outraged Michael Busby "saw it all"

In the course of reviewing and updating parts of this book, we came across an account[2] of what happened that day at Riyadh airport by Michael Busby, an expat, who says he was watching from his villa nearby. He claims the reason the captain of the L1011 went down the full three thousand seven hundred foot (four thousand meter) length of runway, leaving the emergency vehicles standing-by halfway down far behind and moved off onto a taxiway was that the Saudi king's 747 was about to take off and was already taxiing. According to Busby, whenever the king's aircraft was in motion, the protocol was that all other aircraft should stop. Any Saudi not complying would be given a severe prison sentence. Foreign aircrew would be summarily dismissed.

Though this would explain a lot, it is difficult to believe that such a fact, if true, would not have surfaced before or been given more prominence when Busby suggested it. One could understand, however, companies such as Lockheed, hoping perhaps for major contracts in Saudi Arabia, not wanting to go down that avenue.

Busby notes that the passengers were mostly poor Pakistani pilgrims, renowned for bringing stoves with liquid fuel with them. Whether or not that had any bearing on the fire or its propagation, their families would not have been in a position to pursue litigation, especially with their government certainly not wanting to risk offending the Saudi royal family.

A Salutary Lesson?

Elsewhere in this book we have mentioned how passengers having had to evacuate an aircraft via the emergency chutes when no fire eventually breaks out often seek damages for the stress and injuries they have suffered. Looking at this tragedy, they should perhaps ask themselves whether they really would prefer that the crew allow them to stay on board when in doubt.

---

[1] Plug doors cannot be opened if the pressure inside is significantly higher than outside and is why passengers should not panic when they see a frightened passenger trying to do so at 35,000 ft.

The L1011 had a system to relieve the pressure on the ground but apparently, this was not used, perhaps because of damage in the fire. Later extra measures were taken to prevent such an eventuality at all airlines.

[2] http://www.scribd.com/doc/38040625/Death-of-An-Airplane-The-Appalling-Truth-About-Saudia-Airlines-Flight-163 Dated 2010

737 Stops with Fire Upwind (Manchester, UK, 1985)
Breeze blows flames onto aircraft

> A simple matter of how the aircraft stops with respect to the wind, albeit so slight a wind as to be insignificant from a flying point of view, can determine whether passengers live or die.
>
> *British Airtours Flight 28M, August 22, 1985*

The first production Boeing 737 short- to medium-range airliner was delivered to Lufthansa in 1968, and after a start that was so slow Boeing even considered abandoning production and selling the design to the Japanese, the aircraft became the most prolific airliner in the Western world.

By August 22, 1985, British Airtours, a subsidiary of British Airways, was using one for a routine charter flight from Manchester, in England, to Corfu, Greece. Virtually full, with 131 passengers and 6 crew members, the 737 was already engaged in its takeoff run, with the first officer as the handling pilot, when there was a loud thud. Assuming it was a tire blowout, the captain, Peter Terrington, ordered "Stop!" and at the same time pulled back the throttles and engaged reverse thrust. After having reached a maximum speed of 126 knots, the aircraft began to slow, with Terrington checking that the spoilers had deployed.

Terrington told First Officer Brian Love not to hammer the brakes, in order to limit the damage to the landing gear in the event of a blowout, and there was plenty of runway left anyway, as the decision to abort the takeoff had been taken well before V1. The first officer, who had been applying maximum braking, duly eased up on them.

As the groundspeed fell to 85 knots some nine seconds after the thud, Terrington called the tower to inform them that they were abandoning the takeoff. Almost immediately there was a fire warning for the left-hand engine. The tower then informed them there was a "lot of fire," and that the fire appliances were on their way.

With their speed below 50 knots, Terrington queried the tower as to whether an evacuation seemed necessary. The controller replied, "I would do via the starboard side." This was merely twenty seconds after the thud and twenty-five seconds before the aircraft came to a final stop. Some six seconds later, and fourteen seconds before the aircraft eventually stopped, Terrington turned the aircraft to the right so it could exit the runway via the Link Delta taxiway. Then, just before the aircraft came to a complete halt, he told the cabin crew to evacuate from the starboard side.

However, pooled fuel on the ground was burning, and flames were already lapping the rear fuselage. When the rear right-hand door was opened, no one was able to escape from there because of the flames and, worse still, flames soon penetrated through there into the cabin. What had at first seemed to be a minor incident was quickly turning into a disaster.

Difficulty in opening other emergency doors and obstructions of one sort or another resulted in two crew members and fifty-three passengers dying, and fifteen passengers sustaining serious injuries. Sixty-three passengers and one firefighter had minor or no injuries. Had the cabin been better designed, had materials for furnishings produced less toxic smoke, and had the evacuation been conducted in a more orderly fashion and the aircraft not stopped with the fire upwind, perhaps everyone could have escaped.

Training material for US firefighters even cites the disaster in stressing the danger hydrogen cyanide (HCN) given off by burning plastics represents, saying: "The fire killed fifty-four people, of whom forty-seven had possibly lethal cyanide levels, while only eleven had possibly fatal levels of carbon monoxide.

Painstaking studies of the disaster led investigators to make thirty-one recommendations, many of which were at the time deemed too expensive or not worthwhile on a cost-benefit basis, partly because the cost of a passenger fatality was not the $2.5 million or more it is in the United States today.

The thirty-one recommendations included the need to bring aircraft to a stop in such a way that the wind helps rather than hinders, and modifications of the air-conditioning system to prevent the spread of fumes and flashover fires. Allegedly, a passenger sitting next to an emergency exit was incapable of opening it, thus delaying exit from there for a crucial minute.

The late Professor Helen Muir, of the UK's Cranfield University, studied the incident and replicated the conditions in the cabin. When she offered the participants money for getting off quickly, she found they behaved as the passengers on the disaster flight had, climbing over seats and pushing others out of their way. The test suggested that the melee caused by those with a strong will to survive, with them jumping over seats and so on, resulted in far fewer people escaping overall. It seems more people escape if everyone follows the cabin crew's orders.

The underlying cause of the fire was improper cold fusion welding of a defective casing in the left engine. This was in part due to poor collaboration and poor exchange of information between the engine manufacturer (Pratt & Whitney) and British Airways.

737 Pilots Shut Down Wrong Engine (Kegworth, UK, 1989)

Classic case cited in pilot training

The disaster described below was once the one most often cited by pilot-training schools as an example of the dangers of precipitous action.

British Midland Flight 92, January 8, 1989

Pilots around the world were incredulous; some even thought Boeing might have connected the instruments the wrong way around. If not, how could experienced pilots make such a mistake, confusing right with left? In addition, the pilots had believed there was a fire when there was no fire.

The early-evening British Midlands Airways 737 flight from London to Northern Ireland was at 28,300 feet and climbing under rated power toward its cruising height of 35,000 feet. Then, thirteen minutes after takeoff, there was a loud bang, followed by a thumping noise and vibration. Passengers in the rear of the aircraft were disturbed to see flashes issuing from the tailpipe of the number one engine on the left—very apparent in the wintry darkness. Smoke or something like it then started coming in through the air-conditioning.

Pilots can observe events in front of them but not nearly so easily see what is happening to the aircraft and engines behind them. There are plans to install mini–TV cameras everywhere, both for technical reasons and as an antiterrorist measure, but for the moment pilots have to rely on their instruments in the first instance, and perhaps later on reports from cabin crew and passengers. The pilots felt the shudder and vibration and heard the noise. The captain later said he smelled and saw smoke coming in through the air-conditioning; the first officer just noticed the smell of burning.

The captain immediately disengaged the autopilot and took over control, according to standard procedure. He then asked the first officer which engine was giving trouble.

The first officer replied:

It's the lef—. It's the right one.

To which the captain replied:

Okay, throttle it back.

The captain later said he thought the smoke was coming in from the passenger cabin, and, based on his erroneous knowledge of the way the air-conditioning on that model of the 737 was designed, concluded the smoke must be coming from the right-hand engine. In fact, he could not have been sure where the smoke was coming from.

Whatever the facts, this meant the first officer was confirming what the captain already thought. The captain gave his order to throttle back the right engine nineteen seconds after the onset of the vibrations, so we are talking about a short time frame and little opportunity for reflection or study of the instruments.

When questioned later, the first officer could not say which instrument indication made him conclude it was the right engine. From the exact words of the first officer, it would seem that the first officer was not sure but felt obliged to give an answer. Could he have subconsciously sensed the answer the captain expected? Otherwise, why would he switch from "lef—" to "right"?

With the autothrottle disengaged, the first officer duly throttled back the right engine. One or two seconds later the vibrations seemed to decrease, seemingly confirming that the right engine had been the one producing the vibrations. Actually, the vibrations coming from the left engine decreased because it was no longer operating at full rated power, as the autothrottle was disengaged when the captain took over control manually.

The captain's and first officer's preoccupation with communications with ATC and contacting the company retarded the envisaged complete shutdown of the right engine, as the pilots were required to complete the checklist procedure together. In fact, the shutdown procedure was not initiated until two minutes seven seconds after the initial major vibration. During that time the vibration indicator showed no abnormal vibration for the right engine, which was not true for the left engine, where the vibration remained high, but not as high as initially. Indeed, just before the shutdown, the first officer remarked:

Seems we have stabilized. We've still got the smoke.

By adding the remark about the smoke, was he trying to hedge his bets with the captain? Whatever the case, it must have lessened the impact of the statement that they seemed to have stabilized. To be fair, after the shutdown of the right engine, the captain did try to review with the first officer what they had done and what symptoms they had seen, saying, "Now what indications did we actually get? [It's] just rapid vibrations in the aeroplane, smoke . . ."

Unfortunately, communications from ATC cut short this review. (One of the conclusions of the subsequent inquiry was that ATC should not overload pilots in emergencies.)

They were diverting to the airline's home base, East Midlands Airport, which happened to be nearby. Unfortunately, the airport's proximity meant they had no time to reassess the situation or even monitor the functioning of the left-hand engine over an extended period of level flight. The fact that ATC told them to change frequencies only served to increase their already high workload.

They increased power on their one and only engine[1] as they leveled out for a moment and made a turn to line up with the runway, and descended through three thousand feet thirteen miles from the runway.

Descending with gradually increasing flap, they put the landing gear down and set the flaps to 15 degrees. Then at nine hundred feet, with only 2.4 miles to go to touchdown (on the runway), the thrust provided by their remaining engine suddenly fell away.

The captain called for the first officer to restart the number two engine, but they were going much too slowly for a windmill start, and the failing number one engine could not provide the pressure required to restart the good engine. The flight manual did have details of how to relight using the auxiliary power unit (APU), but this explanation only applied to the number one engine.

Given time and a very good knowledge of the system, a pilot might be able to work out a way to restart the number two using the APU. This would involve switching off the air-conditioning and other procedures, and even if the first officer had been able to accomplish this, he could not have brought the engine up to speed in time to save the aircraft. Thus, the fact that the manual lacked such an explanation did not make any difference to the outcome.

Some seventeen seconds after the loss of power from the number one engine, its fire warning system activated. However, at that juncture, lack of airspeed and height, not fire, were the captain's principal concerns. Switching on the PA, he warned the passengers and cabin crew of the imminent contact with the ground by repeating the words: "Prepare for crash landing!"

The captain hardly needed the aural ground proximity warning system (GPWS) to tell him they were below the glide slope as he raised the nose in a vain attempt at least to get over the M1 motorway (freeway), which unfortunately lay transversely in a cutting along the airport perimeter. With the stick shaker[2] indicating they were about to stall, and their airspeed down to only 115 knots, the aircraft grazed the top of the hill lying just before the cutting.

With all lift virtually gone, it lopped off the tops of the trees on the nearside face of the cutting and plunged onward and downward, so that the nose struck the foot of the incline on the far side face. Taking into account the downward component of the trajectory, the encounter with the roughly thirty-degree-upward slope of the opposite face was equivalent to encountering an obstacle sloping at some fifty degrees head-on—at a ground speed of eighty to one hundred knots.

Though the aircraft did slither some way up the far side of the cutting, the deceleration, both vertically and horizontally, was many times more than for a crash-landing on level terrain. Indeed, the stresses imposed were so great that the rear section broke off and ended up lying upside down on top of the fuselage. Partly because the central fuel tank was empty due to it being such a short journey, and partly thanks to the rapid arrival of the airport fire services, there was no major fire.

Kegworth impact sequence. (Courtesy UK AAIB.)

In total, forty-seven passengers perished, sixty-seven passengers and seven crew members were seriously injured, and four passengers and one crew member had slight or no injuries. Though considered the classic case of what not to do, there were a number of contributory factors to what came to be called the Kegworth Air Disaster, in view of its proximity to the village of that name. These include the following:

1. False positive
   Safety expert Professor Peter Ladkin says this is the only case he is aware of where a false positive features in an air accident. That is to say the mistaken corrective action (shutting down the good engine) seemed to be solving the problem, thus making the pilots think they had done the right thing. This is unlike in medicine, where the long period of time over which recovery or improvement of the patient for any unrelated reason can be attributed to action by the doctor or surgeon means such false positives are well known.
   The cessation of the vibrations was one thing, but to confirm things by saying that the smoke disappeared when the pilots shut down the number two engine was, with hindsight, rather dubious thinking, since smoke would not normally disappear immediately.

2. Engine instrument system (EIS) difficult to read
   Before the introduction of two-man flight crews, the primary instruments, showing the performance of the engines were in front of the pilots, and the secondary instruments, indicating the condition of the engines, such as oil temperature and pressure and vibration, were in front of the flight engineer. However, with the sidelining of the flight engineer, these secondary instruments had to be in front of the pilots. In the earlier version of the aircraft, the B737-300, this was done by having traditional cockpit dials with mechanical hands, as in traditional clocks, there being two panels side by side, one with the main flying

instruments, and the other showing the condition of the engines. These earlier ones with needles were easy to read at a glance.

However, as anything mechanical is liable to go wrong and anyway requires costly maintenance, LEDs[3] were used instead of mechanical hands. However, rather than redesigning the panels to take full advantage of the virtues of an electronic display, the designers wanted to maintain the same general layout so pilots could switch from one model of the aircraft to another without expensive recertification.

In reality, LEDs could not simulate the previous clocklike hands, because those available at that time could not be bunched up at the center of the dials to look like a continuous line. Instead, the designers placed three rather pathetic-looking LEDs at intervals around the perimeter of the dials.

These could still be read by pilots with good eyesight when looking for a particular reading but made comparison and noticing anything unusual more difficult. In addition, Boeing had reduced the size of the secondary engine display relative to that for the primary display instruments.

The captain and first officer had very little experience (twenty-three and fifty-three hours respectively) on the 737-400 version, and the airline did not yet have a simulator where they could have practiced using the new engine information system (EIS), with its diodes. In addition, the captain said his considerable experience with other aircraft had led him to distrust vibration readings in general, and he did not include them in his usual scan of the instruments. His conversion training had not included instruction that technical improvements meant that spurious vibration readings were very unlikely.

3. Training and checklists

In the training of the BMA 737 pilots, the need to think or check things out before taking precipitous action was stressed, but as already mentioned there had not been training on a flight simulator with the new hybrid EIS display. There was a checklist for what to do in case of vibration from the engines and one for what to do when smoke occurred, but not one for when they happened simultaneously.

At the time pilots at BMA had not been made fully aware that there was no need to shut down engines completely because of vibration, nor that engine fans which are vibrating or not properly aligned could have their fan tips touching the rubber seals on the periphery and that this could produce smoke and a smell of burning but did not mean the engine was on fire. Thus, as the investigators said, the situation was outside the pilots' experience and training.

4. Workload and stress: Fear of fire

   In many emergencies, airlines usually insist that captains take control. Captains also tend to take control in difficult situations when it is not quite an emergency. Doing something physical makes the captain feel he is coping and relieves stress. The trouble with this is that the captain is concentrating on the physical task of flying the aircraft, or, as in the case of SQ006 at Taipei, maneuvering it over the slippery taxiway in bad visibility and heavy rain, and misses the larger picture.

   The flight data recorder (FDR) revealed that when the captain disengaged the autopilot, the aircraft yawed sixteen degrees to the left, a sign that the left engine was producing less power than the one on the right, but he did not seem to notice, as he did nothing to correct it. The fact that the first officer reported to ATC early on that they had an "emergency situation like an engine fire" shows they were concerned about fire, even though up to then none of the engine fire alarms had triggered.

   It is an interesting psychological point that a smell can instantly transport one mentally to a certain place, and the shaking of the aircraft followed by the smell of burning may have caused the pilots to react more instinctively and precipitously than they would have done in the event of a fire-warning light coming on. Anyway, a fire warning would have immediately indicated which engine had the problem.

   The official report made the additional point that having another pilot take over the handling of the aircraft—as PF (pilot flying)—meant monitoring of the instruments was less consistent than it might have been.

   Up until the onset of the vibration, the first officer had been flying the aircraft and would have been concentrating on the main instruments, not the engine vibration indicator, it being the role of the PNF (pilot not flying; in this case the captain, who did not believe in scanning vibration readings) to do the general monitoring. The captain must have thought the first officer had good reason to say it was the right engine that was giving trouble, and this only confirmed his own opinion, based on how he believed the air-conditioning system worked.

5. Unfortunate timing

   It is almost impossible for pilots to do anything to save the situation when trouble occurs when there is insufficient height and speed at certain critical moments, such as just before landing. They did not have the height or speed to restart the good engine, and not enough height to choose a flat place to land. As said, had the airport been farther away, they would have had a better chance of discovering problems with the number one engine when still high enough to restart the other engine.

6. Passengers and three cabin crew knew which engine had the problem
    Passengers at the rear who had seen the "sparks" from the left engine when the initial trouble occurred were somewhat perplexed when the captain said he had shut down the right engine, but did not inform the cabin crew because the captain sounded supremely confident.

    The three members of the cabin crew who had also seen the sparks apparently did not notice the captain saying the right engine had been shut down. They knew the purpose of the announcement was to reassure the passengers and were no doubt extremely busy with their own duties as they got ready for the unexpected landing.

A retired British Airways flight attendant has suggested to the author that the failure to pick up on the captain's mistake might have come about because cabin staff themselves often get confused about left and right, as they face backwards when addressing the passengers.

Just after shutdown of the number two engine, the captain called the flight service manager (FSM) to the flight deck to tell him to clear things for landing, and at the same time asked him, "Did you get smoke in the cabin back there?" He got the reply "We did. Yes."

This perhaps only confirmed the captain's mistaken view that the right-hand engine must have been at fault. The FSM departed but returned a minute later to say the passengers were panicky, and it was only then that the captain announced to the passengers that a little trouble with the right engine had produced some smoke, but it would be okay, as they had shut it down, and would be landing about ten minutes thereafter.

Cause of the Engine Problem

Subsequent studies showed a vibration harmonic at certain speeds of rotation caused the fan blades to rub against the rubber seals on the periphery, producing the burning smell and causing one blade to break. The manufacturer rectified the harmonic problem and allowed extra space between the blades and the seals to give a greater safety margin, should such vibration reoccur.

Other Similar Incidents

Before these modifications, other 737s with the same engine, including a BMA 737, had similar engine problems when climbing at maximum rated power. Benefiting from the lessons learned from the previous incident, the crews immediately studied the vibration indicator and made sure they shut down the correct engine before landing safely. After those incidents, pilots safely used a lower power rating, until the manufacturer modified the engines as described above.

This was not an in-flight fire at all. Though the first officer had cited the possibility of an engine fire, the instruments did not give a fire warning until the number one engine failed just prior to the crash. As so often when pilots have been partially responsible for a crash, the airline first defended them and then let them go.

Rearward-Facing Seats?

The high number of fatalities and injuries in what was a crash at relatively low speed led the authorities to examine the safety aspects of the seating, the strength of the flooring, and the locking of the overhead lockers. Some asked whether rear-facing seating might have offered greater protection. At first sight, rear-facing seating, as in most British Royal Air Force Transport aircraft, would seem to offer greater protection.

The comprehensive official report said rear-facing seating could impose too great a stress on cabin flooring. In addition, the report said other arguments raised against rearward-facing seats were:

1. They could be less effective in accidents in which the main deceleration force is not along the longitudinal axis of the aircraft.
2. They could expose occupants to the risk of injury from loose objects in an accident.
3. Great improvements had been made in the design and construction of forward-facing seats.
4. They may not be suitable for use in modern jet transports, with their high climb-out angles, and they could be "psychologically less attractive to passengers."

---

[1] With the increase in power, the FDR then began showing maximum vibration for the operating engine, but the pilots did not notice it.

[2] To warn pilots that they are about to stall—flying too slowly to stay in the air—the joystick or control is made to shake.

[3] LEDs are light emitting diodes—a type of indicator lamp used in all sorts of equipment because they consume little current and do not burn out.

Fierce O₂-fed Fire in ValuJet Cargo Hold (Everglades, 1996)

While great attention is paid to what passengers bring onto the aircraft, this is not always true for cargo

> The following case received much publicity, in great measure because passengers' voices on the CVR recordings spelled out their predicament as the metal-melting fire took hold.
>
> *ValueJet Flight 592, May 11, 1996*

The subtropical Everglades National Park, in Southern Florida, is said to be the only place in the world where alligators and crocodiles cohabit. Much of it consists of swamp interspersed with strips of shallow water filled with vegetation, with mud at the bottom. A worse place for recovering the bodies and debris from an air crash would be difficult to find. Not only were the hot and humid working conditions particularly difficult for people in protective suits, but it was also hard to locate the evidence and body parts in the ooze. The only good news for the investigators was that the shock wave, while attracting alligators, scared away poisonous snakes.

ValuJet Flight 592, a DC-9 with 105 passengers, 3 cabin crew, and 2 pilots, had just taken off in the early afternoon from Miami International Airport for Atlanta and climbed through ten thousand feet when the pilots heard a strange sound. No sooner had they concluded it was from an electrical bus than other electrical problems manifested themselves. Shortly afterward, shouts of "Fire! Fire!," presumably from the passenger cabin, could be heard on the CVR.

The thirty-five-year-old female captain of the DC-9, Candalyn Kubeck, did not declare a Mayday but radioed the Miami International Airport controller, saying an immediate return to Miami was required. The controller, not grasping the situation, twice gushed on with the instructions for the routine handover to Miami Center. Once her words had sunk in, he lost no time giving her a heading and a height (seven thousand feet) to return to Miami and asked for the nature of the problem. Kubeck told him it was smoke in the cabin and in the cockpit. Shortly afterward she asked if there was an airport even nearer but continued to be vectored to Miami, perhaps because there was not deemed to be one suitable. Anyway, it became academic, as radio communication with the aircraft was then lost.

Captain Kubeck had more than two thousand hours' experience on the aircraft type, and nine thousand flying hours in total. Neither her ability nor that of her first officer, Richard Hazen, a fifty-two-year-old ex–air force pilot, could have made any difference to the subsequent outcome, an event sometimes referred to as the "ValuJet situation," where an extremely fierce fire in a vulnerable part of the aircraft is only detected a few seconds before the aircraft becomes uncontrollable. In this case, *only forty-nine seconds* had elapsed from the first hint of trouble to the moment someone opened the flight deck door and said:

> *Okay, we need oxygen; we can't get oxygen back there.*

The following extract from the CVR, with some standard acknowledgements omitted for brevity, gives a good idea of how the situation evolved.

Elapsed time	Local time	
−6:54	14:04:09	Takeoff from Miami (V$_R$)
0:00	14:10:03	Strange noise heard in cockpit: probably that of tire bursting in the hold.
0:04	14:10:07	Captain: What's that?
0:05	14:10:08	FO: I don't know.
0:09	14:10:12	Captain: [??] 'bout to lose a [electrical] bus.1
0:12	14:10:15	FO: We got some electrical problem.
0:14	14:10:17	FO: Yeah. That battery charger's kicking in. We gotta . . .
0:17	14:10:20	Captain: We're losing everything.
0:18	14:10:21	Departure control gives frequency for handover to Miami Center.
0:19	14:10:22	Captain: We need to go back to Miami.
0:20	14:10:23	Shouts from passenger cabin.
0:22	14:10:25	Female voices shout: Fire . . . Fire . . . Fire . . . Fire . . .
0:24	14:10:27	We're on fire! (repeated)
0:26	14:10:29	Departure control repeats instructions for handover to Miami Center.
0:29	14:10:32	R/T first officer to departure: *592 needs immediate return to Miami.*
0:32	14:10:35	Departure control: Roger, turn left heading 270, descend and maintain seven thousand.
0:38	14:10:41	Departure control asks what kind of problem they are having and is informed it is smoke in the cockpit and cabin.
0:49	14:10:52	Sound of cockpit door moving.
0:55	14:10:58	Third cockpit microphone: *Okay, we need oxygen. We can't get oxygen back there.*
1:04	14:11:07	Departure control:

		When able, turn left heading 250, descend and maintain five thousand. (592 acknowledges.)
1:09	14:11:12	Third cockpit microphone: *Completely on fire.*
1:11	14:11:14	Shouting from passenger cabin subsides, perhaps because occupants are overcome by smoke.
1:18	14:11:21	Loud sound like rushing air—cockpit window opened?
1:35	14:11:38	R/T FO to departure control: *592, we need the closest airport available.*
1:39	14:11:42	Departure control does not propose another airport; says they (Miami) will be standing by and 592 can plan for Runway 12.
	14:11:45	One-minute-fifteen-second interruption to CVR.
	14:12:48	[CVR stops recording.]
2:54	14:12:57	[CVR restarts.] Loud sound of rushing air?
2:55	14:12:58	Departure controller tells 592 to contact Miami approach, correctly telling them to remain on his frequency.
3:08	14:13:11	CVR recording interrupted for unknown period.
		CVR restarts for a moment. Apart from a radio transmission from an unknown source, one can only briefly hear the loud sound of rushing air. Recording ends.
	14:13:34	Total loss of control is evident from the trajectory of the aircraft. (Pilot incapacitation/aircraft uncontrollable.)
	14:13:42	Impact with ground (radar data).

For the passenger cabin to be completely on fire (and not just filling with smoke) after one minute, nine seconds' elapsed time shows just how quickly the situation got out of hand. In fact, after melting the electrical cables, the fire was eating away at the steel control cables needed to maneuver the aircraft. The aircraft then keeled over, plunged downward, seemed to try to right itself for a moment, perhaps due to action by the pilots or autopilot, and then continued nose-downward into the swamp below.

The force of the impact was such that both the aircraft and passengers ended up as a dispersed collection of difficult-to-recover pieces. All 110 people on board would have died instantly on impact if they had not already succumbed due to the fire and noxious smoke.

With a stroke of luck—a searcher stepped on it—the flight data recorder (FDR) was soon found and, with the cockpit voice recorder (CVR), was retrieved from the swamp. Together with the evidence from the Miami ATC recordings, they confirmed that fire had brought about the crash. In addition, the debris showed the fire had been limited to the forward cargo hold and the area immediately above, thus making the usual painstaking piecing together of the entire mud-encrusted aircraft irrelevant.

For a fire in a hold to set the cabin above on fire, it had to have penetrated the cabin floor. In fact, some of the aluminum used in the construction of the passengers' seats was found to have melted, showing the temperature even there must have reached 1,200°F (650°C) before the aircraft cooled down on hitting the watery swamp.

Investigators later established that the temperature in that hold could have reached 3,000°F (1,600°C). Not only had these high temperatures deformed and melted the floor separating the cargo hold from the passenger cabin, but they had first fused the electrical cables (the first anomaly the pilots noticed) and then the steel cables essential for controlling the aircraft. This would have made it impossible for the pilots to pull the aircraft out of its precipitous dive into the swamp, assuming they were still able to function.

Oddly, the hold in question was a Class D hold and virtually airtight, so that any fire ignited there would normally soon consume the available oxygen and die down, and in the absence of a fire-extinguishing system would only smolder, at least until someone opened the hatch, which would be on the ground. This led to the following two questions:

1. What had ignited the fire?
2. Where had oxygen come from?

As study of the cargo manifest soon revealed, it must have been the five boxes of discarded oxygen generators loaded there. There were also three tires, at least two of which were mounted on wheels and hence liable to burst violently if overheated in a fire.

These oxygen generators came from three secondhand MD-80 aircraft that were being modified and checked out at Miami by SabreTech, a large contractor responsible for ValuJet's line and heavy maintenance. ValuJet had purchased these aircraft to upgrade and improve its fleet.

Oxygen generators supply oxygen to passengers should the cabin depressurize. These consist of canisters placed *at intervals* in the fascia above passengers that produce oxygen by an exothermic (heat-producing) chemical reaction triggered by a spring-loaded firing pin striking an explosive percussion cap. A lanyard pulls out the retaining pin when emergency oxygen is required.

On firing to provide oxygen, the temperature of the canister shell can reach between 475 and 500°F (246 to 260°C), but this heat soon dissipates if the canisters are placed individually with clearance for air around them. However, having dozens of them piled together in a cardboard box and surrounded by Bubble Wrap with nowhere for the heat to go other than into neighboring canisters could produce a chain reaction should one fire accidentally.

The canisters loaded into that hold should have been fitted with safety caps to prevent such accidental firing. Furthermore, the ramp agent present in the cargo hold as the boxes were loaded heard a clink and objects moving around inside one of the boxes, which were stowed on top of the tires without restraints to ensure they stayed in place. Possibly that box could have fallen off.

Probable Cause

The National Transportation Safety Board (NTSB) determined that the probable causes of the accident, which resulted from a fire in the airplane's Class D cargo compartment, initiated by the actuation of one or more oxygen generators being improperly carried as cargo, were:

1. The failure of SabreTech to properly prepare, package, and identify unexpended chemical oxygen generators before presenting them to ValuJet for carriage.
2. The failure of ValuJet to properly oversee its contract maintenance program to ensure compliance with maintenance, maintenance training, and hazardous materials requirements and practices.
3. The failure of the Federal Aviation Administration (FAA) to require smoke detection and fire suppression systems in Class D cargo compartments.

Contributing to the accident was:

1. The failure of the FAA to adequately monitor ValuJet's heavy maintenance programs and responsibilities, including ValuJet's oversight of its contractors and SabreTech's repair station certificate.
2. The failure of the FAA to respond adequately to prior chemical oxygen generator fires with programs to address the potential hazards.
3. ValuJet's failure to ensure that both ValuJet and contract maintenance facility employees were aware of the carrier's no-carry hazardous materials policy and had received appropriate hazardous materials training.

Incidentally, those horrifying words on the CVR so shocked Congress, as well as the public, that criticism of the FAA became especially harsh.

Had the FAA made the installation of fire detectors and fire-suppressant systems mandatory for such Class D cargo holds, the pilots might well have been able to bring the aircraft back before things got out of hand. An interesting point was that the maintenance workers doing twelve-hour shifts were concerned about the safety aspect, but their focus was on the safety of the MD-80s they were working on!

William Langewiesche has covered this organizational accident in detail in the March 1998 edition of the *Atlantic Monthly*, referring to the work on the prevention of such accidents carried out by academics. Incidentally, he adds that some academics claim the extra safety features incorporated into systems can in themselves lead to accidents, as in the Chernobyl nuclear disaster and here in the ValuJet case.

----

[1] Fundamental electric power distributor. (See note for Swissair on-board fire.)

The TWA-800 Controversy (JFK Outbound, 1996)

How could so many witnesses be wrong?

> No other air crash has engendered such a volume of conspiracy
> theorizing—books, articles, and Internet blogs. A cover-up on the sheer
> scale suggested would inevitably have, over twenty years, resulted in
> leaks via whistle-blowers. We therefore include the incident in this
> chapter rather than classing it as a mystery.
>
> *TWA Flight 800, July 17, 1996*

The Boeing 747 for Trans World Airlines Flight TWA-800 to Paris had arrived at New York's JFK International Airport from Athens at 16:31 on July 17, 1996. After cleaning and servicing, it was due to depart for the French capital at 19:00, but because of a delay, caused mainly by a search for a missing item of luggage, it did not depart until 20:03.

The air conditioners were not needed when cleaners and mechanics started servicing the aircraft, but the extra hour's wait with passengers on board at the end and the hot conditions meant they were working hard for about two and a half hours prior to departure and pumping out a great amount of heat. The fuel in the center wing fuel tank would normally have absorbed some of this heat, but as that was virtually empty, since relatively little fuel was required for the short hop across the Atlantic to Paris, the temperature of that tank would have risen significantly.

At 20:18 the aircraft was cleared for takeoff. The climb-out from JFK over the sea proceeded as usual. Then, passing under the control of the Boston Air Route Control Center, it received various instructions regarding its flight level. Up until then the only thing of note was the captain saying:

Look at that crazy fuel flow indicator there on number four . . . See that?

One minute later Boston Center told them to climb from Flight Level 130 to Flight Level 150. As they were complying, there was a loud sound, after which the CVR stopped working (20:31:12). The aircraft had broken up, with the pieces falling into the sea.

With many aircraft in the general area, Boston Center received several reports from pilots about witnessing an explosion, with the most detailed being that of the captain of an Eastwind Airlines Boeing 737:

Saw an explosion out here . . . Ahead of us here . . . About 16,000 feet [4,900 meters] or something like that. It just went down into the water.

The debris had fallen into the sea in a busy area a few miles off the coast, and it was only a matter of minutes before people in all sorts of craft, both military and civilian, were on the scene. As feared, there were no survivors, making the death toll 230.

There was much suspicion that the aircraft had been blown up by a terrorist bomb or shot down by a US missile, or even some secret ray gun under test. Perhaps no other air crash has received the attention this one has, especially as regards conspiracy theories. A respected CIA agent (now deceased) told someone the author knows personally that on reviewing the TWA 800 radar tracks, he right away thought it was distinctly possible that it had been shot down by a US missile. Nothing sinister in this, he would have thought, as accidents do happen.

Much of the NTSB's work involved disproving that such and such, say a missile or bomb, could have been the cause. It concluded that an explosion of the center fuel tank was the cause and that though the equipment in the tank for measuring the fuel level used voltages that were too low to cause ignition, a short circuit in wiring outside might have led to an overvoltage in wires to that equipment.

The NTSB had not been happy with the idea of only relying on avoiding ignition sources in the fuel tank in the early 747-100s as the sole means of protection and thought having the air-conditioning dissipating so much excess heat into the tank was undesirable. One must wonder whether the overheating of the tank, while the suitcase was being looked for, was a factor.

The approach to the investigation was somewhat peculiar, because having an aircraft blow up on its own with no other aircraft that could have collided with it seemed to indicate something suspicious, and hence the FBI played a prominent role from the start, taking over parts of the investigation for a while. Finally, the NTSB concluded it was due to an explosion in the central fuel tank, but it could not find conclusive evidence regarding what initiated it.

Seven years after the disaster, technicians were readying many pieces of the reconstructed jumbo jet, which were to go to George Washington University in Ashburn, for training air safety investigators. The NTSB had spent four years collecting these pieces and putting together a twenty-seven-meter (ninety-foot) section of the aircraft. Salvaging them from the sea had been a major task in itself.

Swissair 111 Cockpit Fire (JFK Outbound, 1998)

Was the flight entertainment system responsible?

> In this second disaster to befall an aircraft departing JFK, investigators
> suspected, but could not prove, the wiring of the entertainment system
> was responsible for starting a fire in the cockpit wiring. Investigators
> recommended that aircrew be made aware of the need to land as quickly
> as possible in the case of potentially serious onboard fires. Any delay
> incurred due to the desire to dump fuel is liable to close the narrow
> window of opportunity to save the aircraft and the lives of those on
> board.
>
> *Swissair Flight SR111, September 2, 1998*

Swissair Flight SR111 had taken off from New York's JFK Airport en route for Geneva at 18:17, with 215 passengers and 14 crew members. The aircraft was a McDonnell Douglas MD-11, a trijet that had evolved from the DC-10, with the addition of much automation to dispense with the need for a flight engineer. A more logical name might have been the DC-11, but one can understand the manufacturer's hesitation to suggest it was a DC-10+. Actually, the famous DC designation stood simply for the rather unglamorous-sounding "Douglas Commercial" and MD fitted the company name similarly.

Following the great circle route to Western Europe, skirting Canada's eastern seaboard, the MD-11 was some sixty-six nautical miles from Halifax and at thirty-three thousand feet when the crew alerted Monckton Center (Monckton high level controller) that a serious situation was developing, but not one that could be classed as an emergency. They did this by using the words "Pan, Pan, Pan" instead of "Mayday, Mayday, Mayday."

Swissair 111 heavy is declaring Pan, Pan, Pan.

We have smoke in the cockpit. Request deviate immediate right turn to a convenient place—I guess Boston. We need first the weather so uh we start a right turn here.

As Boston was then three hundred nautical miles away, and Halifax International Airport only sixty-six nautical miles away, the Monckton controller asked whether they would prefer to go to Halifax, to which SR111 agreed, and began descending from thirty-three thousand feet. The controller informed them that the active runway at Halifax was 06 and went on to ask whether they wanted a vector for it. SR111 answered in the affirmative and was told to turn left onto the north-northeast course of thirty degrees.

There were a number of exchanges with controllers and even a British Airways aircraft listening-in regarding the weather at Halifax. This indicated the pilots thought they had adequate time, and the situation was not critical.

When the Halifax controller subsequently informed them they had thirty miles to go to the runway threshold, SR111 surprisingly demurred, saying:

We need more than thirty miles . . .

The controller instructed them to turn left and to lose some altitude, and SR111 confirmed they were turning left.

SR111 called the controller:

We must dump some fuel. We may do that in this area during descent.

The controller said "Okay," whereupon SR111 told him they could turn left or right toward the south to dump fuel. The controller told them to make what was in effect a U-turn to the left and to inform him when they were ready to dump. They duly made the U-turn.

There were various exchanges regarding the fuel dump and the range of altitude over which it could be done.

Before they could begin the dump, the situation suddenly became desperate.

Swissair SR111:

Swissair one eleven heavy is declaring emergency. We are between uh twelve and five thousand feet we are declaring emergency now at ah time ah zero one two four.

Confirming their clearance to commence dumping fuel on that track, the controller asked to be informed when the fuel dump was completed. He called again, repeating his authorization to dump.

Some six minutes after that, the aircraft struck the water after having made a 360-degree turn. All 229 persons on board lost their lives.

How did the situation deteriorate so quickly? Or was it more serious than thought right from the beginning?

The Final Report

The 352-page final report of the Canadian Transport Safety Bureau can be found via their website, http://tsb.gc.ca/. It has much general technical information about wiring, highly automated aircraft systems, and measures to prevent and deal with onboard fires. Anyone especially interested in this accident should consult the report directly. To sum it up rather inadequately, it made the following points:

1. Aircraft manufacturers, regulators, operators, and pilots do not normally consider smoke or fumes issuing from the air-conditioning system to represent a serious and immediate emergency requiring an immediate landing. For example, in the British Midland accident in which the pilots shut down the wrong engine, the "smoke" exiting the air-conditioning came from the engine fan blades rubbing the rubber seals and did not represent a fire at all.

2. The Swissair crew initially thought it was something affecting the air-conditioning and went through the checklist for that and, reasonably, did not think it was a dire emergency; hence, the use of "Pan, Pan, Pan" rather than "Mayday."

3. The fire started in an inaccessible place in the cockpit, where there was wiring, and involved wiring for the entertainment system. Instead of being connected to a bus dedicated to passenger facilities, the entertainment system, which used a considerable amount of power, was connected to a main bus. When the pilots eventually realized the problem must be electrical in nature, various instruments and equipment were already failing, and they had to fly manually.

4. Contrary to early reports, the pilots could not have reached Halifax even had they opted to come straight in from the moment they declared "Pan, Pan, Pan."

5. Though there was a sign of arcing (spark between wires) on the cable for the in-flight entertainment system cable in the area where the fire started, there was no proof that it was this that actually started the fire.

NTSB Recommendations:

We recommend that the Safety Regulation Group reiterates its advice to airlines that the priority in certain emergencies, such as in-flight fires, is to land rather than to seek to dump fuel in order to avoid an overweight landing.

It should ensure through its inspections that airlines are passing on that information to aircrews.

Furthermore, we recommend that air traffic controllers be given similar advice so that they are able to respond appropriately to such emergencies.

Cockpit Fires

While any onboard fire is potentially a great hazard, those in the cockpit can be particularly pernicious as they can incapacitate the pilots as well as compromise the controls.

One may have been a factor in the mysterious disappearance in 2014 of Malaysia Airlines Flight MH370 that took off from Kuala Lumpur for Beijing only to deviate from its course shortly after departure and end up somewhere in the sea off Australia with no communication from the pilots.

In 2016, the crash of an EgyptAir A320 may also have been due to a cockpit fire with some suggestions it was started by the copilot's mobile phone or tablet, though others think the fire more likely started in the avionics bay just below.

Fragmenting Tire Dooms Supersonic Concorde (Paris-CDG, 2000)
Five-star airliner crashes on two-star hotel

> The Anglo-French supersonic Concorde, the most glamorous and most beautiful airliner ever, stirred the hearts of the public in France and the UK, who had paid out so much in taxes to support it and its mostly wealthy passengers.
>
> *Air France Flight 4590, July 25, 2000*

Concorde's History

The world's first supersonic airliner, the Concorde 001, rolled onto the tarmac in 1967, but according to CNN it took two more years of testing and fine-tuning of the powerful engines before it made its maiden flight over France on March 2, 1969. The original plan was for a production run of three hundred, but in the end it was limited to just fourteen. Air France cannibalized one of those for spare parts in 1982, and another crashed, leaving five for Air France and seven for British Airways when the two airlines withdrew Concorde from service in 2003.

These days the authorities would never certify such a noisy bird, so Concorde was surviving on the certificates issued in the 1970s. Even then the United States would not approve inland flights, so its regular scheduled flights to the States were mostly to the East Coast, and New York in particular. Some said it was sour grapes at being beaten to the post on the part of the United States.

When Concorde finally entered airline service, it was a Pyrrhic victory for its makers or, rather, the two nations of taxpayers who provided such generous funding. Key countries on major overland routes refused to allow it to pass over their territories. For timesaving reasons, it was essential to fly overland, as the aircraft did not have the range to fly the long routes over the Pacific Ocean. Finally, the fuel-guzzling Concorde had the misfortune to come on the scene just as fuel prices were skyrocketing.

The majestic Concorde mainly benefited the rich and famous, though many found first class on a conventional aircraft to be preferable to the cramped seating on Concorde. Some celebrities, such as the British TV personality David Frost, were virtually commuting between London and New York on Concorde. Not so well known is the good use courier companies, such as DHL, put it to in delivering documents and financial instruments for major companies when time really was money. Even so, not enough demand existed to employ fully even that tiny fleet.

Air France had more difficulty filling seats than British Airways, who made a healthy profit out of it. Not only was this because more top business people and celebrities fly to London but also partly because London is nearer than Paris to New York and Washington—quite significant with an aircraft operating near the limit of its operating range. Air France used charter flights and excursions to help fill those seats.

The Fatal Flight

Indeed, the flight that was to last only a minute or two before ending in disaster just after taking off from Paris's Charles de Gaulle Airport in July 2000 was just such a charter flight. It carried elderly Germans to New York to join a luxury Caribbean cruise.

Concorde spooled up its engines to take off with its long beak tipped five degrees downward and pointing just in front of another Air France aircraft, which had brought the French president and his wife back from an official visit to Japan. (The nose usually droops down five degrees for takeoff, and twelve-and-a-half degrees on landing, so the long nose does not obscure the pilots' view.) Though slim and elegant, the Concorde was almost certainly overweight, with fuel representing just over half her total weight.

The captain *thought* they had an all-up weight (AUW) of 185,100 kilos, placing them, he said, at the aircraft's structural limit. He did not know this was an underestimate[1] and did not include nineteen items of baggage loaded at the last minute. It is also likely that the unburned fuel in the rear tank and the nineteen extra items of luggage had moved the center of gravity too far aft.

Much to the surprise of the BEA (Bureau d'Enquêtes et d'Analyses, the French Air Accident Investigation Bureau), the announcement by the control tower of an eight-knot tailwind did not elicit any comment from the aircrew. At the very least, the captain should have considered taking off against the wind. Although a wind of eight knots may not seem much, taking off in the opposite direction would mean sixteen knots less groundspeed. Because of limitations on the maximum speed for the tires, a tail wind of eight knots meant they were too heavy.

An article in the British Sunday newspaper the *Observer*, written by David Rose shortly after the crash, with comments by veteran Concorde pilot John Hutchinson, said they could no doubt have "got away" with being overweight had other things not gone wrong. The crew had decided 150 knots should be V_1, the speed at which they would be committed to continuing the takeoff, and 183 knots should be V_R, the speed at which they would rotate and expect to soar into the sky.

A supersonic delta-wing aircraft like Concorde differs from other aircraft in that the wings do not provide any real lift before rotation. In consequence, the tires continue to bear the entire weight of the aircraft throughout the takeoff run. This in turn means they are particularly vulnerable just before rotation, as there is not only the weight of the aircraft to consider but also the tremendous centrifugal forces acting on their treads due to the wheels' high rate of rotation. In addition, at high speeds sharp objects on the runway are far more likely to cut into them.

Shortly after Concorde had reached V_1, the right-hand tire on the left-hand main landing gear ran over a curved strip of metal that had just fallen off a DC-10 that had taken off shortly before. Curved like a loop, the titanium strip was lying sideways up, with its concave side facing the oncoming tire, thus ensuring the tire would trap it rather than roll it over and bend it flat. The tough metal sliced into the tire, causing the tire to break up under the enormous centrifugal forces. Later, investigators found a 10 lb (4.5 kg) piece of rubber from the tire on the runway near that point.

After complex studies the BEA investigators concluded the lump of rubber had forcefully struck the underside of the wing, pushing it and the fuel tank so rapidly inward that it induced a shock wave in the fuel of such intensity that, with the tank virtually full and no air-filled spaces to absorb the shock, the tank ruptured elsewhere, a phenomenon that had never been seen before in a passenger aircraft. Had the piece of rubber just pierced the tank, as in a previous instance in Washington in 1979, the damage to the tank would doubtless have been less.

From the amount of fuel on the runway and other evidence, it is estimated that kerosene was pouring out from under the wing at a rate of sixty kilos a second. In the Washington incident mentioned above, the fuel did not catch fire, and the aircraft was able to take off and return safely. Perhaps because of the much greater amount of fuel, and very likely because a damaged wire in the landing gear was producing sparks, the leaking fuel then caught fire.

The events leading to the ultimate disaster all happened in some three seconds and must have been very confusing to the crew. The control tower told the Concorde's crew they had flames coming out behind them. Debris, and more likely hot gases from the fire, caused the performance of the two engines on the left side to fall off, resulting in a yaw to the left. The captain rotated the aircraft early as they deviated to the left of the runway, but no sooner had they lifted off than the fire warning for the number two engine sounded and the flight engineer shut it down.

The number one engine seemed to be recovering and able to help them reach the 220-knot speed to fly at least horizontally with the landing gear down. They tried to raise the landing gear but were unsuccessful, no doubt due to damage caused by the fire or debris. They attained a height of two hundred feet, still with insufficient airspeed. The number one engine began to fail, and with the first officer constantly warning them about the lack of airspeed, the aircraft went slightly nose-up, and the left wing dropped down to 115 degrees. The yaw and sideslip meant air was no longer properly entering the good engines on the right-hand side, and they too lost power, though it is possible they were throttled back at the last minute in an attempt to straighten the aircraft's trajectory.

Virtually upside down, the pride of France and Britain crashed on a two-star hotel some six kilometers from the end of the runway and exploded in a fireball. The crash was so violent, with debris scattered widely, that there was no hope of survivors. The death toll was a hundred passengers, six cabin attendants, and three aircrew.

Fortunately, the three-story hotel was almost empty, and on the ground "only" four people were killed and six injured. Forty-five Polish tourists staying at the hotel later returned from sightseeing to find a surprising sight.

Conclusions

Although it does not say so in so many words, the lengthy BEA report into the crash gives the impression the Air France Concorde team of officials, mechanics, and aircrew were an exclusive lot, like those associated with expensive racing cars, and had a somewhat cavalier attitude. With such cars, drivers are liable to take risks when overtaking in the belief that the tremendous reserve of power will always get them out of difficulties, and with the Concorde allow taking off in a tailwind. As already mentioned, the extra distance to New York from Paris meant that Concorde would often be operating just within its safety envelope as regards fuel. To give that little bit extra and lessen the likelihood of being unable to reach New York, the Concorde would often be loaded with excess fuel on the assumption (or pretext) that taxiing would use it up before takeoff.

Just after the accident, maintenance staff who had replaced part of the thrust reverser mechanism on one of the engines became so distressed at the thought they might have been responsible for the disaster that they had to go to the doctor for medication. One can imagine the anguish they endured until it was discovered that the accident was unconnected with their work, carried out properly. However, as mentioned the BEA investigators did find other maintenance workers had forgotten a spacer when reassembling the landing gear during earlier servicing.

An article appeared in the British *Observer* newspaper suggesting that omission of the spacer had caused the aircraft to deviate to the left and that the lack of it only made matters worse when the tire failed, causing the aircraft to veer to the left "like a supermarket trolley with a jammed wheel." The BEA disputed this, saying the aircraft only started to deviate when the thrust from two engines on the left-hand side fell away.

While saying the lack of the spacer had not played any role in the accident scenario, the BEA was unhappy about this evidence of maintenance failures and lack of written material showing procedures had been signed off on completion. It noted that the Air France maintenance people justified working twelve-hour shifts by claiming this avoided wasting time briefing others.

The BEA inquiry took extra time because it was run in parallel with, but separately from, the judicial inquiry, which in the end did not come up with anything dramatic. Some of the investigators from England complained it took a long time to see the evidence, and the BEA report has a note saying that this was due to the judicial inquiry.

The French insurers were very pleased to have come to an amicable albeit expensive agreement with the relatives of the German victims, as they wanted to avoid litigation in US courts. However, the lawyers for the other victims' relatives were not as happy with this quick settlement and asked for extra compensation for their own services.

The direct cause of the accident was running over the titanium strip. There were suggestions that the crew might well have been able to save the aircraft had it not been overloaded, with the center of gravity too far aft, and had the flight engineer not shut down the number two engine. However, the intensity of the fire was so great that the left wing and associated control systems would have failed before reaching the nearby Le Bourget Airport. Indeed, there is evidence that failure of the left elevon was what caused the left wing to drop at the end. Aluminum loses much of its strength when heated to a mere three hundred degrees Celsius and melts at six hundred degrees, and investigators found molten aluminum under the flight path.

One point that the BEA investigators did make about safety in general was that when something is used so little, improvements do not tend to be made. This applied to the tires, for if there had been hundreds of Concordes in operation, developing much safer tires would surely have been thought worthwhile.

When Concorde returned to service, it had new NZG[2] tires, developed by Michelin, which had new materials making them more resistant to foreign bodies and designed so that only small pieces would fly off if they did happen to fail. In addition, the fuel tanks were lined with Kevlar, used for body armor, impregnated with Viton, an expensive synthetic rubber able to withstand high temperatures. Thus, even in the unlikely event of a large piece of rubber flying off the newly developed tires, hitting the underside of the wing, the shock wave produced in a virtually full tank would not create a gaping hole elsewhere.

There had been a close shave, not widely publicized at the time (late 1970s), when the engine on a British Airways Concorde caught fire. The passengers escaped via the emergency slides without too much difficulty, except that a passenger wearing shoes with high heels slit the fabric of one of the slides on her way down. A male, and presumably heavier, passenger following her went right through the aperture and found himself with a sore spine on the ground below.

More troubling was the discovery later that the titanium shield above the engine, destined to protect the fuel tanks, had begun to bubble due to the intense heat of the fire. Had more time elapsed, even titanium would have failed, and there would have been an enormous conflagration. Strangely, there is no mention of this in the individual histories of any of the British Airways aircraft.

The criminal investigation run in parallel with the BEA investigation also concluded that the training of Air France's Concorde aircrew had "weaknesses," which led among other things to the shutdown of an engine before necessary. However, these various failings did not amount to gross negligence or criminal responsibility.

On July 4, 2008, a French judge agreed with a prosecutor's submission that John Taylor, the Continental mechanic who allegedly fitted the nonstandard strip, Stanley Ford, a Continental maintenance official, and the airline itself stand trial for involuntarily causing death and injuries. Also cited for trial were Henri Perrier, seventy-seven, the director of the first Concorde program, and Claude Frantzen, sixty-nine, a former director of technical services at the DGAC, the French civil aviation authority. They were alleged to have known that the Concorde's wing with its fuel tanks was fragile and vulnerable to damage from the outside. A spokesperson for Continental, the only accused with real money, stated, "These indictments are outrageous and completely unjustified," and said any charges would be rebutted with vigor.

This court case did not really reflect the overall situation, as the causes of the accident were many and complex.

In the end the indictments led nowhere.

---

[1] After the accident, the BEA estimated her actual takeoff weight including the 19 extra items of luggage to have been 186,451 kg. This was likely to have been an underestimate, as 20.7 kg was the average for the 103 items on the load sheet, and 84 kg was taken to be the average passenger weight including carry-on luggage. Both averages seem low for this atypical passenger profile.

[2] NZG stands for Near Zero Growth, meaning they hardly grow at all at very high rotational speeds. Michelin's work was not wasted as their new tires are being considered (along with Bridgestone's) for use with the giant Airbus A380.

China Airlines 737 Catches Fire at Stand (Okinawa, 2007)
Loose bolt punctures fuel tank

> Once an aircraft comes to a halt at the stand, passengers assume all they risk thereafter is falling down the stairs as they disembark. However, brakes and engine components may well be still hot enough to ignite any fuel that should fall on them.
>
> *China Airlines Flight 120, August 20, 2007*

China Airlines Flight 120, a regularly scheduled flight from Taiwan, landed at Naha Airport, Okinawa, Japan, at 10:26:52. The aircraft was a Boeing 737NG, with 157 passengers and 8 crew, and only five years old.

As it taxied to the stand, where three buses waited to ferry passengers to the terminal, the pilots retracted the flaps and slats that had been extended during the landing.

The driver of the bus parked farthest out on the apron watched as the 737 reached the midpoint, about fifty meters short of the end, of the line leading to the spot where the 737 would finally stop. To his surprise, he saw fuel in the form of a mist mixed in with the exhaust outflow (blast) of the engine on his side.

When the aircraft came to a stop at the set spot, the fuel continued to spray backwards as a mist, but when the engines were powered down, instead of flying backward as a mist, the leaking fuel began flowing in a stream down along the engine cowling.

Ground staff waiting at the stand noticed a liquid coming off the front of the wing, with one even catching some of it in his hand to see whether it was fuel or hydraulic fluid. But before they could issue a warning, it ignited.

Having completed the parking checklist, the captain was waiting for the ground crew member's instructions when he heard him shouting, "Fire!" Looking out of his side window, he saw thick, black smoke at the far left and behind. Even though the instruments showed no indication of fire, he actioned the extinguisher for the number one engine, and then the one for the number two engine on the right. He also ordered the cabin crew to go to their stations and prepare for evacuation. After running through the procedures, he ordered an evacuation, using Chinese, as most of the passengers were of that nationality.

The fire was burning under, and on, both wings, making use of the over-wing escape hatches out of the question, but luckily, the wind, though recorded as nine knots, was perhaps less adjacent to the terminal and at an angle of ninety degrees, blowing the flames and smoke across and not onto either the rear or front exits.

The last flight attendant to evacuate after all the passengers had was bowled to the ground just after reaching the bottom of the slide by an explosion as the fuel inside the tanks caught fire. The pilots in the cockpit had to evacuate by climbing out of the cockpit window and down an escape rope. The first officer, who was first, was halfway down when he was blown off by the blast from the explosion. The captain followed him. Everyone had escaped, with only four people suffering relatively minor injuries. A fellow passenger had helped a woman on crutches. The cabin crew had performed well, though not many understood the instructions.

Videos (see YouTube and so on) taken from the terminal show the conflagration in its entirety, except for the moment the fire started, and demonstrate the importance of evacuating a burning aircraft as quickly as possible and not dawdling at the feet of the slides—in theory, people come down quickly, and when they slow at the bottom they end up standing up and able to run away. Everyone was evacuated from the cabins in 1.42 seconds, with both pilots out as well 14 seconds later.

The fire services only went into action six minutes after the fire was first reported to the control tower, by which time it had gained a real hold, with the back of the aircraft broken. A fearsome sight. The delay was partly due to communication failures and the distance of the fire station from the stand. Also, drivers of the appliances had to be wary of crossing areas where aircraft were operating.

Aircraft Destroyed for Lack of a Washer

The sophisticated digital flight recorder showed no electrical anomalies, there was no evidence an overheated tire was the source, and the fuel lines showed no signs of having ruptured. Therefore, the fuel the ground handler had seen coming from the right wing had to have come from the tank.

A borescope with a camera inserted into the tank soon revealed the cause of the fuel leak was a hole made by a loose bolt. It had been pushed into the thin but strong wall of the fuel tank as the powerful hydraulics retracted the slats—the bolt could be seen still lodged in the hole!

Investigators found that during a maintenance task to make the slat end stop mechanism safer by putting glue on the bolt threads, carried out ten days before, a large-diameter washer subsequently found inside the wing had been omitted. Because of the narrow diameter of the head, there was nothing to hold it in place, and the jolt when the aircraft landed made it jump out into the confined space. Twenty-one 737s in the US alone were found to have the same problem. The system was eventually redesigned by Boeing to prevent a reoccurrence, and existing mechanisms were replaced worldwide.

Chapter 8
PILOT ISSUES

Uptight BEA Trident Captain's Heart Attack (Staines, 1972)
Combination of fear-instilling captain and diffident rookie first officer

> Lack of a cockpit voice recording made it impossible to determine who did what in the following incident. Graffiti critical of the captain who had been involved in a furious argument in the crew room before departure was found in the cockpit.
>
> *British European Airways Flight 548, June 18, 1972*

The British European Airways trijet Trident had taken off for Brussels from London's Heathrow airport in cloudy and somewhat turbulent weather. Only 114 seconds into its flight, with the aircraft traveling at 162 knots at about 1,500 feet, someone moved the lever controlling the leading-edge droops, causing them to retract.

Droops, as their name implies, are like upper lips on the front (leading edges) of the wings that can be made to droop forward and downward to change the configuration of the wing and enable the aircraft to fly at low airspeeds. Their function is similar to that of the much-better-known slats.

Theoretically, the airspeed would have had to be at least 15 to 20 knots higher than the 162 knots at which they were flying for the aircraft to be able to fly without the droops deployed. In fact, to allow a safety margin, the pilots would only retract the droops once the airspeed had reached 225 knots, and never when the aircraft was in a turn, as it was—an aircraft is more vulnerable in a turn, especially if the turn involves turning away from a strong headwind, as the drop in airspeed can be dramatic. Every Trident pilot would have had this minimum droop retraction speed drummed into him or her and could hardly forget it, as it was indicated next to the droop lever.

Even so, retracting the droops by mistake need not prove immediately fatal, as even in those days airliners had good systems to warn pilots of an imminent stall. Sensors on the wings would detect development of abnormal airflow over the leading edge, and a stick shaker system would shake the control columns to warn the pilots. According to the flight data recorder (FDR), on that BEA flight the stick shaker operated less than two seconds after retraction of the droops.

In addition, aircraft like the Trident, with the engines at the rear and high tails (high horizontal stabilizers), are equipped with a stick pusher system that pushes the nose down in a stall. This is to compensate for the fact that aircraft with engines at the rear and high tails go *nose-up* rather than nose-down in a stall, thus making the stall worse and often irrecoverable if anywhere near the ground.

Apart from in cases of extreme wind shear, an aircraft's airspeed changes progressively, so that in normal circumstances the stick shaker warning (that the airspeed is falling to such an extent that a stall is imminent) occurs well before the last-resort stick push. However, as this was a change-of-configuration stall—that is, not due to a progressive drop in airspeed but due to a change in the shape of the wing—these events took place so quickly that the stick shake warning and stick push must have seemed to have occurred almost simultaneously.

The virtual concurrence of these two events may have confused the pilots, for when the stick push mechanism duly caused the nose to pitch downward, and so much so that the aircraft gained almost sufficient speed to recover, *someone* fought against the stick pusher and pulled the nose up. Once more the stick pusher operated to push the nose down, only for *someone* again to use considerable force to resist it, before finally disabling it. As at no point was there any attempt made to increase engine power, one must conclude that whoever was at the controls was unaware of the true situation.

With the nose yanked up again without redeployment of the droops and no application of additional power, the Trident did what the stick pusher was trying to prevent—it went into a deep, unrecoverable tail-down stall.

Below them a thirteen-year-old boy walking with his younger brother along a footpath saw the Trident with its wings level drop out of the cloud above with the tail, weighed down by its three heavy engines, striking the ground first and breaking off. The rest of the aircraft then belly flopped onto the field with the sound of crumpling and tearing metal as the fuselage and wings separated into various pieces. Despite the large amount of fuel carried at takeoff spilling onto the ground, there was no fire—no doubt because of the lack of forward speed.

With air traffic controllers not even having noticed the aircraft's disappearance, the boy ran to a house some four hundred yards away for help. The occupant rushed to the scene, but her skills as a hospital emergency room nurse were of little avail, as only one person showed any sign of life, and that sole survivor later died in hospital, making the total death toll 118. The official report into the crash praised her highly.

There have been survivors in crashes, even with intense fires, where the wreckage must have looked even worse. However, with the wings providing no lift at all and the forward speed dissipated by the drag produced by the nose-up/tail-down attitude, the Trident had belly flopped almost straight down, with virtually nothing between the occupants and the ground to cushion the shock. Had the aircraft belly flopped on the nearby trees, they might have absorbed enough of the shock to enable some to survive.

With little immediate information about the mysterious cause of the terrible disaster, newspapers resorted to carrying stories about morbidly curious bystanders and gawkers hampering the rescue efforts by blocking the roads. (The official report later denied their presence had affected the death toll.) The absence of fire facilitated the work of the investigators, in particular with respect to the postmortems of the crew members. They established these key points:

1. Someone had retracted the droops before the aircraft had reached sufficient airspeed to stay in the air without them and, contrary to company regulations, when the aircraft was in a banked turn and more vulnerable.

2. The built-in stall warnings (stick shaker) were ignored, and the automatic recovery (stick pusher) resisted several times before being disabled (switched off).

3. Inexplicably, at almost every stage of the short flight, prior to and including the fatal retraction of the droops, the airspeed had been lower than it should have been, and this unusually low airspeed had made the retraction of the droops especially lethal.

4. The captain, who was handling the aircraft, had had a heart attack, and medical experts concluded from the color of the blood in his heart that he was probably dead before the aircraft hit the ground.

The investigators had to resort almost entirely to supposition, as the aircraft did not have a cockpit voice recorder (CVR), even though CVRs had been made mandatory for US-registered airliners much earlier in 1965. (Opposition from the UK pilots' unions had thwarted their introduction in the UK.)

It was impossible to know who did what and why. From terse communications with ATC, it was evident that the captain had been handling the aircraft, but it was not clear who had moved the droop lever.

The initial investigation was followed by a public inquiry, with lawyers arguing for days without producing anything of significance that was not apparent from the original investigation. What were apparent, as in so many crashes, were the human personality factors that experience has shown increase the likelihood of a disaster.

On the flight deck there were actually four pilots:

1. Captain Key, an uptight captain, who supported the company in the ongoing industrial dispute.

2. Twenty-two-year-old Second Officer Keighley. He had only very few hours' line experience on the Trident. He was in the first officer's seat next to the captain.

3. Twenty-four-year-old Second Officer Ticehurst, who was in the monitoring seat.

4. A Captain Collins in the jump seat behind Captain Key. This second captain was deadheading with his own crew to Brussels. Although a freighter captain on another aircraft type, he too was qualified to fly Tridents.

Though technically qualified, Keighley was thought by instructors, both before and after joining the airline, to lack initiative and to be rather diffident. While they thought he would eventually make a good pilot, his relatively young age, unassertive character, and inexperience meant he would have trouble coping with a captain known to be uptight and who had shown just before the flight he could get very angry.

Industrial action at the airline meant that some junior pilots had not had the extra training required to permit them to legally occupy the third seat (behind the captain and first officer), which involved monitoring the flying pilot's actions and performing certain technical functions that used to be done by a flight engineer and necessitated the extra paper qualifications. This quirk in the regulations in turn meant that the least qualified pilot was likely to be at the controls next to the captain.

Furthermore, prior to the flight, Captain Key had been involved in a violent argument in the crew room with another pilot about the strike, and Keighley had witnessed the captain's violent outburst of temper. Thus, there would have been good reason for him to be very circumspect in his dealings with him and hesitate to do anything that might provoke him. In addition, it was said that the considerable air turbulence must have raised Keighley's certainly already high stress level.

The investigators thought the presence of Captain Collins might have distracted the junior pilots. The official report notes that when Captain Collins was found in the debris, he was still holding an air freshener can in his hand. These air fresheners were standard issue for freighter captains, but one wonders what he was doing with it so early in the flight. Was Captain Key sweating so profusely that use of the air freshener was deemed necessary? An awful possibility that no one hitherto has suggested is that Captain Collins was standing up with his air freshener, and this had been a further distraction. He might even have fallen forward when the initial stick push occurred, blocking the view of Ticehurst.

In his book Disaster in the Air, Jan Bartelski, who has worked in senior positions for the International Airline Pilots' Association, says an official accident report should not publicize details such as being found dead grasping an air freshener. However, in the author's opinion it does seem very relevant in that many people have said that Captain Collins's body being found without earphones, slumped away from his seat, might mean he was trying to get to the droop lever to reset the droops. Would he have done this with the air freshener can in his hand?

Jan Bartelski does postulate very logically that an erroneous airspeed reading on the captain's side might have led him to retract the droops because he thought they were going too fast rather than too slow. This would account for the consistently low airspeeds at the various stages of the short flight, which no one can explain, apart from suggesting this resulted from the captain being in pain.

Possible Contributory Causes

1. The airline's training.
2. Reports of similar incidents being overlooked because of bureaucratic muddle, staff absence (holidays), and bad luck.

Outstanding Questions

1. Why was the aircraft flying more slowly than it should have been at various stages of the climb-out?
2. Why was the droop lever moved?
3. In addition, why, even if the captain was incapacitated, didn't the other two pilots, and especially the monitoring pilot, realize what was happening?

Aftermath

The British government belatedly faced down opposition from pilots and made the installation of cockpit voice recorders mandatory, as in the US.

Colonel on Korean Air Cargo Flight Banks Too Far (Stanstead, 1999)
Copilot did not dare offend captain even though in danger himself

---

This crash had something in common with the disaster just described: a crusty senior captain, and a lowly, inexperienced copilot.

Korean Air Cargo Flight 8509, December 22, 1999

At the time of this incident, Korean Airlines had a culture quite the opposite of that practiced through cockpit resource management in Western airlines. Firstly, Korean culture meant that younger people obeyed their elders even when not agreeing. Secondly, most pilots were ex-military, and in the military disobeying orders was not tolerated, and the more senior the officer, the more difficult it would be.

Korean Air Cargo Flight 8509 took off from London's Stanstead Airport in the early evening of December 22, 1999. The captain was a former colonel in the Korean Air Force with 13,490 flying hours, of which 8,495 were on the Boeing 747. As was the custom, senior officers such as he were catapulted to the rank of captain and had little understanding of flying as part of a team.

His copilot had only 195 flight hours on the 747 out of a total of 1,406. The flight engineer, however, was very experienced.

On the leg to Stanstead, a fault with one of the aircraft's inertial guidance systems disabled the attitude indicator on the captain's side. This, an artificial horizon, indicates whether the nose is pointing up or down and whether the aircraft is banking to the left or right.

When the aircraft lifted off from the Stanstead runway, the computer warned the pilots that there was a discrepancy between what the attitude indicator on the captain's side was showing and what the one on the copilot's side was. When it had happened in daylight on the incoming flight, the pilots easily grasped the situation and switched to a second guidance system. Anyway, they could see the actual horizon. At Stanstead, the maintenance people thought they had solved the problem, not realizing it was due to a faulty inertial guidance system.

The captain banked, and with no bank showing banked further and further despite the warning. The flight engineer called out "Bank" to no avail, and with the copilot, who had been brought up to not offend superiors, especially be they military, saying nothing, the wing snagged the ground fifty-five seconds after taking off, and the aircraft crashed with no survivors.

Korean Air had a rethink and, with the help of Delta and a cultural change, became one of the safest airlines.

Overcontrolling Copilot Swishes off Airbus Tail (NYC, 2001)
Why him?

> Though wake turbulence was a key factor in this incident, it has been
> included in this chapter, as it was a propensity of the first officer,
> perhaps accentuated by his training, that was most likely the cause.
>
> *American Airlines Flight 587, November 12, 2001*

The sixteen-thousand-foot runway at Denver International Airport in Colorado is the longest commercial runway in North America. It owes this distinction to the city literally being a mile high and thus needing an exceptionally long runway to cope with heavily laden aircraft that accelerate less quickly at the lower atmospheric pressures found at high altitudes. The previous record holder, the slightly shorter 14,572-foot (4,441-meter) Runway 31L at New York's JFK airport, probably still holds the record for the best takeoff prospects, as it is at sea level, as well as being exceptionally long.

A drawback of New York's 31L is that the city lies straight ahead, albeit in the distance. To avoid areas that are sensitive for noise and other reasons, including terrorist action, aircraft taking off in that northwesterly (310-degree) direction usually have to make a sharp left turn shortly after takeoff so as to avoid the city and fly mostly over water.

On November 12, 2001, almost two months to the day after 9/11, with the images of the two hijacked aircraft slamming into the World Trade Center's twin towers still vivid in people's minds, Japan Airlines Flight JL47 to Tokyo took off from 31L at JFK. It was just after 9:00 a.m. The shorter, parallel runway, 31R, on the other side of the terminal buildings was being used for incoming aircraft, including the soon expected British Airways supersonic Concorde Flight 1 from London.

JL47 acknowledged the takeoff clearance from the tower but waited perhaps fifteen seconds longer than usual before beginning its roll. In addition, with the large amount of fuel required for the thirteen-hour-plus flight to Tokyo, it would have climbed out more slowly than most other 747s just hopping across the pond to Europe. The tower had already warned a police helicopter to exercise caution about wake turbulence from several heavy jet departures over a suburb to the left of the runway extension[1] called Canarsie.

Once well into the air, the Japan Airlines pilots would retract the flaps and commence a high-power climb. It is precisely when climbing under high power with a clean configuration—that is, no flaps or slats sticking out—that a large aircraft, such as the 747, produces the most wake turbulence.

This consists of vicious vortices of swirling air produced at the wingtips, where the air squished under the wings meets the lower-pressure, relatively fast-moving air above them. These hang behind the aircraft, and, spinning like tops, gradually descend, until their energy either dissipates or they are broken up by other turbulence. In fact, turbulent or stormy weather conditions help protect following aircraft from wake turbulence, as the turbulence breaks up the spirals. It is in calm conditions that wake turbulence is the most treacherous. That day, aircraft taking off on JFK's 31L were encountering a light headwind of just nine knots.

The next aircraft in line to take off was American Airlines Flight 587, an Airbus A300 bound for the Dominican Republic with 260 people on board, a flight expected to take some four hours but which in fact was to last only 104 seconds from liftoff. First Officer Sten Molin was the pilot flying (PF), with Captain Edward States in command. The tower had warned the A300 of the danger of wake turbulence, but this was routine in the circumstances. In fact, many considered it a phrase used to cover the controllers' backs, with the effect being somewhat like someone crying wolf too often. Nevertheless, Molin was obviously concerned, for he said to the captain:

You happy with that distance?

The captain replied:

Aah, he's . . . We'll be all right once we get rollin'. He's supposed to be five miles by the time we get airborne. That's the idea.

First officer:

So you're happy . . . Lights?

Captain:

Yeah, lights are on.

Preparations for takeoff continued.

First officer:

Takeoff checks complete. I'm on the roll. Thank you, sir.

The aircraft began its roll and reached the takeoff decision speed, V_1.

Captain:

V_1!

The aircraft reached rotation speed, V_R.

Captain:

V_R!

Captain:

Rotate!

The captain announced they had reached the safe takeoff (climb) speed, V_2, and then V_2 +10 knots, giving them an extra safety margin.

First officer:

Positive rate. Gear up, please.

With the aircraft five hundred feet off the ground, the aircraft was banking left as the control tower ordered the crew to turn left and fly the bridge climb, like JL47 ahead of them. However, they had commenced their left turn earlier than JL47, with the result that they were cutting the corner downwind of the Japan Airlines craft rather than following in its precise tracks.

Also, in taking off with a just about acceptable gap between them and the Japan Airlines 747 ahead, they had not taken into account that they would be climbing much more quickly, as that aircraft was weighed down by the great amount of fuel required to fly to Japan.

The control tower signed off, telling the A300 to contact New York Departure (Control). They duly informed departure they were at thirteen hundred feet and climbing to five thousand. Departure confirmed they had them on their radar and said:

Climb. Maintain one three thousand.

They acknowledged this. The aircraft passed through fifteen hundred feet at roughly 220 knots, and the bank eased as it completed its turn to the left. The Japan Airlines 747 was ahead of them but upwind and farther out on the curve and somewhat higher, but not so much higher, due to its load of fuel.

As Captain States and First Officer Molin passed through 1,769 feet at 239 knots, departure called, telling them to turn left again.

587 heavy, turn left . . .

The cockpit area microphone (CAM) picked up a squeak and a rattle.

Captain:

Little wake turbulence, huh?

[The wake turbulence from one of the JAL 747's wingtips.]

First officer:

Yeah!

The aircraft was in a steep but not excessive bank as it continued to make the second left turn as ordered. Fourteen seconds later, and traveling at 238 knots, it encountered the wake turbulence from the JAL 747's other wingtip.

In a strained voice, Molin asked for maximum power, and the following exchange took place between the two pilots, with Captain States speaking first:

You all right?

Yeah, I'm fine.

Hang on to it. Hang on to it.

This time the buffeting seemed more serious, and the flight data recorder (FDR) shows First Officer Molin continuously making rapid full-sweep, to-and-fro rudder inputs in addition to the aileron inputs that were normal for such a situation. In fact, the rudder is not used very much in general apart from compensating for a crosswind on landing with the aircraft traveling relatively slowly.

The cockpit area microphone (CAM) recorded the sound of a thump.

CAM:

> *[Loud thump.]*

CAM:

> *[Loud bang.]*

Molin:

> *[Human grunt?]*

CAM:

> *[Roaring noise starts and becomes louder.]*

Molin:

> *Holy s***!*

There followed warning chimes, which were probably first ECAM chimes and then stall warning chimes.

> *[Roaring noise decreases and ceases.]*

Molin:

> *What the hell are we into? . . . We're stuck in it.*
>
> [Possibly more chimes, lasting some four seconds overall.]

Captain:

> *Get out of it! Get out of it!*

[End of recording.]

When Captain States was saying "Get out of it!," neither he nor Molin knew the vertical tail fin had broken off some fourteen seconds earlier and was falling into Jamaica Bay below.

The aircraft, minus its vertical stabilizer (tail fin), continued a little farther and crashed onto a narrow band of land separating Jamaica Bay from the Atlantic Ocean, which is part of Queens. None of the 260 people on board the Airbus 300 survived. The death toll on the ground could have been considerable, as it was a residential neighborhood, but was limited to five persons, since most pieces hit unoccupied property or open spaces.

The big question was why Molin had swished the rudder back and forth in such a manner. The investigators had some difficulty determining just how extreme Molin's movements had been, because of the sampling rate—that is, the frequency with which the FDR measures the angle of the rudder—which was then four seconds, during which time it would have been possible for him to move the rudder back and forth (without it showing up).

Another problem was that the readings, as are many others on FDRs, are filtered to avoid confusing spikes. This averaging would not matter when dealing with a rudder used in virtually one position, as would usually be the case, but obscured the magnitude of the extremes that investigators so much wanted to know in the present situation.

One major line of inquiry was obviously to find out whether the tail fin had some inherent weakness, either due to poor design or previous overstressing that had left undetected cracks. Since it was made of composites rather than the traditional metal, there was much talk that composite tail fins were inherently dangerous, partly because, unlike metal tail fins, cracks are so difficult to detect. Some American Airlines pilots even talked of boycotting the Airbus A300.

In the end, and after exhaustive examinations and tests, the investigators concluded the accident had not resulted from an inherent weakness but from Molin's aggressive use of the rudder, and in particular his rapidly moving it back and forth.[2]

The next question posed was why Molin had done this. Should the manufacturer have warned that doing so was dangerous? As large sums of money and prestige were at stake, there then ensued a battle between Airbus and American Airlines. Investigators looked into Molin's personal life so extensively that his father, also a pilot, complained. Inquiries almost got to the point of ascertaining how Molin performed in bed. He was said to be a little immature socially, and his deferential manner in his exchanges with Captain States do hint at this. However, investigators found nothing dramatic. On the other hand, American Airlines maintained Airbus knew of the dangers and had done too little to warn pilots.

Airbus maintained that the airline had trained Molin to use the rudder in recovery situations, and being nervous about wake turbulence even before takeoff, he expected it and overreacted when it occurred. Following this line of thought, one could say the first encounter keyed him up so that he was more aggressive the second time.

Some experts said that as he reacted mildly the first time, the turbulence the second time might have been rather different and produced the yawing movement to which he reacted instinctively by using the rudder.

Perverse though it might seem, had First Officer Molin been captain rather than first officer, the subsequent joust with wake turbulence might have never taken place. When Molin asked the captain just before takeoff whether he was happy with the distance between them and the JAL aircraft ahead, he was evidently thinking it would be better to wait perhaps thirty seconds more.

This expression of serious doubt to seniors in the form of a question brings to mind the worst-ever aircraft disaster, at Tenerife, where the flight engineer used the words "Did he not clear the runway then?"

Had Molin been less polite and said forcefully "I think we should wait a moment," the captain of Flight 587 might have done just that and not tried to justify himself. Admittedly, the KLM flight engineer at Tenerife would have had to be very assertive to halt that fatal takeoff roll.

Conversely, had Captain States been flying the aircraft, his laid-back attitude would probably have meant he would have let the aircraft have its head and not have used the rudder so disastrously, if at all. However, his telling Molin "Hang on to it. Hang on to it!" might have made Molin even more determined to keep the aircraft on a leash, though "hang on" was not literally telling him to swish the rudder, but it could have been interpreted as that.

With hindsight, one can also see that had Flight 587 followed the track of the Japan Airlines 747 more closely instead of cutting the corner, as ordered by ATC, it would have missed the turbulence altogether, as it would have been blown downwind. In a sense, the required separation was not maintained once the corner was cut. Of course, the premature turn over water would subject built-up areas to less noise.

An article on a now-terminated website called mywiseowl half seriously said, "The primary purpose of flight data recorders (FDRs) is to reduce the manufacturer's liability by correctly assigning blame to the failing assemblies or persons, with the famous secondary purpose being to prevent future accidents."

Had FDR evidence not been available in this accident, would anyone have thought the first officer would have made those rudder movements? Thus, the FDR did indeed help save the day for the manufacturer. The airline put considerable pressure on the investigators, making new points until they were told not to make new representations unless based on new facts. Whether or not this pressure was a factor in the change made to the order in which the contributory factors were listed, it is notable that board member Carol J. Carmody, with the concurrence of board member Richard F. Healing, added a dissenting statement to the report.

While they agreed with its conclusions as regards probable cause, namely the excessive use of the rudder by the first officer, they objected to the fact that the vice-chairman had changed the order in which the contributory causes were listed from that in the original draft. This gave the impression that more weight should be given to the characteristics of the A300 rudder system[3] than to the nature of the airline's Advanced Aircraft Maneuvering Program (AAMP).[4] They pointed out that Molin had not been known to use the rudder abnormally before partaking in the program, and when questioned by a captain about excessive use of the rudder in another incident after taking part in the program, Molin had insisted the AAMP directed him to use the rudder in that manner. Carmody went on to say, "To elevate the characteristics of the A300-600 rudder system in the hierarchy of contributing factors ignores the fact that this system had not been an issue in sixteen million hours of testing and operator experience—until the AAMP-trained pilot flew it."

This puts Molin in a rather different light. Known as a hands-on pilot, with some saying his flying was as smooth as silk, maybe his only propensity or quirk was to follow things too much to the letter.

In 2012, and in the light of other cases where the A300 tail assembly had been subjected to excessive stress but not enough to cause it to break off, the NTSB said the A300 should have greater built-in protection against such events, which suggests they no longer think another pilot doing what Molin did is inconceivable. Though the initial report was not revised, it does suggest, as some investigators felt, Airbus bore some responsibility, though the airline's Advanced Aircraft Maneuvering Program (AAMP), involving application of significant rudder inputs, was surely a factor.

Interestingly, the latest advice to pilots, admittedly often destined for smaller craft, is to do absolutely nothing on encountering wake turbulence. If pilots follow their instinct and immediately try to correct a vortex-induced sharp bank, they in practice end up doing so just as the vortex sends them in the opposite direction, making matters much worse. Only when they are out of the vortex should they make corrections.

Added Safety Feature Sometimes Responsible for Disasters

Though there is absolutely no direct link, it is sad to think that in the case of the Asiana A320 that crashed into the Java Sea in 2014, an intermittent fault with the rudder travel limit unit (RTLU), designed specifically to stop pilots doing what Molin did, ultimately led to the disaster.

Due to the fault, that had been occurring on and off for over a year, the captain had to repeatedly reset the system. Exasperated, using a technique a mechanic had taught him that should not have been used in the air, he momentarily switched off a couple of the computers. The autopilot disengaged, and then due to a whole series of blunders and the captain using baffling language the copilot misinterpreted, the aircraft finally stalled and crashed.

---

[1] As explained in connection with the collision between an airliner and a Cessna at Milan's Linate Airport, "runway extension" means the imaginary line that would be drawn if the runway continued.

[2] The NTSB presented a simulated video for the inquiry, showing a side view of the aircraft and the movements of the controls made by Molin.

[3] Four years earlier, the A300-600 pilot of American Airlines Flight 903 lost control at 16,000 feet on approach to Miami and moved the rudder from side to side to recover, almost tearing off the tail in the process.

Subsequently, an Airbus internal memo noted, "Rudder movement from left limit to right limit will produce loads on fin/rear fuselage above ultimate design load." If that memo had been generally circulated, that fact not only might have been incorporated in the AAMP program but mentioned in the manual.

[4] The AAMP program to enable pilots to right aircraft after upsets included aggressive use of the rudder.

Weary Colgan Air Pilots Gossip Before Stall (Buffalo, NY, 2009)

Few people know 50 percent of flights in the United States are by regional carriers

> The feeder airline pilots were tired. One was also not very well, and they allowed gossiping to encroach on the time they should have been using to prepare the landing.
>
> *Colgan Air Flight 3407, February 12, 2009*

Though not a major disaster in terms of the number of lives lost, the Colgan Air Continental Airlines Connection Flight 3407 crash [1] described here received much attention because fatal air crashes in the States had become rare. Another reason was that it made the American public realize how many flights seemingly flown by the majors (with their flight numbers and even their livery) are actually operated by regional airlines.

Though the Canadian-built Bombardier C4 turboprop airliner [2] used for Flight 3407 had Continental livery, employees of Colgan Air [3] operated and serviced the flight exclusively. Following the accident, legislators passed laws to make purchasers of airline tickets more able to determine on which airline they are actually flying.

Salaries and working conditions at commuter airlines, as opposed to the majors, are less than generous. The forty-seven-year-old male captain and the twenty-four-year-old female first officer, whose gross salary had been $15,800 during the previous year and was less than that of a bus driver, spent much of the flight discussing the difficulties they had in relation to this. (The cockpit voice recorder also recorded the captain stating that he earned a gross salary of about $60,000.)

The captain lived in Florida, and the first officer in Washington State, both on the other side of the country from their New Jersey operating base. Both had at one time considered having proper crash pads in New Jersey, but for financial reasons neither had one—there was always the option of resting or sleeping in the crew lounge.

The first officer had come in from Seattle early that morning. She had managed to get a reasonable amount of rest in the crew lounge and on the flight from Seattle but was suffering from a cold and had stated in an exchange prior to takeoff that had circumstances been different she would have called in sick.

She said, "I'm ready to be in the hotel room," to which the captain replied, "I feel bad for you."

Continuing, she went into detail:

This is one of those times that if I felt like this when I was at home, there's no way I would have come all the way out here. If I call in sick now, I've got to put myself in a hotel until I feel better . . . we'll see how . . . it feels flying. If the pressure's just too much . . . I could always call in tomorrow. At least I'm in a hotel on the company's buck, but we'll see. I'm pretty tough.

The CVR recorded her sniveling and occasionally yawning during the flight. The captain himself suffered from a sleep deficit built up over several days and could be heard yawning on the CVR. The rest that he had had that morning in less than ideal conditions in the crew lounge could hardly have made up for that deficit.

The flight was to be a short fifty-three-minute hop across New York State to Buffalo, on the edge of Lake Eerie. Near the border with Canada, Buffalo is the second-largest city in New York State, and consequently they were over built-up areas as the aircraft approached the airport.

The CVR recorded the Newark tower controller clearing the airplane for takeoff at about 21:18:23 EST. The first officer acknowledged the clearance, and the captain said:

All right, cleared for takeoff. It's mine.

The intended cruise altitude for the flight was sixteen thousand feet mean sea level, and the flight data recorder showed that, during the climb to altitude, propeller and airframe deicing were turned on (pitot static deicing had been turned on prior to takeoff) and the autopilot was engaged.

The turboprop reached its cruising altitude of sixteen thousand feet at about 21:34:44. The cruise part of the flight was routine and uneventful, and the only thing remarkable about it was the rapport between the two pilots, which resulted in them indulging in almost continuous nonessential personal conversation. Although not in conflict with the sterile cockpit rule, which prohibits nonessential conversations within the cockpit during critical phases of flight, as they were above ten thousand feet it did mean that certain tasks, such as checklists, were delayed until the very end of the flight, when they would be busy.

At about 21:53:40 the first officer briefed the captain on the airspeeds for landing, with the flaps at fifteen degrees (Flaps 15), giving 118 knots as the reference landing speed (V_{REF}), and 114 knots as the go-around speed (V_{GA}). The captain acknowledged this information.

Determining the so-called reference speed (for landing) is a key aspect of any landing, as it is the airspeed needed to ensure a stabilized, safe landing. However, the airspeed that the first officer gave was that for normal conditions, not icy conditions, under which, say, an extra twenty knots is added to allow for wings, stabilizers, and control surfaces performing less than optimally.

The flight continued to descend under approach control, with the captain confirming in his briefing the airspeed for the fifteen-degree flaps landing. At about 22:10:23 the first officer asked whether ice had been accumulating on the windshield, and the captain replied that ice was present on his side of the windshield and asked whether ice was present on her windshield side. The first officer responded, "*Lots of ice.*" The captain then stated, "*That's the most I've seen—most ice I've seen on the leading edges in a long time. In a while anyway I should say.*" About ten seconds later the captain and the first officer began a conversation that was unrelated to their flying duties.

At about 22:12:18 the approach controller cleared the flight crew to descend and maintain 2,300 feet, and the first officer acknowledged the clearance. Afterward, the captain and the first officer performed flight-related duties but also continued the conversation concerning other matters. At about 22:12:44 the approach controller cleared the flight crew to turn left onto a heading of 330 degrees. At about 22:13:25 and 22:13:36 the captain called for the descent and approach checklists respectively, which the first officer performed. At about 22:14:09 the approach controller cleared the flight crew to turn left onto a heading of 310 degrees.

The altitude hold mode became active as the aircraft was approaching the preselected altitude of 2,300 feet, which was reached at about 22:14:30, with the airspeed approximately 180 knots at the time. The captain began to slow the aircraft less than three miles from the outer marker to establish the appropriate airspeed before landing. According to FDR data, the engine power levers were reduced to around forty-two degrees (flight idle was thirty-five degrees) at around 22:16:00, and both engines' torque values were at minimum thrust at about 22:16:02. The approach controller then handed them over to the BUF ATC tower controller. The first officer's acknowledgement of this instruction was the last communication between the flight crew and ATC.

When the first officer, at 22:16:21, told the captain that the gear was down, the airspeed was roughly 145 knots. The reduced airspeed meant the autopilot had to apply additional pitch trim in the aircraft-nose-up direction to maintain the 2,300-foot altitude. An ice-detected message appeared on the engine display in the cockpit. About the same time the captain called for the flaps to be set to fifteen degrees and for the before-landing checklist. The airspeed was then about 135 knots.

At 22:16:27.4 the CVR recorded a sound similar to the stick shaker. (The stick shaker warns a pilot of an impending wing aerodynamic stall through vibrations on the control column, providing tactile and aural cues.) The CVR also recorded a sound similar to the autopilot disconnect horn, which repeated until the end of the recording.

When the autopilot disengaged, the airspeed had dropped to 131 knots. FDR data showed that 0.4 seconds later the control columns moved aft and that the engine power levers were advanced to about seventy degrees (the rating detent was eighty degrees) a second later. The CVR then recorded a sound similar to increased engine power, and FDR data showed that the engine power had increased to about 75 percent torque.

The stall warning (stick shaker) activates seven knots or so before an actual stall. However, as the reference speed switch on the icing panel had been turned to "Increase," the stick shaker would activate at about twenty knots higher than it otherwise would have done. In addition, the airspeed indicator would have shown the low-speed red markings coming up on the band. Surprisingly, in view of all the talk about icing, the pilots had not even considered increasing their reference landing speed.

The captain had not called out "Stall" when the stick shaker activated, nor had the first officer called out "Positive rate" when the aircraft climbed in response to the captain (presumably) pulling back on his control column. While the increase in power (not to the maximum) could not immediately influence the airspeed, putting the aircraft into a climb certainly did, and it declined rapidly.

After pitching up, the aircraft rolled to the left, reaching a roll angle of forty-five degrees left wing down, and then rolled to the right. As it rolled to the right through wings level, the stick pusher[4] activated at about 22:16:34, at which point Flaps 0 was selected. At about 22:16:37 the first officer told the captain that she had put the flaps up, even though she had not been ordered to do so and in so doing had made recovery from the stall even more difficult, as the aircraft would have to fly even faster in a no-flaps configuration.

FDR data confirmed that the flaps had begun to retract by 22:16:38; at that time the aircraft's airspeed was about a hundred knots. FDR data also showed that the roll angle reached 105 degrees right wing down before the aircraft began to roll back to the left and the stick pusher activated a second time (at about 22:16:40). The aircraft's pitch angle was minus one degree.

At 22:16:42 the CVR recorded the captain making a grunting sound. The roll angle had reached about thirty-five degrees left wing down before the aircraft began to roll again to the right. Afterward, the first officer asked whether she should put the landing gear up, and the captain said, "*Gear up*" and an expletive. It was too late. The aircraft pitched about twenty-five degrees nose down with a roll angle of a hundred degrees right wing down respectively, with the aircraft entering a steep descent. The stick pusher activated a third time (at about 22:16:50), and two seconds later the FDR data showed that the flaps had retracted fully. About the same time the CVR recorded the captain stating, "*We're down*," and the sound of a thump.

The aircraft had come down on a single-family home, and a post-crash fire ensued. The death toll was fifty: one person killed in the house, and all forty-nine occupants of the aircraft. The fire, partly augmented by escaping gas from a gas main to the house, destroyed much of the aircraft, making recovery of the bodies and evidence very difficult.

The investigators, as usual, first sought the extremities, such as the nose, wingtips, and tail, to see whether the aircraft had crashed in one piece or broken up before hitting the ground, which it had not. They quickly recovered the FDR and the CVR and found them to be in good condition.

Probable Cause

The NTSB ultimately determined that the probable cause was the captain's inappropriate response to the activation of the stick shaker, which led to an aerodynamic stall, from which the airplane did not recover.

Contributing to the accident was:

1. The flight crew's failure to monitor airspeed in relation to the rising of the low-speed cue.
2. The flight crew's failure to adhere to sterile cockpit procedures.
3. The captain's failure to effectively manage the flight.
4. Colgan Air's inadequate procedures for airspeed selection and management during approaches in conditions where the aircraft would be subject to icing.

There were differing opinions at the NTSB as to how much fatigue was a factor, with some believing it should have been included in the contributing causes. After publication of the official report, a lawyer for families suing over the crash said that newly released e-mails showed Colgan Air doubted the captain's ability to fly the plane six months before it crashed. These e-mails were not made available to the NTSB in the course of its inquiry. Colgan Air had admitted that it might well not have employed the captain had it been aware of all the checks that he had failed before it engaged him.

The NTSB recommended that there should be a limit to the number of failed checks a pilot is allowed before his or her dismissal. However, the FAA considered this to be too arbitrary and impractical. Yet there have been a number of cases, including a recent Nigerian one, where pilots responsible for crashes have been found to have a checkered history as regards check rides, and with hindsight should perhaps have been let go. Instead of being fired, such pilots are generally sent for remedial training, but one wonders whether any amount of remedial training can make up for a lack of innate aptitude or ability to analyze while under stress.

At one time many commercial pilots in the US came from the military, where the weak ones were quickly weeded out.

Ironically, it was the extra safety feature, namely the special switch for increasing the reference speed setting for icy conditions, that caused the stick shaker to operate and the captain to react perhaps in panic, which bears out the contention of Charles Perrow that safety systems can cause accidents.

The first officer, who was considered to be above average for her level of experience, did not help matters by raising the flaps. Her failure to note the rising low-speed cue with its red markings on the band of the airspeed indicator was attributed to the fact that she was preoccupied with other tasks.

Gossiping had delayed tasks such as checklists. Then the lowering of the flaps and undercarriage would have diverted her attention from the altimeter to the central console. Of course, the captain should have been monitoring the airspeed himself and picked up on the rise of the red low-speed cue. Could his alleged tendency to rely more than usual on the autopilot to do the flying have got him out of the habit of monitoring airspeed? His failure to call out "Stall" possibly led to the first officer reacting incorrectly by raising the flaps (to which the captain did not object). To what extent her bad cold and fatigue impaired her monitoring and decision making is difficult to determine.

Sterile Cockpit Rule

This is sometimes cited as an example of what can happen when pilots fail to obey the FAA's sterile cockpit rule, requiring pilots to refrain from nonessential activities, such as idle gossip or unnecessary interaction with cabin crew, during critical phases of flight, in general defined as heights below 10,000 feet (3,048 meters). Passengers will usually hear a chime when the aircraft passes through that 10,000-feet altitude.

---

[1] The previous disaster had similarly involved a commuter airline, the Comair Flight 191 crash in August 2006, where the aircraft with 50 people aboard had taken off on the wrong (much too short) runway. Only the first officer survived, having suffered terrible injuries.

[2] Designed to carry 74 passengers and 4 crew members (2 pilots, 2 cabin crew).

[3] Colgan Air was a regional carrier that flew routes under contract for US Airways, United, and Continental. Pinnacle Airlines Corp, the parent company of Pinnacle Airlines and Colgan Air, filed for bankruptcy protection on April 1, 2012. Pinnacle said it would rework contracts with Delta and end flying for United and US Airways. The company also said it wanted to cut its labor and operating costs.

[4] As a last resort, it pushes the nose down so the aircraft can gain speed. It can be countered and overridden by the pilot.

Chapter 9
PILOT CRASHES AIRCRAFT INTENTIONALLY?

SilkAir Flight 185 in Near-Vertical Dive (Indonesia, 1997)
NTSB convinced it was suicide; Indonesians and others not so sure

> A US lawyer later managed to sue the makers of the 737's rudder power control unit (PCU), with the jury not permitted by law to consider the NTSB findings.
>
> *SilkAir Flight 185, December 19, 1997*

The Boeing 737 was the newest in SilkAir's fleet and had been delivered to the airline only ten months before. With ninety-seven passengers and seven crew, it had taken off at 15:37 local time from Jakarta's main international airport, bound for Singapore, a short flight of some eighty minutes.

The captain was forty-one-year-old Tsu Way Ming, a Singaporean, and the first officer twenty-three-year-old New Zealander Duncan Ward. Before flying commercially, Captain Tsu had been a Skyhawk pilot and Skyhawk squadron instructor in the Singapore Air Force, attaining the rank of major before taking voluntary retirement.

Everything about the flight that day seems normal on the cockpit voice recorder (CVR), with the exchanges between the pilots, and with the cabin crew, nothing out of the ordinary. Neither was there anything strange about the captain's routine address to the passengers over the PA system.

At 15:53 the pilots reported reaching their thirty-five-thousand-feet cruising height, and twelve minutes later, at 16:05, Captain Tsu handed over control to the first officer, saying, "You have the controls" as he absented himself. There is the sound of a seat belt being unbuckled, after which the cockpit voice recorder ceases recording.

Although there was no CVR, a routine exchange between the air traffic controller and First Officer Ward was recorded at Jakarta at 16:11, with no indication of anything untoward. However, just over a minute later the controllers' radar showed the aircraft dropping 15,500 feet in just thirty seconds before disappearing from the monitor.

The controllers did not have to wait long for confirmation of the crash, as though in an isolated country area, the scene of the crash was inhabited. There were even eye witnesses who saw the aircraft, having shed some parts, dive almost vertically at terrifying speed into the wide Musi River, near Palembang, Sumatra. There could be no survivors.

Though wreckage was scattered over several kilometers, most of it was in a 200- by 260-foot (60- by 80-meter) area at the river bottom. Because of the force of the impact, the wreckage was in tiny pieces, much of it embedded in the mud.

Not a single complete body or body part was found, since the aircraft and passengers disintegrated on impact. Only six positive identifications were later obtained from the few recovered human remains.

Unbelievably, the whole trajectory, from cruise height at thirty-five thousand feet to impact with the river, took less than a minute.

Powerful dredgers were brought in to scoop up the mud containing the fragments of wreckage in the hope they would include one that could provide a clue as to why the accident had happened. Despite the murkiness of the water, fumbling divers found the flight data recorder only five days after the crash. That, together with the discovery sixteen days later of the bright-orange cockpit voice recorder in the material being dredged up, made the investigators believe the case might be solved relatively quickly.

The Investigation

Apart from the Indonesians, investigators from several countries were involved, and notably the NTSB, from the US. Since the 737 was used in great numbers all over the world, finding any inherent flaw was of great importance.

Unbelievably, 73 percent of the aircraft, albeit in tiny pieces, was recovered. They revealed that the engines were at full power and the controls set to push the nose down rather than up. That alone suggested a pilot might have flown it into the ground on purpose.

However, there had been crashes involving the Boeing 737, described in Chapter 6, where the aircraft had suddenly banked and dived into the ground. It had taken the NTSB years to show that it was due to a malfunction of the rudder power control unit (PCU).

Senior executives of airframe makers, component makers, and engine makers must live in constant fear of waking up one morning to learn something they are responsible for has an inherent flaw, and worse, has caused a disastrous crash. In a scenario that was later to work against them in a US court, Parker Hannifin, the world's largest maker of hydraulic valves, was alerted that radar tracks suggested the accident had something in common with the 737 ones just mentioned, and immediately sent its people out to the site to try to ensure its rudder power control unit (PCU) was recovered.

Retrieved intact, the unit was sent to the US to be checked, and returned, having allegedly been found to be working perfectly—the PCU on the SilkAir 737 was one of the new redesigned ones.

The US's NTSB concluded, after a great deal of painstaking work, that even in the unlikely event of the PCU having caused the upset:

1. The pilots could easily have recovered;
2. The aircraft could not have descended as rapidly as it did without input from a pilot.
   Examination of the debris had shown the engines were at full power, the horizontal stabilizer trim was set at fully forward, and no devices, such as air brakes, were employed to slow the descent.

Other pilots, independently of the NTSB, simulated the descent, and none could achieve such a rapid one without manual input.

Also, the captain could have disabled the CVR—there was a click suggesting he had done so manually on getting up after releasing his seat belt—and later found a pretext for asking the first officer to go back to the passenger cabin, then locked the cockpit door before finally shutting off the FDR. Or he could have struck the first officer on the head. All this would have happened in an ATC no-man's land between that last radio exchange with Jakarta and calling Singapore at PARDI.

On leaving the Singapore Air Force as a major, Tsu had been fast-tracked to the rank of captain at SilkAir, and then to that of line instructor pilot. There were, however, some reservations about his character, and it was thought he needed to be monitored in this respect. In fact, these reservations proved justified, for in the year of the accident he was involved in two incidents.

During a routine approach, which the first officer thought necessitated a go-around because he was too high and too fast, Tsu started weaving to lose height. The "violent rolls, left and right" were, according to the first officer, very disturbing and scary. In the end, they were unable to lose enough height and had to go around. Instead of doing a regulatory conventional one, Tsu did a tight visual circuit at 170 knots within the hills surrounding the airport, with the undercarriage and five units of flap still extended.

As an ex-aerobatics pilot, Captain Tsu probably did not think it amounted to much and failed to report the incident, as required.

Rumors that he was unsafe circulated, and the first officer was asked to submit a report. On the next flight, the two of them were rostered together. Tsu said how displeased he was to hear the rumors about his nearly crashing and about not being suitable for command. The first officer told him he was not the source, which seemed to please Tsu. But just as they began to taxi, he swiveled round in his seat and pulled the CVR circuit breaker, located on a panel behind him. When the first officer objected, Tsu said he had done so to prevent the conversation being overwritten. He wanted to use it as evidence. With the first officer refusing to fly with the CVR disabled, Tsu then said they would return to the terminal to download the recording but changed his mind when the tower asked him to explain why. Tsu then reset the circuit breaker, allowing them to depart. This embarrassing incident, of Tsu's own making, resulted in him being demoted to a mere captain.

There was evidence that Captain Tsu had financial difficulties—his stock exchange trading account had been blocked for nonpayment. However, having sold a couple of properties, he was actually solvent. What was first seen as a sure sign that it was suicide, namely a life insurance policy starting the day of the flight was later found to have been done to comply with the terms of a mortgage he had taken out. Also, his father was willing to lend him money.

However, the day of the flight did coincide with the anniversary of the day he was involved in an incident where three of his fellow Air Force pilots were killed.

That said, the captain seemed so normal to colleagues, and even in the way he conversed in the cockpit and briefed the passengers. Also, would one ever kill oneself to claim insurance for one's wife and children on the day the policy started? Doing so would look highly suspicious. However, the behavior in the cockpit leading to his demotion suggested an impetuous streak. The fact that he had asked his wife to meet him at the airport on his return from that flight is seen by some to be proof he was not premeditating suicide.

When the Indonesians produced their final report, they said it was impossible to determine the cause of the crash. This conclusion so infuriated the NTSB investigators, who had worked so hard, that they added their comments to the report. The investigation had taken three years and left the relatives of the victims angry, and with no closure and a sense of disbelief. To sum up, the logical conclusion was that it was suicide, but there was nothing to prove it in a legal sense or from Tsu's demeanor in talking to his copilot or more notably the passengers.

The Private Investigation and Los Angeles Trial

A US lawyer launched an independent investigation of the recovered rudder power control unit (PCU), and in 2004 a Los Angeles jury ruled that was the cause of the crash. The jury could not be told of the NTSB's findings due to a law forbidding it to help ensure the NTSB would come to conclusions without factoring in how they might be used in litigation. Parker Hannifin, the manufacturer of the unit, was ordered to pay US$44 million to the plaintiffs' families. It then appealed the verdict, resulting in an out-of-court settlement for an undisclosed but significant amount.

Among the many arguments that helped sway the jury were Parker Hannifin being immediately alerted thereby suggesting the PCU was potentially suspect, the old flight data recorder that had been installed in place of the solid state one had on that and earlier flights worked intermittently, and if Tsu had switched off the voice recorder why not the data recorder at the same time?

Was EgyptAir Flight 990 a Matter of Revenge? (JFK Outbound, 1999)

Revenge for being taken off US flights for sexual misconduct

> The US investigators concluded the pilot crashed the aircraft
> deliberately; the Egyptian side disputed this. Video cameras in the
> cockpit would have made it possible to prove beyond doubt, in this case,
> and other contentious ones, that aircraft were crashed on purpose.
>
> *EgyptAir Flight 990, October 31, 1999*

On board the Boeing 767 EgyptAir flight from New York to Cairo was EgyptAir's chief 767 pilot, Hatem Rushdy. The night before he had told First Officer el-Batouty, who was on board as number two in the relief crew due to take over midflight, that it was to be el-Batouty's last flight on routes to the United States.

According to the UK's *Guardian* newspaper dated March 16, 2002, el-Batouty faced ruin following allegations of sexual misconduct, including exposing himself to teenage girls, propositioning hotel maids, and stalking female hotel guests. Flights to the US came with extra pay besides prestige, so his downgrading and loss of face would be obvious.

Having come from Los Angeles, the EgyptAir 767 lifted off at 1:20 a.m. EST from New York's JFK Airport, where it had stopped to refuel and pick up passengers before continuing on to Cairo. There were 217 persons on board, including 14 crew, of which four were pilots in two groups—the primary crew and the relief crew.

In the half hour after takeoff, the aircraft climbed to its initial cruising height of thirty-three thousand feet, with flight attendants, and finally Hatem Rushdy, paying a visit to the cockpit. After Rushdy left, el-Batouty got up from his seat in the passenger cabin, where he was meant to be resting in readiness for taking over as copilot for the midpart of the flight, and entered the cockpit. He then pulled rank, intimating to the much younger primary copilot that they should swap shifts. The latter strongly objected, as it would complicate matters later in the flight and mean he could not rest as planned. Anyway, it was contrary to regulations.

Probably sensing el-Batouty's single-minded determination, the primary captain, who was a personal friend, persuaded the younger primary copilot to cede to the older man's demands. There then followed a routine exchange with the en route air traffic controller, Ann Brennan, who had eight years' experience.

01:47 a.m. ATC:

EgyptAir Nine-ninety, change to my frequency one-two-five-point-niner-two.

Captain of EgyptAir 990:

Good day ... [Pause while switching frequency.] ... New York, EgyptAir nine zero heavy, good morning.

ATC:

EgyptAir nine-ninety, roger.

Having confirmed they were on the right frequency, Brennan, with her headset still on, moved away from her monitor to deal with paperwork. When she looked again six minutes later, the echo indicating where EgyptAir 990 should be according to its flight plan was freewheeling, because the computers could find no correlation with its transponder. In vain she called them to get them to reset their transponder and tried to contact them through other aircraft and facilities, even asking an Air France aircraft in her zone to fly over them. With it becoming evident something had gone terribly wrong, she, in consultation with her supervisor, set search-and-rescue operations in motion.

The fact that Brennan had not been watching what happened to EgyptAir 990 did not make any difference other than to slightly delay the rescue, when there was no one to rescue, and what she would have seen was recorded anyway and not lost.

What had happened in those six minutes?

At 01:48:03, and shortly after the exchange with air traffic control, the captain announced he was going to the toilet. The aircraft was flying on autopilot at its assigned cruising height of thirty-three thousand feet with no indication of a problem, so it was a reasonable time to do so, for, as he explained, the toilets would be free while the passengers ate their meals. There was the whirring sound of the captain's seat being moved back, followed by the sound of the cockpit door closing.

Then, at 1:48:30, there was a muffled sound that sounded like three syllables, with emphasis on the middle one. Despite repeated analysis, the NTSB was never able to grasp what was being said or even determine who was speaking. Some thought it sounded like "Control it" or "Hydraulic." At 01:48:39, el-Batouty said in a clear but soft voice:

I rely on God! (Tawakkalt ala Allah!)

The significance of these words in Arabic was later to be hotly disputed, with some saying they meant he was going to do something dramatic, and others saying they are words that slip off the tongue easily—when starting your car, for instance.

This was followed by the sound of the whirring of an electric seat and disengagement of the autopilot. If it had disengaged on its own, there would have been an audible warning to alert the pilots.

For four seconds the aircraft flew on, with nothing happening.

"I rely on God," said el-Batouty just as the thrust levers were sharply pulled back to idle. The elevators in the tail that control the pitch of aircraft then moved to the three-degrees-down position. As a result, the nose pitched downward and the aircraft did what is called a bunt, whereby it suddenly pitches so sharply downward that it creates near-zero or negative g. It is used to throw hijackers off their feet or to get astronauts used to the weightlessness of space flight.

Meanwhile, el-Batouty was repeating "I rely on God" again and again.

The elevators were moved even farther downward, with the pitch of the aircraft attaining forty degrees downward. In the twenty-five seconds from the pulling back of the thrust levers and the movement of the elevators, the aircraft had plunged ten thousand feet, from thirty-three thousand to twenty-three thousand feet, and it was not over.

Almost unbelievably considering the conditions, the captain got back to the cockpit sixteen seconds after the initial pitch down must have made him realize something was wrong. The cockpit voice recorder transcript gives some indication of the situation:

Captain:

What's happening? What's happening?

El-Batouty:

I rely on God.

For about fifteen seconds there were various thumps and clicks, together with a continuous aural master warning, followed by another "I rely on God" from el-Batouty.

Captain:

What's happening? [unidentified voice]: [unintelligible]

Captain:

What's happening? What's happening?

El-Batouty:

I rely on God.

Captain:

What's happening?

Captain:

What's happening, Gamil? What's happening?

Captain:

What is this? What is this? Did you shut the engine[s]?

Captain:

Get away in the engines. Shut the engines.

El-Batouty:

It's shut.

Captain:

Pull. Pull with me. Pull with me. Pull with me.

The recording ends at 01:50:38 due to lack of electrical power.

Analysis of the flight data recorder data while still working showed the captain, probably standing, had been pulling back on his control column, while el-Batouty had been pushing forward on his, with the unusual result that the left elevator was up and the right one down.

The captain must have had some success in his struggle, for the radar tracks show the 767 leveled out at about sixteen thousand feet and climbed to nineteen thousand, and possibly went as high as twenty-two thousand feet. The recovery and climb only prolonged the agony of all on board, for with el-Batouty having cut the fuel supply to the engines, it would have been impossible to restart them in time even were it possible to subdue him. The cockpit was in darkness, with the sophisticated displays dead and the only light coming from the basic instruments running on battery power. In the passenger cabins, there would have been light from the emergency lights, including those set in the floor showing the way to the emergency exits.

A major search for survivors was undertaken, but it soon became clear no one could have survived. In a decision they were to regret, the Egyptian authorities handed over responsibility for the accident investigation to the US, and notably the NTSB, even though the accident occurred over international waters and they had the right to lead the inquiry.

Nine days later the flight data recorder was recovered from the depths and the cockpit voice recorder shortly afterwards. What they revealed was not what the Egyptians expected; instead of an inquiry pointing the finger at a rich foreign entity, such as Boeing, that could be sued, one of their own pilots was to blame, bringing shame to their country, and el-Batouty's references to God precipitated a poisonous religious angle, which was played up in the Western press.

What the NTSB thought would be a simple open-and-shut case came to test their patience and waste their experts' time. The Egyptians sent over a team, which installed itself near the NTSB headquarters in Washington, and spent a year coming out with one theory after another as to why, apart from el-Batouty's actions, the aircraft crashed. Each of these had to be checked out. They at one point said it was a cover-up because the US was ashamed at the poor performance of its air traffic control. It was Air Traffic Controller Brennan's fault!

The Egyptian side closed ranks, evidently under orders, making it impossible to get anything other than robotic answers from individuals, whether concerning el-Batouty or the views of their experts.

Though at first sight some of the Egyptians' points may have seemed valid, it beggared belief that the fault in the aircraft they suggested should occur the precise moment el-Batouty found himself alone at the controls. There can be hardly any doubt he crashed the aircraft intentionally.

It seems that revenge, following the conflict with the head 767 pilot, was the principal motive. One might bear in mind that the desire to lay the blame on someone is an element in many suicides.

An article on the ABC News website after the EgyptAir crash said, "Although the actions of the Flight 990 copilot may never be clearly understood, the allegations [that the pilot crashed the aircraft on purpose] raise [general] fears about whether pilots may be suffering anything from depression to anxiety, or even a death wish. But are the fears founded? There are, in fact, very few instances in which a pilot's or crew member's deliberate actions are known to have brought down an airplane:

1. In 1982 a Japan Airlines DC-8 approaching Tokyo's Haneda Airport plunged into Tokyo Bay. It was determined that the pilot had pushed the nose of the plane down after fighting off the first officer for control. Twenty-four people were killed and 141 injured. The pilot, who survived and pretended to be a passenger when picked up, was later put into a psychiatric institution. [Author's note: After this incident, Japanese passengers flying with Japan Airlines would scan the faces of the airline's pilots at the airport for any tic, suggesting mental problems, resulting in them being twitchier than usual.]

2. On August 21, 1994, an Air Morocco jet fell from the sky after its autopilot was disconnected, and fifty-one people were killed. The cockpit voice recorder suggested that the pilot wanted to commit suicide and deliberately crashed the plane.

3. On December 19, 1997, a SilkAir Boeing 737-300 at cruising altitude went into a steep dive after its flight data recorder and cockpit voice recorders had been turned off. It crashed in Palenbang, Indonesia, killing 104 people. The pilot had recently been demoted and was suspected of committing suicide/murder.

4. On October 11, 1999, an Air Botswana pilot, recently grounded for medical reasons, took off in a company plane and threatened to crash it into company headquarters. He ultimately crashed into two other Air Botswana jets on the ground, destroying all three craft and perishing.

5. The closest such American incident involved an off-duty Federal Express pilot, who, while traveling on a FedEx cargo flight on April 7, 1994, attacked the flight crew with a hammer and spear gun in an attempt to take control of the DC-10 and crash it into Federal Express headquarters in Memphis, Tennessee. The injured flight crew managed to gain control and safely land the airplane."

He had chosen the hammer and spear gun as weapons in the hope that the crash he planned would look like an accident and his family would benefit from the insurance, which would not be the case were a gun found in the wreckage.

To those we must add the Germanwings case described next.

Whatever one may think of el-Batouty, his risking his career through his publicly outlandish behavior in the US brings into question his judgment, which is all-important for a pilot. He should have been dismissed or given non-flying duties on those grounds alone.

He was a large, popular man three months shy of his mandatory retirement age. Many senior people around him were no doubt aware of his risky behavior but laughed it off, pleased to receive the Viagra he would buy in the US as presents.

Lessons might be learnt from financial institutions, where vast sums of money rather than lives are at risk. There, a dismissed staff member is immediately escorted from the premises. While that may be over the top, when a pilot might feel demeaned on losing his flying privileges, measures might be taken to ensure that he or she is never left alone in the cockpit.

Suicidal Germanwings Pilot Locks Out Captain (French Alps, 2015)
The conundrum of medical confidentiality

> Within two days of the incident there came the shocking news that the aircraft had been crashed intentionally by a pilot with known mental problems.
>
> *Germanwings Flight 9525, March 24, 2015*

Modern technology in this case allowed some relatives of those on board to follow the flight in real time on Flightradar24, and they became aware something was wrong as soon as, if not before, the air traffic controllers.

Indeed, one of the first things the French air accident investigators did on finding no transponder returns from Germanwings Flight 9525 and repeated calls to unanswered was to ask Flightradar24 to provide them with its data, which showed the A320 at its cruising height of thirty-eight thousand feet, staying on course toward Germany, but gradually descending, until disappearing sixty-two miles (one hundred kilometers) northwest of Nice.

The air traffic controllers alerted the local mountain rescue team. Almost as if fate had willed it, the first man from the team to be lowered onto the crash site from a helicopter caught sight of an orange box as he, having been told not to touch anything, made his way carefully through the carnage to see whether there were any survivors. The orange box was the cockpit voice recorder, an item that can take days to find, if ever, and it was to enable investigators to solve the mystery without painstaking examination of every scrap of debris.

Together with the flight data recorder, which was recovered later, it revealed that the first officer, Andreas Lubitz, exploited the captain's bathroom visit by locking him out of the cockpit. He then set the autopilot to the lowest altitude possible, a hundred feet, and calmly waited for the inevitable.

The captain could be heard banging on the cockpit door, pleading to be let in, and even trying to smash it, while Lubitz breathed steadily and was chillingly quiet. Passengers could be heard screaming just before the impact with the mountains.

The flight was a regular Germanwings flight to Dusseldorf that had taken off from Barcelona at 10:01 a.m. local time with 144 passengers and 6 crew. It reached its cruising height of thirty-eight thousand feet at 10:27 while still over the Mediterranean, but as it passed over the French coast near Toulon four minutes later, at 10:31, it began descending and finally hit the mountains ten minutes later, at 10:41.

The inquiry found that Lubitz had similarly toyed with the autopilot height setting on the outbound flight. Whether that was to practice, or whether the captain returned before Lubitz felt fully resolved, is uncertain. On the return leg, the toilet next to the cockpit was out of order, forcing the captain to go all the way to the back of the aircraft. It is thought that Lubitz might well have given the captain a diuretic to make him absent himself. Descending even onto mountains from thirty-eight thousand feet takes time, and at various stages Lubitz hastened the descent.

Just before getting up to go to the toilet, the captain told Lubitz to get ready for the landing at Dusseldorf, and he did not realize the significance of his copilot's subsequent comment, "Hopefully... We'll see."

Mental Condition

All attention became focused on Lubitz, and when it was found he was being treated for mental problems and had had to take time off from his training at a Lufthansa facility in the US due to depression, anger grew. Doctors knew, and a torn-up sick note for the day of the flight was found in his flat. Germany has very strict laws regarding medical confidentiality, and it was claimed doctors could not have broken it. Even a chit saying someone needs to be off work does not have to indicate the nature of the illness.

In 2009, Lubitz halted his pilot training with Lufthansa, telling friends he was suffering from stress. He was also found to have lied on a FAA form to be approved for flight training in the US, usually a serious offense, but was given a second shot. A former girlfriend said he admitted receiving psychiatric treatment. He allegedly told her, "One day I'll do something which will change the whole system, and everyone will know and remember my name."

He had studied modes of suicide on the Internet, and shortly before the flight had gone online to learn how to lock the cockpit door definitively.

According to a doctor treating him, he not only suffered from depression but also from a psychosis that he was losing his sight, when there was nothing wrong with his eyes, and he lived in constant fear of losing his coveted pilot's job, which was everything to him.

Aftermath

The US and China, for instance, already had a rule that there must always be two crew members in the cockpit, and some European countries and airlines immediately followed suit. Lufthansa initially did not but reversed that policy under pressure.

One recommendation was that there should be insurance so that pilots should have some compensation in the event of losing their license and not face what may seem to them ruin.

Chapter 10
UNCONTAINED
ENGINE DISINTEGRATION

"Uncontrollable" DC-10's Miracle Landing (Sioux City, 1989)
Impossible to replicate feat in simulator

> Engines are made of the most advanced materials and their parts are x-rayed for defects so that even twin- as opposed to four-engine airliners can fly miles away from possible landing places, and also because engines can disintegrate and their parts fly out with tremendous centrifugal forces. This chapter describes two cases where the latter happened, and exceptional airmanship more or less saved the day.
>
> *United Airlines Flight 232, July 19, 1989*

United Airlines Flight 232 from Denver to Philadelphia, with a stopover in Chicago, was more than halfway through its first leg. On board the three-engine DC-10 were the two pilots—Captain Haines, aged fifty-seven, and First Officer Records, aged forty-eight—and the flight engineer, Second Officer Dvorak, aged fifty-one, as well as 8 cabin attendants and 285 passengers. It was midafternoon, and they were flying in almost perfect weather at the economical height of 37,000 feet and at their designated cruising speed.

Just as they were effecting a slight change of heading, the aircraft juddered, and passengers heard a thud at the rear. The instruments indicated something had happened to the number two engine, mounted in the tail, and that it was spooling down. It would have to be shut down in accordance with the engine shutdown checklist, which should be no big problem, as it was the center engine and the aircraft could fly easily on two engines, and especially if the thrust on each side was equal.

While losing one engine was not much reason for concern, much more worrying was the fall in pressure and loss of fluid in all three hydraulic systems. First Officer Records informed the captain that the aircraft was no longer responding to the controls, and on taking over, the captain was able to confirm the desperate situation for himself. Even by deploying the ram air turbine (RAT) to power the emergency generator to supply the backup hydraulic pump, he found it impossible to restore hydraulic pressure. The aircraft was descending slowly, with a gentle rolling and pitching motion.

Minneapolis Center proposed they divert to Des Moines International Airport, which though a hundred nautical miles away had the advantage of wide, long runways. However, the local controller suggested they try Sioux City, as they were already heading in that direction. The DC-10 pilots agreed this was the best option.

With some difficulty they had established direct contact with the company's maintenance department in San Francisco, but no one could come up with any helpful suggestions. Meanwhile, Captain Fitch[1], a forty-six-year-old off-duty company DC-10 check and training captain, who had been sitting in the first-class passenger cabin, made his presence known and offered his services, which were to prove invaluable as events unfolded.

Captain Haynes asked Fitch to go back to the cabin and check on the damage, and on his return, gave him the task of working the throttles—a key role, as it was the only way they could maneuver the aircraft, and one that would be continuously hands-on. This would leave Haynes and First Officer Records free to deal with the other tasks and, hopefully, find a solution with the help of the airline's experts.

The situation was very similar to that of the Japan Airlines 747 just described in that the DC-10 had what is called a phugoid motion, like a sine wave. On pitching downward, it would progressively gain speed, until at a certain point it would pitch upward and progressively lose speed, until at a certain point it would pitch downward, for the whole cycle to be repeated. The problem was that after each cycle, the aircraft ended up lower, and it would not be long before it struck the ground.

Experimenting with a throttle lever in each hand, Fitch did achieve some measure of control, and finally, by being proactive and anticipating the phugoid motion, was largely able to iron out the highs and lows. The fact that the aircraft had an inherent tendency to yaw to the right and pitch downward, with the right wing tending to drop, made his task even more difficult, as he had to solve the two problems simultaneously. However, one feature that makes the DC-10 easier to control using the engines is that, with the center engine high up at the tail pushing down, the two engines under the wing are very low slung, so that when power is applied to them the nose is levered upward.

Realizing the situation could deteriorate at any moment, Fitch told Captain Haynes they needed to get down as soon as possible. Haynes concurred and asked the first officer to give him the V-speeds for a no-slats, no-flaps landing—they would be able to lower the landing gear using the backup method. The clean—that is, no slats, no flaps—maneuvering speed would be two hundred knots. Furthermore, the speed would have to be above that for Fitch to be able to control the aircraft.

They had dumped fuel, but fifteen tons remained, as the DC-10 has an automatic dump shutoff valve to ensure that dumping cannot reduce the amount of fuel below that level.

The Sioux Gateway Airport[2] controller told them one runway was closed but they could land on either of the other two active runways. Haynes intended to go for Runway 31, which was the longest and widest. However, as they got nearer, Haynes realized that as they could only manage to make right turns and just about manage to fly straight ahead, they could not turn left to line up for an approach to Runway 31. He asked for, and obtained, permission to go for the closed Runway 22, which by then lay straight ahead. The controller reassured them by saying there would be no trouble with the wind and that the presence of a field at the far end would make up for its shorter length.

The Touchdown and Aircrew Performance

They had no micro-control—that is to say, no means to make last-minute adjustments, such as for the crosswind component—and their speed and rate of descent were extreme to say the least. Maneuvers made by means of control surfaces—that is, ailerons, elevators, and rudder—take effect almost immediately, while control inputs via the throttles have a delay of twenty seconds or even more, depending on what is required. For example, to get into the desired position for touchdown would necessitate preplanned inputs performed some forty seconds beforehand. If something unforeseen occurred, there would be insufficient time to make a correction.

The normal landing speed for a DC-10 is 140 knots, and they were doing 215 knots and accelerating. Worse still, their rate of descent was 1,850 feet per minute and increasing, compared with a normal touchdown rate of descent of no more than 200 to 300 feet per minute. Besides adding an extra ten knots to their landing speed, the quartering tail wind was causing them to drift away from the runway centerline.

Through dexterity and/or good luck, Captains Fitch and Haynes had brought them in almost exactly as intended. An amateur video (the professional photographers had been expecting the aircraft to come in on the main runway) of the event gives the impression that it is going to be a perfect routine landing. Captain Haynes[3] has said this video is very deceptive. Though it appears to show them steady at 300 feet, they were losing control and did not have sufficient height to regain it.

The reduction in speed meant the aircraft was commencing a down phugoid and reasserting its tendency to bank to the right. Haynes has also said Captain Fitch, who was at the throttles, added power to correct that bank but that unexpectedly[4] the left engine spooled up faster than the right, making the situation only worse.

The bank increased to twenty degrees, at which point the right wingtip hit the ground just short of the runway and to the left of the centerline, with the nose pointing somewhat downward. The starboard landing gear came down just left of the runway, with the engine on that side striking the ground also to the left of the runway almost at the same time. Accounts then differ.

According to most news reports, the right wing sheared off, and the aircraft flipped on its back, slid down the runway, and then off to the right of it, over a total distance of almost a kilometer. According to Captain Haynes, the aircraft did not cartwheel, as many news reports maintained, but slid sideways on the (intact) left landing gear and the stub of the right wing for two thousand feet. Finally, the left wing came up, and because there was no weight in the tail, as it had broken off, the aircraft flipped over and bounced.

Examination of the wreckage showed the fuselage had broken into five pieces:

1. The cockpit, where all four occupants survived.
2. The first-class cabin, just behind the cockpit, where, apparently, no one survived.
3. Almost three-quarters of the economy-class cabin, from the front galley behind first-class right back to the trailing edge of the wings, where almost three-quarters survived, albeit with injuries.
4. A short section comprising six rows of seats from the trailing edge of the wings to just before the tail section, where only one person survived, with injuries.
5. The tail section behind that, with just two rows of seats, in which seven people, surprisingly, survived.

Incredibly, 185 people out of 296 survived, making a death toll of 111. That so many survived was to some extent due to the sterling efforts of the cabin crew and their rigorous training in a simulator, which made a crash landing and fire seem real.

Even so, their contribution would have probably been in vain, supposing they were even still alive, had not the high rate of descent and the 215 + 10 knot ground speed been absorbed in some manner. This is a prime example of the fact mentioned in this book's preface that the more horrendous-looking crashes can be the most survivable, due to the fracturing and crumpling absorbing the shock.

Unlike the Staines "heart-attack" crash near London's Heathrow, where the aircraft belly flopped with the landing gear up and the wings level and ended up virtually intact, but with everyone dead or dying, the DC-10's right wing and right landing gear and right engine struck first and progressively absorbed the downward momentum as they crumpled and sheared off. Then the 3,000-foot skid decelerated the aircraft even more, before the breaking up of the fuselage into five pieces absorbed almost all of what speed remained.

The official report into the accident mentions that the crop of high corn where the main part of the fuselage ended up upside down hampered rescue efforts by firefighters and recommended that having agricultural crops in proximity to runways should be reconsidered. Thinking of how the rain-sodden ground helped slow Qantas 1 on its 100 mph overrun at Bangkok, one cannot help wondering whether the high corn helped slow the aircraft.

Looking at the feat from a piloting perspective, the main lesson does seem to be the importance of cockpit resource management (CRM), though, according to Captain Haynes, it was first called cockpit leadership management. Each person performed his allotted task and "ideas were thrown around."

With hindsight, one idea that could/should have come to mind was that the pilots transfer fuel from the tanks in the right wing to those on the other side to correct the tendency to bank to the right. Haynes admitted in his presentation at NASA's Dryden Flight Research Center that people would be likely to think of that nowadays, but they did not then, perhaps because they were too preoccupied with immediate problems. One wonders why people back at maintenance and operations did not think of it either. One problem was that those on the ground took some time to grasp fully what was happening, as the situation was so unbelievable. Certainly, the retention of so much fuel because of the DC-10's fuel dumping restriction was unfortunate—another example of a measure deemed to increase safety doing the opposite.

The Mechanical Causes

The underlying cause of the incident was failure of the engine's fan blade disk due to a casting fault that had remained undetected for seventeen years. The engine had a protective girdle designed to arrest the loss of a single blade flying out prior to the engine being shut down (or spooling down on its own), not to parry multiple blades and heavy fragments of the disk retaining them.

The accident led to much more stringent checks during the manufacture of engine components and during maintenance. In recent years the need to ensure engines are reliable enough for ETOPS has spurred progress in checking such components. The accident also prompted action to find ways of maintaining some measure of control even in the event of such a multiple hydraulic failure.

An interesting point later made by Captain Haynes when giving a talk at NASA's Dryden facility was that modern computers could be made to adapt to flying an aircraft by juggling engine power alone—the pilot would then move the controls in the normal way, and the computer would hopefully do the rest!

Indeed, the FAA and NASA have been looking into this in view of the possibility that airliners will be struck by shoulder-launched missiles, such as happened to the DHL A300 at Baghdad.

Apart from being famous as an exceptional feat of piloting, the Sioux City miracle landing was captured on video and is often shown in documentaries to demonstrate that many survive crashes one would think cannot be survived.

---

[1] Dennis Fitch died of brain cancer on May 7, 2012, aged 69. Many obituaries said how much he was admired as a person and portrayed him as the hero of the Sioux City DC-10 landing.

[2] Sioux City Gateway Airport has the unfortunate three-letter code SUX.

[3] Slide show presentation by Captain Haynes at the NASA Ames Research Center, Dryden Flight Research Facility, Edwards, California (1991).

[4] Just before touchdown, the first officer is telling Fitch to apply left throttle—there seems to be some inconsistency.

Masterful Handling of Stricken A380 (Singapore, 2010)

Shrapnel damaged the wing, fuel lines, wiring, and control systems

> Some in the aviation world said it was not difficult for Qantas to be the "safest airline" (with no hull losses in the jet age), because, unlike the US majors, it only flew long-haul routes to safe airports. The edge-of-your-seat incident we now describe shows it perhaps really does deserve the title.
>
> *Qantas Flight QF 32, November 4, 2010*

Twitter reports said pieces of the aircraft were raining down on Indonesia's Bataan Island. Australians later watching TV even saw footage of locals holding up aircraft pieces, including a large section of an engine cowling with the red markings typical of Qantas livery. Fortunately, no one on the ground was injured or killed, though one piece, still hot, narrowly missed a child at primary school.

Qantas CEO Alan Joyce traveling by car in Sydney with his investment advisers received a phone call asking why the airline's share price was tumbling. His heart sank on being told it was in response to Twitter reports that a Qantas A380 superjumbo had crashed.

Qantas staff raised his spirits somewhat by assuring him that though the aircraft could not be contacted by satellite phone, they were certain it was still airborne, as there had been no report from Singapore air traffic control that it had disappeared from their screens, and Qantas maintenance was as usual receiving telemetric ACARS data (used to report faults automatically before an aircraft lands) from the aircraft. However, the data was so voluminous that maintenance had initially thought the system itself might be "playing up." Never had they seen anything like it.

QF32

The A380 superjumbo Airbus in question was the first one in the Qantas fleet and named Nancy-Bird Walton after the famous Australian aviatrix. With the flight number QF32, it had taken off on schedule at nine thirty that morning from Singapore, where it had stopped to refuel and allow some passengers to embark on its almost twenty-two-hour journey from London to Sydney, or disembark.

The flight was unusual in that in addition to the usual two pilots plus another to allow them to take turns to rest over such a long leg, there were two extra ones, and very senior ones at that. One was an under-training route check captain there to carry out the rostered captain's mandatory annual check, himself being checked by an even more senior check captain.

Too many management pilots in the cockpit in emergencies can spell disaster, and it was good that the rostered captain, Richard de Crespigny, arranged to meet the other two operating pilots beforehand to explain the presence of the check captains and to make clear that he was ultimately in charge, though, of course, following the modern concept of crew resource management (CRM), the captain no longer plays God and considers all views.

The crew and the pilots' flight times were as follows:

1. Richard de Crespigny, captain, and the pilot in command, fifteen thousand hours
2. Matt Hicks, first officer, eleven thousand hours
3. Mark Johnson, second officer, eight thousand hours
4. David Evans, senior check captain, seventeen thousand hours
5. Harry Wubben, route check captain, twenty thousand hours
6. Michael Von Reth, the exceptionally experienced cabin service manager

As customary, the captain sat in the left-hand seat, and the first officer in the right-hand one. Harry Wubben, the route check captain under training himself, insisted on sitting in the middle just behind them, saying that otherwise he could not check on what the captain was doing. De Crespigny was not happy with this arrangement, as to be any use, his second officer should be there. Some tension built up, which Harry Wubben astutely defused by stating he would also assume the role of second officer when required.

On board were 469 people: 440 passengers, 24 cabin crew, and the 5 pilots mentioned.

The leg to Singapore, at almost eight hours, was shorter than the twelve-hour one from London, so the comparatively light A380 did not need a very high thrust setting to lift off easily from Singapore's 4,000-meter (13,123-foot) center runway. This meant the engines were less stressed than they might otherwise have been and were to be later when climbing under higher power.

It was almost a perfect day as they executed a slight left turn over the sea to take them toward Bataan Island and continued to climb. Apart from having to deviate slightly to avoid ash from a volcano, it looked as though they were in for an easy flight.

Having retracted the flaps and slats to put the aircraft into the clean configuration they would maintain all the way to Sydney, they passed through seven thousand feet on their way to their much higher cruising height, and still gently accelerating.

Moments later there were two bangs in quick succession, which de Crespigny said sounded like a car backfiring. From their instruments, the pilots could see right away it was trouble with the number two engine. De Crespigny immediately set the autopilot to "Altitude Hold," a maneuver he had learnt from a pilot whose 767 suffered an engine disintegration.

It was the quickest way to make the aircraft fly horizontally with reduced thrust in the least demanding way possible, though in fact, with the autothrust surprisingly not working, he had to move the levers himself. He then made a "Pan, Pan, Pan" call to air traffic control in Singapore to signify a serious problem requiring aircraft in their vicinity be kept away, and that as far as possible they not be troubled with ATC demands while they dealt with it.

The passengers were nearer the engine that disintegrated and had heard the bangs and muffled noises more loudly. Not only had the aircraft shuddered, it seemed to them to be falling, but this was an illusion, as by reducing the thrust and leveling off, the captain had abruptly stopped the high-power climb, to which the passengers, with their seatbacks continually pressing firmly against their backs, had become accustomed and which they had come to take as normal.

Pilots are trained to deal with engine failures and even engines falling off, which they are designed to do in certain circumstances, and it is not usually such a big deal, since the aircraft can still fly on those remaining. However, dozens of faults were coming up on the ECAM (electronic centralized aircraft monitor) in addition to the first one, which had been "Engine 2 turbine overheat."

In both training on the simulator and operational flying, the most alerts the monitor would ever show at one time would be two or three. The cabin service manager, who had an unobstructed view of the damage, wanted to tell the pilots how serious it was but found the intercom inoperable.

According to the ECAM, the failed number two engine's thrust had to be reduced to idle and the engine given thirty seconds to settle. It then showed "Fire" before reverting to "Overheat." It was finally shut down and the fire suppressant control selected, but no confirmation that it had functioned was received because the wiring was severed.

These operations were carried out by the first officer with the approval of de Crespigny, who had assumed manual control. The aircraft was flyable but responses to inputs were sluggish. Could the aircraft be gently turned to point back to Singapore's Changi Airport, with its long runways and good emergency services? Gingerly, they executed a wide turn with little bank and found to their relief it could.

Second Officer Mark Johnson was sent back to the passenger cabins to investigate. In first class, right behind the cockpit, he recognized a Qantas second officer traveling back to Sydney, who wasted no time in drawing Johnson's attention to the last thing on his mind, the inflight entertainment system! There on the screen, for all the passengers to see, was a view of the aircraft from a camera high up on the tail, with the left wing punctured by holes and fuel streaming out of the back.

Moving over to the window inset in the left-hand door, Johnson could see what the camera could not, namely the rear cowling of the inside left engine missing and that the inner workings of the engine were exposed. A closer look at the wing revealed a hole so large a man could pass through. Johnson then moved through the cabins, stopping from time to time as "well-placed" passengers drew his attention to disturbing damage. After consulting and briefing the cabin service manager, who needed some info to help him reassure the passengers, he returned to the cockpit, trying to exude confidence he did not feel.

What he had not been able to see was that the fuselage had been penetrated by shrapnel that had severed the two bundles of control wiring which ran up and down it like two spinal cords, making the aircraft paraplegic, as they were to later realize. Fortunately, no passengers had been struck, and at that low altitude the cabin pressure had been virtually equal to the outside pressure, so there had been no explosive decompression, with oxygen masks dropping down.

Now the stream of ECAM warnings made some sense, though the pilots could not know the damage included that to the "spinal cords." The exhausting tasks of dealing with each fault thrown up on the ECAM, and taking appropriate action in consultation with the others, fell to First Officer Matt Hicks.

However, a thought had to be given to the state of the passengers, whom Cabin Service Manager Von Reth had been trying to reassure with the little information he had. Passengers coming off an aircraft safely after a crisis often complain that though the cabin crew did their job, barking orders and telling them to keep calm, the pilots did not tell them anything, not realizing they were too busy saving their lives.

The presence of the extra pilots enabled de Crespigny to delegate that task to Senior Check Captain David Evans, whose calm, measured voice and convincing-but-not-quite-frank words did much to reassure them, and was, according to the quietly confident cabin service manager, Michael Von Reth, the game changer in keeping them under control. He himself was going through the cabins looking for what are called "ringleaders," people who stand up, go into the aisles or to the doors, and shout "We are going to die!" or similar. Panic is infectious and leads to pandemonium if not nipped in the bud, something that requires psychology and "presence."

Taking Stock and Resolving ECAM Faults

In an aircraft that the pilots knew might at any moment suffer an extra malfunction rendering it virtually if not completely impossible to fly, it took over an hour and a quarter for them to deal with and assess the many faults as best they could and see whether they would be able to land.

The slats on the left wing had been damaged, together with their control systems, and those on both wings had been locked for protection, so they needed to ensure not only that they would have enough low-speed lift and control to land when very much overweight, but also the ability to stop before overrunning the runway, with the fuel very likely catching fire.

Though electrical supply to the left wing had been cut, a fire might still be smoldering there, which was one reason the other pilots opposed de Crespigny's suggestion they climb to ten thousand feet to give them extra height to play with should the three remaining engines they were relying on fail. Besides putting extra strain on the unsure engines, the alteration to the slipstream could possibly exacerbate a fire.

Fuel and its distribution were major problems, as most of the pumps and fuel galleries were inoperable, making it impossible either to move some of the fuel farther forward to adjust the center of gravity or to jettison any. As a result, they were forty-four tons overweight to land. They could not fly around for hours to burn it up, for not only could something else go disastrously wrong, making them lose control, but the longer they flew, the greater the three fuel imbalances between the amount in the left wing, right wing and horizontal stabiliser tanks became. With the right wing already almost eleven tons heavier than the left, limited control of the aircraft meant a difference of much more than that would make them flip over, and particularly at slow speed on landing.

The hydraulics situation was also very worrying. The pressure in the green channel, fed by pumps in the left wing, had fallen to zero, and the yellow one, fed by four pumps driven by engines on the right, was none too healthy; the ECAM was saying two should be shut down, which the pilots hesitated to do as once shut down they could not be restarted. Finally, after some debate they were shut down, with the system seeming okay.

To realize what was involved, and not least the strain under which the pilots must have been, one should read de Crespigny's book *QF32*, still available on Kindle. It also gives a good insight into the complexities of the latest aircraft, which though taken in hand by the computers should ideally be understood by pilots. In this respect, de Crespigny had the advantage of having studied this while researching a book he was writing, even visiting manufacturers, such as Boeing, Airbus, and Rolls-Royce.

Flying around in a circuit near Singapore, they tried to resolve the almost one hundred problems thrown up by the ECAM, and after they had been solved in one way or another, decided to concentrate on what did work rather than on what did not. Enough was working for the aircraft to fly, but was that sufficient for them to land safely?

They would be coming in faster than usual, and even if they were in one piece on the runway, they might not be able to stop before reaching the end, since there were few spoilers to slow them initially, and they only had about 50 percent braking power, meaning the working brakes would be white hot and liable to set fire to the fuel leaking from the wing. Also, they were considerably overweight, and if they came down hard, the landing gear would collapse, with possibly dire consequences.

They calculated that if all went well and they came in at 166 knots, they could stop with 100 meters (109 yards) to spare. There was a grass overrun area, but there, though the wheels might sink in and slow them, the landing gear might collapse due to their weight, and a fire ensue.

Before making the attempt, and to make it less risky, de Crespigny insisted they perform a control check, something he had experience of in the military, where pilots of aircraft damaged in action had to determine what they could still manage. This in fact simple concept consisted of trying out the landing maneuvers they intended to make high up with enough height to recover—a dress rehearsal of the landing at four thousand feet.

If they found themselves losing control at 166 knots, they would have accelerated and brought the flaps up one step. The point of it was not only to be sure the aircraft would behave under those conditions but also to prevent any speed or stall alarms causing them to react unnecessarily and dangerously. The control checks showed the aircraft could be landed without stalling, though commands would be sluggish.

The Landing and Aftermath

After a long, gentle approach, de Crespigny managed to flare to break their descent and, amazingly, touch down right at the very beginning of the 4,000-meter runway. At the last moment he executed a rarely used counterintuitive maneuver, pushing the sidestick fully forward to make the upwards-facing A380 tip forward. Doing so lifted the low-hanging wheels at the middle and poised the aircraft on a cushion of air, which by virtue of the so-called ground effect helped soften the impact of the overweight superjumbo.

Though four thousand meters of runway lay ahead, the first thousand flashed by with little evidence of slowing, since only limited use of the spoilers and only one engine producing reverse thrust, combined with degraded braking, meant high-speed retardation was poor. Then things started looking better, and they eventually came to a halt just over one hundred meters from the end of the runway.

Being on the ground was in some respects more dangerous than being in the air, for up in the sky the fuel leaking from the wing just disappeared, with nothing hot to ignite it. On the ground it was not only pooling under the aircraft like kindling for a bomb fire, there were also the overheated brakes, with temperatures climbing toward 1,000 degrees Centigrade, that could ignite it at any moment.

With it being the "Kangaroo Route" between the UK and Australia, de Crespigny knew there were many old people, some in wheelchairs, not to mention babies and infants who were liable to be injured coming down the chutes. Once outside, passengers might start a fire with their shoes or mobile phones or might wander to the vicinity of the engine they had been unable to shut down. He therefore did not order an evacuation but told the cabin crew to be constantly ready for one.

Finally, fire crews, who had been holding back due to the unstopped engine, sprayed gallons and gallons of water onto the brakes and covered the fuel with foam. The danger passed as the brakes cooled. Even so, the passengers had to endure an uncomfortably long wait with no air-conditioning before they were finally allowed to leave without luggage via stairs and were bused to the terminal, where de Crespigny debriefed them.

The care de Crespigny took to look after the passengers, with the help of the other pilots, meant when they eventually faced the press pack on leaving the terminal, they were happy and grateful people and well informed enough not to allow the press to put words into their mouths, such as their aircraft was on fire, which was not the case.

The Extent of the Damage

Close examination on the ground revealed that shrapnel had severed more than 600 wires, including all of one of the "spinal cords" transmitting essential data from systems. The wing had been struck in more than one hundred places, and the fuselage in some two hundred. A heavy chunk of metal had missed the top of the fuselage by inches. Fourteen holes were found in the fuel tanks. Perhaps 95 percent of the systems, including the number one and two AC bus providing essential power for many of them, had failed.

Rolls-Royce

Believing BP made a big mistake by being so open about the Gulf of Mexico oil rig disaster, Rolls-Royce refused to comment publicly regarding the likely cause of the disintegration and how that might have come about, much to the annoyance of Qantas and others. The airline's CEO, Alan Joyce, said BP's mistake lay not in being forthcoming but in saying the wrong things.

The lack of information led to the general impression that Rolls-Royce was running a shoddy, primitive manufacturing operation at the plant where the engine had been made, with poor attention to detail and inappropriate quality control. Though that was partly true, and the issues have since been addressed, the manufacturing and quality control there used very sophisticated equipment. The error arose from using the wrong datum (reference point) for drilling into the stub pipe to make room for a filter after the object moved slightly; the control check did not reveal the discrepancy, because that wrong datum was again used as reference. The error would have been obvious to the naked eye had the weld at the end of the stub pipe not obscured the cross-sectional view.

According to an Australian investigator, Rolls-Royce had considered the possibility of such a fire but calculated the engine would power down on its own accord rather than speed up—he jocularly remarked the engine was too good! Perhaps this was borne out by the lengthy struggle the firefighters had to shut down the recalcitrant engine, first with water and finally with firefighting foam.

Besides improving manufacturing and quality control, Rolls-Royce have taken measures to ensure such an event cannot have such dire consequences, and this includes software to shut the engine down in such a case. Nevertheless, it is scary to think an engine maker could spend large sums to make fan disks that never should fail—never happened in forty years—and then find some very minor component on the outside could cause one to spin so fast one would.

With Emirates ordering engines from Rolls-Royce for their next batch of A380s, Rolls-Royce's reputation did not appear to suffer irreparable damage, but the financial cost was enormous. Firstly, they had to replace many of the engines in use by six or so carriers, then not only pay compensation to the insurance company for the A$139,000,000 it cost to repair the aircraft expensively in Singapore over a period of a year and a half rather than in China, but also compensate Qantas for loss of revenue due to the whole Qantas A380 fleet being grounded.

Outstanding Airmanship

We said fortuitous circumstances made the "The Miracle on the Hudson" possible. To be consistent, we should ask what role luck played here.

Unlike in Sully's case, it was not primordial, but bad luck could have jeopardized everything. Had it been raining heavily, as it so often does in Singapore, perfectly judging the delicate landing, where hitting too hard with a way-overweight aircraft could be fatal, would have been nigh impossible. Also, with only limited control via the ailerons, a significant crosswind component would have been impossible to counter. A wingtip and then an engine would have snagged the side of the runway, and the aircraft would very likely have spun around and broken up like the DC-10 did at Sioux City after landing with controls taken out following a similar engine disintegration (described in the previous piece). Gusts of wind, as said, could have blown fuel up onto the white-hot brakes and started a fire.

While they were also lucky to have the two highly experienced check captains and a psychologically astute cabin service manager, no one can dispute how well they did and how exceptionally talented they were individually.

Chapter 11
METAL FATIGUE

The Hidden Flaw of the World's First Jetliner (1952 to 1954)
Front-runner falls; others learn

> It does not always pay to be first with a radically new aircraft concept, and the example of Britain's brief stand on the pedestal with the world's first jetliner rather proves that.
>
> *BOAC Comet 1 G-ALYZ, October 26, 1952*
> *Canadian Pacific Airways delivery flight, March 3, 1953*
> *BOAC Flight 783, May 2, 1953*
> *BOAC Flight 781, January 10, 1954*
> *South African Airways Flight 201, April 8, 1954*

At the end of the Second World War, and before sky-high taxes made it better to be an accountant than an engineer, the British still expected to be in the forefront of aviation technology and aircraft production. The problem was that while the British had excelled in producing fighters, such as the Spitfire and the Hurricane, the Americans had built up an invincible lead in bombers and heavy transport aircraft that they could switch to work as civilian aircraft at little cost.

With no hope of challenging on that front, the British set about designing an airliner that would exploit a domain where Britain had some sort of lead: jet engines. This would enable them to leapfrog the Americans with an airliner able to fly almost twice as fast. This project, envisioned by a government committee even before the war ended, had strong governmental support.

As the key feature was to be the use of a jet engine, the government perhaps unfortunately placed its support behind a distinguished aircraft company that produced its own engines—the de Havilland company, maker of the famous all-wood Mosquito, as well as the later Vampire jet fighter.

The major concern from the design point of view was conceiving an airliner that would handle well on takeoff and landing, as well as at the much higher cruising speeds contemplated. Not enough thought was given to the fact that flying an airliner at thirty-five thousand feet or more put it in a completely different category as regards pressurization and the effects thereof, and notably metal fatigue of which not so much was known at the time.

Passengers do not require cabin pressure to be the same as that on the ground, but for them to be comfortable, it should not be lower than that at about seven thousand feet, a height at which even people with asthma are comfortable.

For post–Second World War piston-engine airliners, such as the very successful and beautiful Lockheed Constellation, cruising at twenty-one thousand feet, having the cabin pressure equivalent to that at seven thousand feet, meant the pressure difference between the inside and outside corresponded to fourteen thousand *feet* (twenty-one thousand minus seven thousand). The difference for a jet, such as the Comet, cruising at thirty-five thousand feet would be double that (thirty-five thousand feet minus seven thousand, which equals twenty-eight thousand feet).

In addition, there is a world of difference between a fighter with no passenger cabin and an airliner destined to carry dozens of passengers. Indeed, it took de Havilland four years to get a prototype into the air and two and half years more to have a certified aircraft. A limited scheduled service between London and South Africa finally began on May 2, 1952.

With airlines worldwide showing keen interest, de Havilland soon realized its jetliner should have been larger and proposed a Comet 2, followed by an even larger Comet 3 for what should have been highly profitable transatlantic services. Affluent passengers were boasting about their flights, not only stressing their speed but also their comfort and smoothness.

The First Accident (No Fatalities)

On October 26, 1952, less than six months after the start of commercial flights, the first mishap occurred. A Comet taking off in rain at night from Rome's international airport with a full complement of passengers started juddering and buffeting just as it was reaching the rotation speed (V_R) of around 112 knots. Sensing the aircraft was stalling and would not gain enough speed, the captain pushed the control column forward but, finding that did not solve the problem, cut the engines and applied maximum braking.

The aircraft exited the end of the runway at high speed and ended up some distance away, with the landing gear ripped off and fuel leaking onto the ground from a ruptured fuel tank. Fortunately, there was no fire and all the occupants survived—very shocked but unscathed. The aircraft was a write-off.

Somewhat like the Qantas overrun at Bangkok, the absence of fatalities or even injuries meant the incident received little publicity. The Italian investigators did not even publish their report. Having found marks on the runway showing the tail had struck it a number of times even before reaching the point where the aircraft would normally rotate for takeoff, the investigators concluded it was not the aircraft's fault but the pilots' for prematurely raising the nose on the takeoff run.

Group Captain John Cunningham, de Havilland's famous chief test pilot, carried out tests that showed raising the nose too early increased drag so much that the aircraft could never reach takeoff speed. One wonders whether this shortcoming was not in part due to de Havilland previously working mainly in designing fighters with surplus power, where a tiresome quirk like that would not matter.

Early jet engines did not have the abundant power they have today, enabling them to soar into the sky at a considerable angle. They were not much different from piston-engine aircraft, such as the Lockheed Constellation, which to the author seemed to take off like swans, with their "flippers flicking over the water's surface." Manufacturers had not yet fully considered exploiting leading-edge devices able to give considerable extra lift without excessive drag at low speeds. Even the early Boeing 707-100s that stepped into the Comet's shoes only had partial slats inboard of the outboard engine pylon. Full-wing slats only came in later versions of the 707.

All Comet pilots were told to keep the nose down until reaching takeoff speed, but with no dedicated attitude indicator showing the attitude in degrees, any abnormal nose-up attitude proved difficult to detect when concentrating on the runway ahead and occasionally glancing at an artificial horizon, which would only give a rough indication. In addition, the hydraulic controls of the day did not simulate a feel, and the pilots could be pushing the tail down with the elevators without realizing it.

The Second Accident

When the first of three Comets ordered by Canadian Pacific Airlines was ready for delivery early in 1953, de Havilland persuaded the Canadians to have it delivered via Australia. This would provide a chance to demonstrate it Down Under and produce worldwide publicity if they could break the record for the time taken to fly from England to Australia in the process.

The aircraft took off from England on March 2 with a five-man flight crew headed by Canadian Pacific's overseas operations manager, Captain Pentland, who in his administrative role had little experience of flying the Comet and had never taken off at night at its controls. Early the next morning, and while still dark, the Comet was taking off from Karachi at its maximum permissible takeoff weight, only to find the Rome scenario repeated but with a less favorable outcome. After failing to lift off either on the runway or on the 180-meter overrun, the Comet exploded on encountering a number of obstacles situated beyond. The five-man crew and six de Havilland technicians, who were traveling as passengers and were going to assist Canadian Pacific with the aircraft, were killed.

Again, witnesses noted the unusually high nose-up attitude maintained for most of the takeoff run. Just as at Rome, the runway marks showed the tailskid had scraped the runway at several points. It seemed the pilots corrected the nose-up attitude at the last moment and almost succeeded in lifting off.

Investigators concluded that fatigue and pressure to beat the record had been contributing factors in what they classed as "pilot error."

Canadian Pacific immediately canceled its order for the two aircraft de Havilland was prepping for delivery. The accident was bad publicity for de Havilland, but it could still attribute it to pilot error and argue that with some modifications and better training there should be no more such silly accidents. It was fortunate that up to then the number of fatalities had been mercifully small.

The Third Accident

The Comet bound for Delhi had taken off from Calcutta midafternoon on May 2, 1953. There had been no problem lifting off, as the aircraft was quite light, with a crew of six and thirty-seven passengers, and only the fuel required for the two-and-half-hour leg to Delhi.

Twelve minutes after takeoff, the crew duly reported to Calcutta that they were climbing to thirty-two thousand feet, before calling the controller at Delhi, who replied without getting a response. Repeated efforts failed to get any response, and nothing more was heard radio-wise from the aircraft.

Witnesses later reported hearing a bang during an exceptionally severe thunderstorm and seeing a blaze in the sky around that time and at the location of the aircraft.

The aircraft had, in fact, disintegrated in mid air. Based on evidence that the horizontal stabilizer had failed due to an exceptional downward force, an Indian High Court judge concluded the accident was due to the storm, and that any aircraft would have broken up in the same circumstances.

Experts reexamining the debris in England drew the same conclusion. They thought it possible the lack of feel in the controls had led the pilots to overcontrol, especially as the captain had for much of his career flown flying boats, where heavy control inputs were the norm. It was thought the explosive decompression (presumed to have followed separation of the tail fin) occurred as the aircraft was climbing through twenty-seven thousand feet.

People were getting apprehensive, but there was no proof that there was anything inherently wrong with the aircraft itself. The eight accident-free months that followed seemed to confirm this.

The Fourth Accident

Then, south of the island of Elba, on January 10, 1954, a Comet inexplicably disintegrated in midair.

The aircraft with six crew—in those days the pilot and first officer were complemented by an engineer and radio officer—and twenty-nine passengers had taken off from Rome for London on the last leg of the run from Singapore. The fact that the aircraft had disintegrated in midair probably while climbing through twenty-seven thousand feet, just as in the Indian case, was thought to be a coincidence.

De Havilland had tested the Comet's cabin under high air pressures and severe conditions prior to production and concluded that it should have been able to withstand fifteen times more flights than the 1,200 that the aircraft had done. The Comets were withdrawn from service for modifications made on the basis that the cause was most likely an in-flight fire or engine fan blade separation.

The Fifth Accident

Eleven weeks after the Elba accident, the Comets were put back into service, only for yet another inexplicable accident to occur two weeks later. Again, the aircraft happened to have taken off from Rome. It was the evening of April 8, 1954. The Comet in question was one leased to South African Airways and crewed by staff from that company. With only fourteen passengers, the flight can hardly have been a profitable one—memories of the Elba accident had no doubt dampened people's enthusiasm.

The pilots reported to Rome they were climbing to its cruising height of thirty-five thousand feet and seven minutes later called Cairo to report their ETA, after which nothing more was heard of them. The aircraft had disintegrated twenty nautical miles off the Italian coast, southeast of Naples. It had occurred at roughly the same time after departure and at a similar height to the previous one off Elba.

With Britain's dream shattered, the aircraft's Certificate of Airworthiness was withdrawn. Until the exact cause of the disasters could be definitively determined, and adequate steps taken to prevent any chance of a re-occurrence, the Comet bandwagon was halted in its tracks.

The Cause is Found

Prime Minister Churchill ordered that everything should be done to resolve the matter, including testing of the fuselage to destruction in a water tank pressurized and depressurized repeatedly to simulate actual flights.

The wreckage of the last disaster had fallen in such deep water that little could be done to recover it with the technology available at the time. However, the Elba wreckage was more accessible, and a crucial piece of evidence was recovered: the tail, with paint marks on it indicating it had been hit by a passenger seat ejected from the cabin, which could well have happened due to a pressurization failure of the cabin.

In late June, the fuselage that was being tested in the water tank finally failed. The failure had occurred at a rivet hole adjoining one of the square escape hatches. More detailed inspection revealed fatigue cracks that could have been the failure point in rivet holes in the vicinity of the cutout for the ADF antenna in the roof of the cabin.

If the failure had not occurred at the escape hatch, which had been recovered intact, it in all likelihood would have occurred in the vicinity of the ADF cutout. Assuming that the failure had occurred there, they tried to work out where the part in question would have fallen, and finally recovered it. The telltale fatigue cracks were duly found, and they had the definitive answer they wanted.

The five remaining Comet 1s had become junk, and a number of Comet 2s in the course of production were modified and bought by Britain's Royal Air Force. They had a heavier-gage fuselage, the safer round, smaller windows, and with the RAF operating them at a slightly reduced cabin pressure gave good service over a number of years without any serious problems.

Too Late

In October 1958, more than six years after the first scheduled Comet service to South Africa, BOAC launched a transatlantic one with the larger, radically redesigned and strengthened Comet 4s. What was to be the Comet's second honeymoon was cut short three weeks later by the arrival of Boeing's much larger 707, with its longer range, greater speed, and greater economy, developed in tandem with the KC-135 air refueling tanker, funded by the US Air Force Strategic Command.

Not only was the 707, as everyone knows, an outstanding success, but so was the KC-135, from which it was developed, with more than eight hundred of both incarnations produced in various versions over the years. In fact, it was the KC-135 that was the money-spinner, for the arrival of a competitor from Douglas meant Boeing had to make expensive upgrades and modifications to the 707.

Boeing and other US makers, such as Douglas, had learned much from the Comet disasters with the help of the British, who made information freely available. De Havilland's last stab at the big time was with the rear-engine Trident trijet. However, there again Boeing spoiled the party, with its similar 727 trijet, not because the Trident had serious technical faults but because it had been designed solely with the needs of British European Airways, or rather its chairperson, in mind.

Flexing Aloha 737 Lands with Passengers Exposed (Hawaii, 1988)

Photo showing passengers exposed to elements as on an open-top bus shown around the world

---

In the 737, Boeing designed the fuselage so in the event of a failure a structural flap would open to release the cabin air without destroying the structure. Unfortunately, when the failure occurred on Aloha Flight 243, the updraft sucked a female flight attendant standing below up into the aperture, plugging it, with the result that a large section of the roof separated. Pictures of the Aloha jet with passengers sitting in the open as if on an open-top sightseeing bus received worldwide publicity.

Aloha Airlines Flight 243, April 28, 1988

U sed for inter-island hops, the eighteen-year-old Boeing 737 had 89,680 cycles to its credit, or rather, debit. Dividing the number of hours it had been airborne by that number, the average flight time worked out as twenty-five minutes. Only one other 737 in the world had as many cycles.

Admittedly, the impact of these cycles on the structure was in theory less than usual in that the aircraft would hardly ever reach great cruising heights. However, this plus was negated by the fact that they were operating in, and parked in, the warm, salty, and humid environment associated with the Hawaiian Islands. Furthermore, the airline operated within tight financial margins and extracted the most from its geriatric fleet by doing important fuselage checks piecemeal overnight, it being imperative to have the aircraft back in the sky early the next day.

Some operations, such as checking rivets and the bonds holding the fuselage together, require not only very diligent and tedious work but also setting up gantries and safety harnesses if upper parts of the fuselage are to be checked safely, all limiting the effective inspection time. Perhaps worst of all, workers suffering from upset sleep rhythms can very easily miss defective areas.

On boarding in the good light of the early afternoon, one of the passengers noticed a small horizontal crack along a line of rivets to the right of the forward left-hand door and had even thought of reporting it to the crew but did not do so for fear of being thought foolish.

The aircraft was already making its ninth flight of the day, this time the thirty-five-minute leg from Hawaii's Big Island (Ito) to Honolulu, on Oahu Island. Its route would take it close to several other islands in the Hawaiian chain, including Maui. Twenty minutes after takeoff, it had leveled out at its cruising height of twenty-four thousand feet, with thirty-seven-year-old First Officer Madeline (Mimi) Tompkins at the controls. She had eight thousand hours' flight experience and hoped soon to be promoted to the rank of captain, like Captain

Schornsteimer, sitting next to her, who had a similar amount of flying experience.

A loud bang behind the pilots signaled the beginning of an explosive decompression. Tompkins later said the decompression was so rapid that the displacement of the small amount of air in front of her was enough to throw her head back. Donning their oxygen masks according to regulations, with the captain taking over control, as is usual in an emergency, the two pilots tried to assess the situation. Glancing behind him, Schornsteimer was shocked to see some blue sky through the doorway where the blown-off flight deck door had once been. Had he been able to see the full extent of the damage, he might not have made such a rapid emergency descent, with the airspeed attaining 280 to 290 knots even with deployment of the air brakes. This is standard practice in the case of a decompression, so the passengers can breathe more easily, but is contraindicated when doing so might lead to the structural breakup of the aircraft.

Having seen the sky through the cockpit doorway, the captain would have realized some passengers would not have oxygen masks, and he felt a rapid descent was necessary for their benefit. He could not see enough to realize that with so much of the upper fuselage missing, he was exposing them to a vicious slipstream and, worse, going so fast could bring about the breakup of the structurally impaired aircraft. Schornsteimer later received many plaudits for his airmanship but was criticized by the investigators on this point.

As the aircraft slowed at a height where breathing was easier for all, the plight of the eighty-nine terrified passengers and the remaining two female cabin attendants became more bearable. Some had seen the shocking sight of Flight Attendant Clarabelle Lansing, who had been working at the airline for thirty-seven years, sucked upward into a hole (the safety flap) that opened up in the left of the roof near row five, before the whole section around it came away.

Another flight attendant, Jane Sato-Tomita, who had been standing three rows farther forward, not far from the flight deck door, had been severely injured by flying debris and, notably, that missing door. The other flight attendant, Michelle Honda, had been thrown about and suffered bruising but, being much farther back, in the part of the cabin that still had a roof, was relatively unscathed. Dragging herself on her hands and knees up the aisle against the slipstream, she had to shout in passengers' ears to communicate. Unable to contact the flight crew over the intercom, she at one moment thought the pilots might be incapacitated and asked likely prospects whether any of them could fly a plane.

Unlike the flight attendants standing in the forward part of the aircraft briskly serving drinks in the little time available before their expected arrival at Honolulu, the seated passengers there had seat belts to save them from serious injury or even death.

The pilots, who were very much alive and the least subject to the elements, found the aircraft was controllable, albeit somewhat abnormally. Apart from the distortion of the airflow over the fuselage caused by the gap, the nose was in fact drooping down about a meter below where it should have been and springing up and down. Some maintained only the seat-fixation rails running longitudinally along the cabin floor were holding the aircraft together—hence the springiness.

The pilots decided to divert to Maui's Kahului Airport, by then only some twenty nautical miles away. This was on the opposite side of the island and involved an approach between two high mountains, which, though not inherently difficult, could have been fraught with danger for an aircraft in its condition, as the violent wind and turbulence could easily break the fuselage apart. Luckily, the wind was much weaker than usual.

Their immediate problem was radio communication, largely due to the wind noise, which made it virtually impossible to hear or be heard. It took some time even for the Maui controller to get their flight number right and then to grasp the seriousness of their predicament, of which even the first officer, handling the radio, was unaware, as she had no intercom contact with the passenger cabin and had to assist the captain. Maui ATC asked them to switch frequencies, as if it were a normal flight.

The investigators criticized this obligation to change radio frequency on the grounds that aircrew dealing with an emergency should not be distracted by this task, which had the added risk that contact with the aircraft might be lost altogether. This reminds one of the criticisms leveled at ATC in connection with the famous crash of a 737 at Kegworth in the UK the following year, where the pilots shut down the good engine instead of the bad one and, incidentally, had to hop from frequency to frequency.

On lowering the landing gear, the pilots found there was no green light to confirm the front nose wheel was down and locked. Fearful that the aircraft would break up if they delayed, they decided to forgo the usual procedure of overflying the airport for the tower to make a visual check, which would entail a circuit to bring them back to the touchdown approach. The captain decided to come in anyway.

As they came in, the control tower was able to confirm the front wheel was down but obviously could not be sure it was locked.

When the amount of flap was increased to fifteen degrees for the landing, the aircraft began to misbehave, and the setting was quickly reset for five degrees, with the first officer having to consult her manual and make quick calculations to work out the V_{REF} (landing airspeed) for that configuration, which was 152 knots. However, finding the aircraft unstable at that speed, the captain opted for

170 knots, but as he moved the throttles the port engine failed and would not relight.

With the other engine he was still able to bring the aircraft in at the higher speed, telling the first officer, who suggested increasing the flaps to forty degrees detent at the last moment, to wait until they were on the ground before doing so. Despite or thanks to the higher than usual speed, the aircraft touched down smoothly, with the nose wheel holding. Observers saw the fuselage flexing as if it might snap at any moment. Using the brakes and reverse thrust on the operable starboard engine, the captain successfully brought the aircraft to a halt on the 6,995-foot (2,133-meter) main runway.

It was only after shutting everything down and moving back to assist with the evacuation that the pilots realized how close they had been to the aircraft snapping in half at any time or tearing itself apart during their rapid descent to a level where passengers could breathe without masks. With no upper fuselage to take the compressive forces, the fuselage could have snapped upward at the last moment had they come down too hard on the nose-wheel. It is not clear whether anyone has mentioned that the lack of a green to confirm the nose-wheel was locked might have made the captain more careful than he would have been to bring the nose down gently and may have been a blessing in disguise.

The only passengers unscathed, apart from eardrum problems, were those sitting right at the rear of the passenger cabin, which retained its roof. Those immediately behind the break were hit by debris and suffering from lacerations, often to the face. Worst affected were those sitting in the area exposed to the elements, some of whom were trapped in their seats in places where the floor had buckled upward beneath them due to the air pressure in the hold below.

Those able to vacate their seats and walk either evacuated via the stairs at the rear or the emergency slide at the main door at the front, with at least one dithering passenger having to be pushed onto the slide by the first officer. The fire crew was able to extract and evacuate the others in less than half an hour and dispatch them to hospital by ambulance.

There were a whole range of major and minor injuries, including lacerations from flailing wires and even electrical burns caused by contact with them. It took time and therapy for some to recover psychologically.

The Investigation and Way the Aircraft Was Constructed

Search of the ocean where the explosive decompression had occurred failed to produce either the body of the ejected flight attendant or any useful material evidence. With key evidence missing, investigators had to work out what had happened from cracks and disbonding on the periphery of the hole.

To understand the conclusions of the investigators, one needs to know how the fuselage of those early 737s was assembled or, rather, riveted and glued together.

To save weight, the skin of an airliner is often only the thickness of a credit card. Even the thickness of the paint or lack thereof can make quite a difference to the overall weight. The fuselage consists of overlapping panels mounted on a frame and riveted together. However, unlike a ship's hull, where the rivets can hold the thick plates together with tremendous force, rivets holding the outer skin of an airliner can only get a limited purchase, as they have to be countersunk to ensure a clean airflow over them. If subjected to repeated stresses, cracks develop in the skin around them.

Therefore, Boeing and other manufacturers decided to bond the overlaps with epoxy resin, in principle using the rivets solely to hold the surfaces together while the resin cured (set) and bonded the joint. Of course, the rivets would provide residual strength to the joint in the event of failure of the bond and allow time to discover the problem during maintenance.

As many who have used epoxy glues know, mixing the two resins together in the right quantities with adequate mixing every time is no easy task. Boeing thought it had solved the problem by putting the mix on a scrim tape. Kept in a refrigerator to prevent curing, the tape would be used as needed and sandwiched between the overlaps before riveting. In theory, it should have produced a perfect bond, but in practice this was not always true for a number of reasons, including dirt and humidity.

If an aircraft is being operated in a hot, salty, and damp environment, corrosion in the poorly bonded areas can push the bonded parts apart, so that they in turn disbond until only the rivets are holding the assemblies together. Then, with all the stress and weakness concentrated along that line, a long crack can result.

Boeing was aware of the problems associated with the bonding in the earlier models of the 737 but largely dismissed the possibility of a failure leading to a catastrophe, as it had devised a safety flap system. This consisted of having the skin crisscrossed by tear straps that would arrest and divert any crack that was actually opening up (called the lead crack) at right angles, thus creating a triangular flap that would blow out and relieve the pressure before further damage was done. These tear straps created ten-inch squares, big enough to relieve the pressure but not so big as to significantly weaken the structure in the process. Not everyone agreed this would work, thinking it might not save the situation in the event of a crack along a line of rivets extending over several potential flaps.

From evidence from the passenger mentioned above and signs of a crack on the same horizontal line of rivets on either side of the gap, the NTSB investigators concluded that this was what had indeed happened.

Theory That Ejected Flight Attendant Plugged Flap

They later accepted as theoretically possible but unlikely a scenario put forward by Matt Austin, an engineer specializing in pressure vessel explosions, which suggested that the safety flap mechanism did initially work as intended but subsequently failed because the lost flight attendant, Clarabelle Lansing, was sucked into it in the process.

Austin was basing this on his experience of pressure vessels, where an initial rupture can be forced wide apart by a fluid hammer effect when debris, for example, blocks the orifice, the powerful hammer effect around the hole being similar to that produced when one replaces the plug in one's bath while water is still flowing out.

According to Austin, Clarabelle Lansing was the plug. There is certainly some evidence that that was what happened, as some passengers saw her being sucked leftward and up into an opening. Signs of blood spatter found (possibly seen in photos) on the remaining part of the fuselage aft of that point seem to be a good indication that she was caught there for a moment. If she had gone straight out as the roof opened up, there would not have been any blood.

However, the NTSB's investigators stood by their initial findings as to the trigger for the failure, and their recommendations remain valid. Adding such grisly details would not change much. Should they have recommended that safety flaps be considerably larger than just ten inches square? Tests might show that the individual would have to be standing up and wearing a skirt to be sucked sideways and up into the venting flap.

Rethink Regarding Aging Aircraft

The Aloha incident had a galvanizing effect in aviation circles, with the realization that many had been complacent regarding aging aircraft. Outside those circles, the pictures of the roof of an airliner blowing away as if it were a tent in the wind produced tremendous pressure from both the public and the media on Congress, which in turn pressured the FAA, with the threat that if nothing was done, aircraft over a certain age would be banned outright.

The Aloha incident and subsequent accidents had the beneficial effect of getting Congress to fund the employment of additional FAA field inspectors, their number having dwindled from 1,672 to a low of 1,331 in 1984 despite the need for more stringent checks to prevent airlines cutting corners to maintain profitability in the face of deregulation.

One problem was that the airlines were not following the directives and advice provided by the manufacturers and the FAA. In their defense, one should point out that the FAA had drafted its directives in something resembling legalese and that a small airline like Aloha would have needed someone to translate these into a form the mechanics could grasp.

Many said some other airlines were as bad, and that Aloha was merely the first to have a failure because their old, high-cycle aircraft were operating in the particularly hot, salty, and humid environment in the Hawaiian Islands.

The incident and the strained relations with John Maple, the principal maintenance inspector, assigned by the FAA to Aloha, made Boeing rethink its confidentiality policy. Boeing, thinking like medical doctors, felt that patients might hesitate to come to them with problems if they thought they would be reported to the regulators.

FAA inspector Maple, who said he had prompted Aloha to get Boeing to review its maintenance procedures, was furious to find himself excluded from the meeting at which Boeing reported the results. Since then Boeing has told airlines to report problems themselves, as it will do so anyway.

Boeing might have been justified in its thinking about confidentiality, but now, with lawyers avid to find any cover-up and sue, the downside risk of that informal approach is probably too great anyway.

Although not in any of the official reports, it is alleged Boeing staff traveling as passengers on the aircraft in question felt it was in a pretty bad way due to the rattles, as did another frequent traveler, who said it felt like an old car.

The excellent article by Martin Aubury in the BASI journal from which some of the information here is drawn finishes by saying the teams setting out to investigate the problems with aging aircraft concluded that checking the obvious was not enough, and that:

1. If signs of corrosion are found, it must be assumed the state of the whole aircraft is questionable;
2. If a problem is found, a proper fix is essential. Repairs are sometimes badly done, with questionable techniques. These probably seem strong at the beginning but may fail later, as happened in the case of the repairs to the aft pressure bulkhead in the JAL 747, which resulted in the worst single-aircraft disaster ever.
3. A repair making one part too strong may produce stress elsewhere. In addition, repairs may hide other defects or make them difficult to see. They therefore need especially careful and thorough checking.

Aftermath

The incident made airlines all over the world and the FAA and other overseers bring in stricter regulations regarding periodic checks of old aircraft. These checks are time-consuming and expensive and are why some airlines prefer to dispose of their older aircraft.

Chapter 12
TURBULENCE, ICE,
AND WIND SHEAR

Sightseeing BOAC 707 Breaks Up (Mount Fuji, Japan, 1966)
With no flight data recorder, American tourist's movie camera was key evidence

> When flying still had something of the frontier spirit about it, a BOAC captain might sometimes offer his passengers a treat. What better treat could there be than giving them a close-up view of Japan's renowned Mount Fuji?
>
> *BOAC Flight 911, March 5, 1966*

Weather conditions that day were so good that from Tokyo one could clearly see the ice-capped cone of 12,390-foot Mount Fuji some eighty kilometers or more away. This prompted the captain of the British Overseas Airways Corporation (BOAC) flight to Hong Kong to change his flight plan at the last minute.

Instead of the usual Jet Green 6 airway, from Tokyo's Haneda Airport out over the sea to Oshima Island and on to Kagoshima in the south, he would take the scenic overland Red 23 airway, skirting Mount Fuji and onward to the industrial city of Nagoya. The seventy-five-strong American tour group he had on board would surely be highly appreciative.

Despite the bright sky overhead, the atmosphere at Haneda Airport was more than somber. Only a month earlier an All Nippon Airways Boeing 727 had crashed fifteen kilometers away from the airport. As if that had not been enough, the previous night, with many aircraft diverting due to visibility just below the one-kilometer minimum, a Canadian Pacific DC-8, whose captain was determined to land there, had circled for forty-five minutes, until visibility improved just enough for him to be allowed to make an attempt.

In their haste to seize the opportunity before the weather closed in again, the pilots of the DC-8 descended too steeply. The tower told them they were too low for the distance they were from the runway and instructed them to abort the landing. Disregarding this, the pilot continued, his aircraft first snapping off a number of frail posts set in the sea to support the approach lights before smashing into the unforgiving concrete of the airport perimeter sea wall.

Only eight of the seventy-two occupants survived the ensuing blaze.

The next day the pilots and passengers of the BOAC Hong Kong flight could hardly have failed to gaze at the charred debris of that DC-8 as they turned onto the runway to line up for their takeoff for Hong Kong. According to some imaginative Japanese newspapers, this horrific sight might have contributed in some way to the deadly accident that was to befall the BOAC flight, in that it must have unnerved the pilots.

Leaving the charred debris behind, the BOAC 707 lifted off and made a wide turn to join the Red 23 airway. On reaching almost seventeen thousand feet, rather than continuing his climb to cruising height, the captain initiated a gradual descent. Then in what must surely have been an intended deviation from the airway that for safety reasons skirted Mount Fuji, the aircraft apparently veered from a heading of 246 degrees to a heading of 298 degrees magnetic, taking it toward the conical peak of the mountain, only some ten miles distant. Possibly the captain intended to "snake" so passengers on both sides would get a view without rushing to one side and possibly risking upsetting the aircraft.

Just as the passengers were about to get the wonderful close-up view of the mountain, a violent force of air pushed the vertical tail fin (vertical stabilizer) so sharply to the left that it broke off, taking the horizontal stabilizer on the left-hand side with it. This, possibly with the help of the violent air turbulence, destabilized the aircraft, so that shock of the airstream striking it obliquely immediately made it break up. People below could see it falling from the sky, trailing vaporized fuel, thought by many to be smoke. Some even took photographs. No Mayday call or any other radio transmission was made.

Authorities diverted a US Navy A-4 fighter from the nearby Atsugi air base to look for signs of wreckage. It encountered such violent turbulence in the area that the pilot almost lost his own craft.

According to Macarthur Job (*Air Disaster: Vol. 1*), the navy pilot said afterward, "I flew into the same turbulence and truly thought the airplane was going to come unglued . . . My oxygen mask was pulled loose, my head was banging on both sides of the cockpit canopy, the instruments were unreadable, and the controls about useless. Somehow, I managed to get the nose pointing up more times than down and eventually climbed out of the turbulence. When I got back to Atsugi, the A-4's g meter had registered +9g and –4g, and the fighter grounded for inspection."

Other aircraft flying nearby, but not quite so close to the mountain, also reported severe turbulence.

Ultimately, searchers found the wreckage of the BOAC 707 scattered in a wide path some sixteen kilometers long. As the aircraft had broken up at a considerable height, there was no possibility of survivors, despite much of the fuselage hitting the ground in one piece.

The Investigation

For the investigators, finding out precisely what had happened was difficult because searing heat from burning fuel had seriously damaged the flight data recorder (FDR), which in those days was located in the nose.

The aircraft did not have a cockpit voice recorder (CVR), despite their installation being made mandatory in the United States the year before (1965). In the United Kingdom the pilots' union, like that of their US counterparts, had long opposed the introduction of a "spy in the cockpit," and their installation only became mandatory for UK-registered airliners eight years later following the crash of a Trident airliner taking off from London's Heathrow airport where a domineering uptight captain, rostered with a novice copilot and a not much more senior first officer, had a heart attack possibly resulting from an extremely heated argument with another pilot a about strike action to which he was opposed. Without a CVR to help them, investigators were unable to determine what transpired in the cockpit.

Thus, with no FDR or CVR, air accident investigators had to try to work out what had happened to the 707 through careful examination of the wreckage and analysis of reports from amateur eyewitnesses, who from the ground could hardly make accurate estimates of the initial height and course of the aircraft. Luckily, the film recovered from a passenger's 8 mm movie camera seemed to answer that question, and it is the basis for the aircraft headings just cited.

As mentioned, it was ascertained that the vertical stabilizer (vertical tail fin) snapped off in a leftward direction and struck the left horizontal stabilizer, causing that to break off too. The ventral stabilizer (an inverted "shark's fin" under the belly beneath the tail) also broke off in a leftward direction, suggesting a powerful leftward gust was responsible rather than overaggressive use of the rudder, as in the case of the Airbus 300 taking off from JFK in 2001.

The passenger's 8 mm movie camera had been filming out of the window when it received a shock sharp enough to skip two frames. Still working, it fell to the floor and showed seats and carpet before stopping. From the footage, investigators estimated that the aircraft had been traveling at a speed of between 320 and 370 knots.

They found some preexisting cracks in the tail area, but tests showed these did not significantly weaken the tail and that some exceptional force must have been at work. (Nevertheless, inspections revealed similar cracks in other 707s, and in consequence design changes were made and more frequent inspections ordered.)

The loss of two Japan Air Self Defense Force fighters due to severe turbulence near Mount Fuji in February 1962 should have been lesson enough, but it was only after this tragic and unnecessary sightseeing accident that the scenic airway option was withdrawn.

Wind Shear, Microbursts, and Wake Turbulence (1982 and 1985)
Dangers that catch pilots, and even passengers, unawares

> After a series of wind-shear disasters in the United States, airports most prone to wind shear were belatedly equipped with extremely expensive devices to detect the phenomenon. They have proved highly effective.
>
> *Pan American World Airways Flight 759, July 9, 1982*
> *Delta Flight 191, August 2, 1985*
> *Delta Air Lines Flight 9570, May 30, 1972*

Wind shear occurs where two bodies of air moving in different directions come together. This can result in a dramatic drop in airspeed.

Not only can an aircraft find itself going too slowly and stalling, but also the air in the second body may be going downward, thus tending to slam the already vulnerable aircraft into the ground.

One particularly nasty form of wind shear is the microburst. It has a life span of about five to ten minutes and affects an area normally no more than two and a half miles (four kilometers) across. It consists of a small, very intense downdraft that descends to the ground and spreads out in all directions from the point of impact, rather like water from a faucet falling on a flat surface.

Thus, a pilot might take off at the "correct" airspeed into, say, a twenty-knot headwind, with it also pushing him upward and, having gained a modicum of height, suddenly find himself in a thirty-knot tailwind that is also pushing him down faster than the wind that pushed him upward. The fifty-knot drop in airspeed could result in, at worst, a stall, and at least a high sink rate, compounded by the effect of the downward draft, giving the pilot little time to recover, especially if obstacles like trees or perimeter fences lie in his path.

Two incidents in a three-year period made the American traveling public very aware of wind shear and somewhat nervous of flying. Many phoned their congressman, insisting that they do something so that pilots could be forewarned.

Pan American World Airways Flight 759, July 9, 1982
ATC had warned of wind shear in all quadrants as the Pan American World Airways Flight 759 prepared for takeoff from New Orleans International Airport. The captain told the first officer to let the speed build up on takeoff and said they would switch off the air-conditioning, so the forward engines would be able to provide more power.

These precautions were to no avail, for after takeoff the aircraft rose to roughly 100 feet (60 meters) and then began to sink. It seemed it might manage to stay in the air, especially as the ground effect would come into play, but at an altitude of some 50 feet (15 meters) it struck some trees 2,376 feet (724 meters) from the end of the runway. Losing further height, it continued, clipping trees and houses, before crashing in a residential area 2,234 feet (681 meters) farther on, destroying six houses and substantially damaging five more. The impact, explosion, and fire that followed resulted in the deaths of all 145 on board and 8 people on the ground, making a final death toll of 153.

The inquiry concluded that the accident was due to microburst-induced wind shear, and that part of the problem would have been the pilot being unaware of what was happening, notably the reversal of wind direction and downward push associated with the microburst phenomenon.

The fact that the total distance from the end of the runway to the crash site was about 4,610 feet (1,405 meters) shows how they might very easily have been able to recover. No doubt the initial strike with the trees, and further clipping of trees and houses, impaired the performance of the aircraft.

This was an example of what can happen when taking off. In an incident three years later, the aircraft was attempting to land.

Delta Flight 191, August 2, 1985 (Dallas/Fort Worth)

There were thunderstorms in the area of the Texas airport as Delta Flight 191, an L1011 jumbo jet with 163 passengers and crew on board, approached Runway 17L (the left of two or even three parallel runways in a 170-degree direction) for landing. There was a rain shaft and scattered lightning coming from a thunderstorm cell in the airliner's final approach path, but the pilots decided the weather was passable and continued the approach.

Some fifteen to thirty seconds after the L1011 entered the weather, however, the rain and lightning intensified, and the airplane was buffeted by a violent series of up-and-down drafts. The headwind increased rapidly to twenty-six knots, and then, just as suddenly, switched to a forty-six-knot tailwind, resulting in an abrupt loss of seventy-two knots of airspeed. The jet was only eight hundred feet above the ground when this happened, leaving the pilots little room to maneuver when the airplane began to lose airspeed and altitude at the same time. Thirty-eight seconds later, Delta Flight 191 crashed into the terrain short of the runway, killing all but twenty-six of those on board.

Following this disaster, improved weather radar was set up at US airports where microburst were frequent, and NASA did research on onboard systems to detect them as notification from the ground can incur fatal delays. They found the most effective system was Doppler radar, now with commercial aircraft around the world equipped with such systems crashes due to microbursts should be something in the past.

Wake Turbulence

One phenomenon, wake turbulence—that is, mini-tornadoes or vortices coming off the wingtips of large aircraft—can endanger following aircraft. Surprisingly, the calmer the weather conditions, the more dangerous these are likely to be, as without turbulence to break them up, they can float for miles.

The worst scenario is generally when a large aircraft heavily laden with fuel and passengers is climbing out after takeoff under high power in clean configuration— that is, with flaps, slats, and undercarriage retracted. Air traffic control uses the word "*heavy*" in the call signs for large jets as an extra warning.

The amount of turbulence generated does not depend purely on size, with the Boeing 767 being well known for nasty turbulence. Airbus is trying to demonstrate that the wake turbulence generated by its superjumbo A380 is not commensurate with its size and that it should not be penalized by being designated as requiring extra separation from following aircraft.

Delta Air Lines Flight 9570, May 30, 1972 (Greater SW Airport, Fort Worth)

The first instance of wake turbulence from heavies—large jets, such as the 747, DC-10, 767, 777, and A380—causing a following aircraft to crash was in 1972, shortly after the first ones entered service.

Fortunately, the airliner in question was being operated as a training flight with no passengers and only three crew and an FAA inspector on board. A McDonnell Douglas DC-9-14, it was landing at the Greater Southwest International Airport in Fort Worth, Texas, just after a DC-10 had performed a touch-and-go maneuver on the same runway.

As the DC-9 passed over the runway threshold, wake turbulence from one of the DC-10's wingtips caused it to rock and then roll so much that the right wingtip dipped almost ninety degrees and snagged the ground. The aircraft caught fire as it inverted and veered to the right, ending up 2,370 feet beyond the point where it first made contact. Three of those on board were killed by the trauma of the impact and the fourth by the fire.

Air traffic control had warned the pilots they could expect turbulence, but as they had already twice landed behind the DC-10, the inspectors believed the cry wolf syndrome operated, whereby repeated ATC warnings lose weight. The investigators thought it likely that the pilots believed wake turbulence only represented a danger to tiny aircraft and not to ones as large as the DC-9.

Though the death toll was small, it could have been considerable had it been a normal revenue flight with passengers on board. The NTSB investigators studied the question of wake turbulence extensively and recommended the gap between aircraft be lengthened, especially when flying under visual flight rules, where the pilots are making the judgments.

So many factors can affect wake turbulence, and it can have no effect except when they all come together. In this incident, the relatively calm conditions and flat ground prevented the vortexes from breaking up, the slight crosswind component prevented the vortex from spinning off to the side of the runway, and the slight tailwind component kept it there.

191 is an unlucky number, like 13

Because there have been an improbable number of crashes with flight number 191 most airlines avoid using it in the US.

They were:

1. Prinair flight 191, Puerto Rico in 1972. Stalled on go-around. Of the twenty people onboard, five were killed including the two crewmembers.
2. American Airlines flight 191, Chicago in 1979. Engine slung under the left wing came off damaging slats and hydraulics. Aircraft stalled. All 271 passengers and crew killed as well as two people on the ground. Described in Chapter 6.
3. Delta Airlines flight 191, Dallas 1985, described here.
4. Comair flight 191, Lexington, 2006. As Delta Connection flight 5191, it mistakenly took off at dawn on a short runway only used for light GA aircraft, and overran, killing all but one of the fifty people on board. There was only one controller in the tower and the talkative crew had not been obeying the FAA's sterile cockpit rule.

    The incident particularly shocked people because it marked the end of an exceptionally long period during which there had not been a single airliner crash in the USA.

Ice Builds Up Behind De-Icing Boots of Turboprop (Chicago, 1994)

Aircraft on hold repeatedly traversed clouds containing supercooled water droplets

> Ice is a major problem in the Northern US, but resultant crashes
> normally stem from inadequate deicing prior to takeoff. In this case, the
> aircraft had an inherent design flaw that only revealed itself in flight
> under very specific conditions.
>
> *Eagle Air 4184, October 31, 1994*

The ATR was a French-Italian turboprop airliner suitable for short routes, such as the one from Indianapolis to Chicago. The flight was operated by Simmons Airlines for American Eagle. On board were sixty-four passengers and four crew.

The twenty-nine-year-old captain, Orlando Aguiar, had almost 8,000 hours' flight time, of which 1,548 hours were in the ATR. The thirty-year-old first officer, Jeffrey Gagliano, had 5,000 flight hours, of which 3,657 hours were in the ATR. Both were highly regarded. For one of the two flight attendants, it was her first day with American Eagle.

The flight took off from Indianapolis at 14:55, having been held waiting for forty-two minutes for IFR clearance to O'Hare. Because of the short distance, the ATR was already beginning its descent from its cruising height of 16,300 to 10,000 feet at 15:13. Shortly after leveling off, they were told by a trainee female controller under supervision, whom the pilots between themselves referred to as the "the controller chick," to enter a holding pattern in the expectation that they would have to continue in their race-track circuit for some twenty-five minutes, subsequently extended for a further fifteen minutes. Chicago's O'Hare is always very busy, and in bad weather there can be long delays, with international flights given preference and humble turboprops at the back of the queue.

With the airspeed reduced as they made the holding pattern circuits, the nose of the aircraft had pitched up, making it uncomfortable for the passengers and even for the flight attendants going up and down. They therefore extended the flaps slightly to push the nose down, unaware that this would affect the way ice built up on the wings.

Just before the time they had been told they would have to hold expired, they were authorized to descend to 8,000 feet. As the aircraft speeded up, reaching 186 knots, the overspeed warning sounded to tell them they were going too fast for their fifteen-degree flap setting. Not needing them, they retracted them completely rather than reducing speed. A difficult-to-place sound coming from the airframe, almost as if it is quivering, can be heard on the CVR. An instant later the flight data recorder shows the control column moving sharply rightward and the aircraft tipping to the right.

Together the pilots try to right the control column but find that impossible. Then, just as it seems they are going to plunge to the ground, the aircraft rights itself, with the pilots seeming to be about to regain control, only for it to happen again, this time with no reprieve, and the aircraft plunges to the ground in a field near a tiny Indiana township called Roselawn, with it beginning to level out, suggesting if the pilots had had two or three thousand feet more height they could have recovered.

All sixty-eight persons on board were killed, and the force of the impact of the partially inverted aircraft was so great that the wreckage consisted of tiny pieces, though it was not damaged by fire, as the fuel had been scattered by the impact. The site was declared a biohazard because of the state of the fragmented bodies.

The Inquiry

Steve Frederick, another pilot working for Eagle Air and a close friend of the copilot of the crashed ATR, started a campaign, distributing leaflets three weeks after the crash saying the aircraft was dangerous, and he even appeared later on a morning TV show to say the same. The airline not only fired him but managed to have his pilot's license revoked, grounding him for seventeen years. Pilots later refused to fly in icy conditions.

The inquiry concluded that the trainee air traffic controller was in no way responsible, and that the pilots, through no fault of their own, by using fifteen-degree flaps to lower the nose exacerbated the tendency for ice to form on top of the wing behind the deicing boots when flying in cloud with supercooled rain droplets. When they retracted the flaps due to excess speed on the descent, the nose went up, causing the wing with the ice on top not only to stall but create a powerful vortex at the trailing edge, sucking the aileron upward with such force that without "power steering" pilots could not overcome it, making the aircraft uncontrollable.

Looking at the records, investigators found there had been previous cases, including one over the French Alps, where the aileron had behaved in exactly the same way in icy conditions. The investigators were not happy with the certification process by the French, which was rubber-stamped by the FAA. Admittedly, it was not easy to foresee the way ice would accrue and make the aileron behave in that way. Overall, the ATR incidents and their history should have been investigated much more thoroughly. The FAA softened an airworthiness directive stopping the AR being used in icy conditions, but for PR reasons the US airlines from there on used them exclusively in the warmer south.

Moment of Turbulence Results in a Death (off Japan, 1997)
Why passengers should keep seat belts on

> Passengers scoffing at their captain's suggestion that they keep their seat belts on just in case do not realize that a smash against the ceiling panels or other solid objects could quickly wipe that know-all look from their faces. Though pilots try to avoid clear air turbulence, it can occur unexpectedly.
>
> *United Airlines Flight 826, December 28, 1997*

Most of the passengers on the United Airlines 747-100 that had taken off from Tokyo's Narita Airport for Honolulu at 9:05 p.m. local time were Japanese holidaymakers planning to spend the extended New Year vacation in the warm climes of Hawaii rather than in wintry Japan.

Two hours out of Narita, they were flying above the Pacific en route to Honolulu at the economical cruising height of thirty-one thousand feet. There were 374 passengers and 19 crew members on board, and some people were moving about the cabin. These included cabin crew with their trolleys, as the passengers had just finished dinner.

Nine hundred and twenty-five miles (1,480 kilometers) east of Tokyo, the aircraft encountered some undulation, and the captain turned the seat belt sign on as a precaution. However, it was not accompanied by a chime, or an announcement in either English or Japanese, so passengers probably gave it less attention than they otherwise might. The captain also radioed Northwest Flight 90, ahead of him, to ask how the ride was, and he was told it was smooth with an occasional ripple of light turbulence at their altitude.

They then encountered some turbulence, followed, about two minutes after the undulation, by severe turbulence. First the aircraft rose, and then it descended so abruptly that heavy food trolleys hit the ceiling, making indentations, before falling back down on people. A purser holding on to a counter found himself upside down with his feet in the air.

One Japanese woman hit the ceiling and fell back in such a contorted way that, despite resuscitation efforts by flight attendants and one of the two doctors on board, she died from her injuries. Japanese TV broadcast a video taken by a passenger showing screaming passengers being thrown about the cabin and debris in the aisles.

A pilot reassured the frightened passengers over the PA, saying, "We have just hit air turbulence and the aircraft descended thirty meters [almost one hundred feet]. There is no danger of a crash."

The drop must have seemed much more than one hundred feet to the terrified passengers. The 747 then turned back to Tokyo's Narita Airport so the injured could receive medical attention. The captain could have landed on Midway but decided Tokyo would be preferable in view of the better medical facilities.

Though 110 out of the 374 passengers were alleged to have been injured, only 10 were kept in hospital. Nine of the 19 crew members were cited as injured, with three of them in a serious condition.

Examination of the flight data recorder showed that at the moment of most severe turbulence the aircraft had experienced a peak acceleration of +1.81 g, pushing the passengers down in their seats, followed by a negative acceleration of –0.82g, lifting them, pushing the aircraft violently down and them upward.

There are an increasing number of cases of passengers being injured by turbulence when not strapped in. We cite this one, as it shows that even where the severe turbulence does not last long and the aircraft does not drop thousands of feet, people can be injured. Perhaps when an aircraft rises suddenly, it is best not to be reassured by the fact that it is going up, and a good idea to hold on tightly in case the aircraft then abruptly drops, with passengers rising to the ceiling.

Chapter 13
CONTROLLED FLIGHT
INTO TERRAIN (CFIT)

Autopilot Disengages as Pilots Fiddle with Bulb (Everglades, 1972)
No one flying aircraft

> The 1972 crash of an Eastern Airlines TriStar into Florida's Everglades is often cited as a classic case of an aircraft flying into the ground while ostensibly fully under control. That is, there was nothing wrong with the aircraft or its instruments, and the pilots even knew where they were—but not how low.
>
> *Eastern Air Lines Flight 401, December 29, 1972*

As the Eastern Airlines Lockheed TriStar with 176 people aboard, and with the captain, first officer, flight engineer, and a company technician on the flight deck, came in to land at Miami, everything was normal until they lowered the landing gear. Instead of three greens, there were only two; the lamp indicating the nose wheel was down and locked had failed to light up.

The crew informed ATC it wanted to abort the landing and investigate the nature of the problem. ATC instructed them to turn off out of the way of other aircraft and proceed northbound at two thousand feet.

With the autopilot set to maintain that height, the two pilots fiddled with the bulb, ultimately trying to remove it from its socket to see whether it had blown. Meanwhile, the flight engineer, who had been told by the captain to go down to the avionics bay below to look at the wheel via the observation port, came back and said he could not be sure, and went back to look again, accompanied by the company technician.

While the two of them were away in the avionics bay, the chime indicating a deviation of more than 250 feet from the set height sounded just near the seats they had vacated. Had he been in his usual position, the flight engineer would have heard it, but the two pilots wearing headsets evidently did not.

ATC was also getting concerned about their height, which was down to nine hundred feet, but instead of warning them straightforwardly, he politely asked, "How are things coming along out there?"

The Eastern Airlines TriStar captain's reply that they were returning to land suggested they were in control of the situation.

The descending aircraft continued on in the dark over the Everglades swamp, with nothing showing up on the ground to indicate how near it was—rather like the Peruvian aircraft over the sea described later in this chapter.

Turning the aircraft southward, the first officer noticed the lack of height even though his autopilot light was on.

First officer:

We did something to the altitude!

Captain:

What?... We're still at two thousand... right?

Hey... what's happening here?

There was a bleep from the radio altimeters, indicating the proximity of the ground, followed by the sound of the left wing hitting the ground.

The ground proximity warning would certainly have sounded earlier had they not been in a left banking turn with the bottom of the aircraft facing sideways.

The left landing gear struck next, after which the aircraft began to break up, with wreckage spread over a swath of flat marshland 1,500 feet long.

It took rescuers thirty minutes to reach the scene in helicopters and airboats (shallow-draft flat-bottom boats with a fan at the back that can skim over the swamp and reeds). Only two sections of the aircraft were intact—the tail and a section of the cabin. The rest was in pieces.

It is remarkable that 77 out of the 176 persons on board survived. The death toll was 99. The two pilots and the flight engineer were among the fatalities, though the company technician in the jump seat, who had absented himself from the flight deck and was still in the avionics bay to try to check visually from there whether the front wheel was down, did survive.

The press noted that this was an accident involving the new breed of wide-body airliners and wondered whether it heralded deadlier accidents in terms of the number of lives lost at one time.

The investigators attributed the disaster to pilot error.

They said, "It seems that in turning around to tell the flight engineer to go down to the avionics bay, the captain must have exerted a strong force on the control column equivalent to fifteen pounds, thus disengaging the autopilot on his side. Unfortunately, the autopilot on the first officer's side was mistakenly set to only disengage if the applied force exceeded some twenty pounds, compared with rather less on the captain's side. As a result, the light on the first officer's side would have stayed on, giving him the impression that the autopilot was maintaining them at two thousand feet."

Preprogrammed to Hit Mount Erebus (Antarctica, 1979)
A question of command responsibility

> This accident, in which a DC-10 full of sightseers flew into a mountain at the South Pole, took place at a time when the advantage of using computers to manage the flight was being exploited, but well before development of the enhanced ground proximity warning systems also taking into account the height of the terrain ahead.
>
> *Air New Zealand Flight 901, November 28, 1979*

Staff at Air New Zealand's Flight Operations Division had a sinking feeling of culpability in their stomachs on learning their sightseeing DC-10 had flown straight into a mountain at the South Pole, for that was precisely what they had programmed it to do.

Without informing the pilots, they had made a correction to the final waypoint the night before, moving it from the end of McMurdo Sound, where the track to it passed over water, to a point behind Mount Erebus, where the track would pass right over the 12,450-foot-high mountain. The airline's chief executive then compounded the scandal by appropriating and shredding documents to try to cover up the mistake.

The scandalous aspect that made the affair a cause célèbre in normally uneventful New Zealand is not why this disaster is of particular interest here. Rather, it is because it highlights three common elements essential for preventing air accidents:

1. Command responsibility
2. Cockpit resource management (CRM)
3. Consideration of all possibilities when something odd happens, and taking conservative action if any doubt remains

"Command responsibility" is the tenet that the captain of a ship or aircraft has the ultimate responsibility for ensuring his passengers' and vessel's safety, whatever the pressures put on him by management or circumstances. There are no excuses. It is responsibility to a higher God—professionalism in the best sense of the word. Something the captain of the *Titanic* apparently did not demonstrate.

CRM is the concept that the crew works as a team, with even the lesser players able to contribute. In this case, that person's contribution came twenty-six seconds before the impact, with a flight engineer registering his increasing disquiet by saying:

I don't like this!

In fact, both flight engineers had been getting increasingly concerned.

"Consideration of all possibilities" relates to the fact that the crew assumed they could not receive the VHF radio transmissions from the US base for some technical reason without thinking that their inability to receive them—they only work in line-of-sight situations, unless the parties are very close—might be because high land, possibly a mountain, blocked their path.

The Disaster

As in the case of the BOAC sightseeing disaster near Japan's Mount Fuji and the Airbus crash at the Habsheim Air Show, jaunts with sightseeing passengers on board are inherently more dangerous than serious commercial flights, as they tend to generate an easygoing attitude and involve unfamiliar routes and circumstances.

Air New Zealand's sightseeing trips to the South Pole were no exception. However, involving a six-thousand-nautical-mile flight lasting up to twelve hours, they represented a major operation. Weather conditions permitting, the aircraft would descend to two thousand feet near the Pole for a better view. This low-level flying would consume fuel at a high rate, and even with a virtually full load of fuel at takeoff, the DC-10 would be operating without much margin for playing around before the long return leg home.

Before the introduction of GPS, navigation on long routes over water with no radio beacons depended upon inertial navigation systems (INS), which were and are also used for guiding intercontinental ballistic missiles (ICBMs).

The idea behind INS is quite simple.

1. A combination of gyroscopes measures changes of direction.
2. Sensors attached to inertial weights detect acceleration and deceleration.
3. A computer integrates the two to calculate the position.

The system is in triplicate, allowing the computer to ascertain if one has failed, as it will be the odd man out. As with ballistic missiles, users can program the desired route into the system so the aircraft flies to the destination automatically.

In airline use, the system works very well provided the pilots put in the correct coordinates. This has perhaps not always been the case (see Korean Airlines 747 Shot Down over Russia).

The INS system is a form of dead reckoning, so, unlike the GPS system, which is based on satellites, accuracy decreases with time and distance. Some versions are referred to as AINS (area INS) in that they are considered accurate over a specific area.

These Air New Zealand trips to Antarctica were popular with airline staff, and in consequence rather than choosing pilots with experience of polar flying, the treat was ill advisedly shared out as a perk. The captain chosen that time was forty-five-year-old Captain Collins. He had eleven thousand hours' flying experience, three thousand of them on the DC-10. A qualified navigator, he was regarded as a very capable pilot who exercised good judgment, but he lacked experience of flying in Arctic conditions.

The rest of the flight crew consisted of two first officers, Cassin and Lucas, and two flight engineers, Brooks, forty-four, and Moloney, forty-nine. Brooks had been on one previous flight to the Antarctic, while Maloney had the most airline experience. As usual on these trips, a celebrity was included, his role being to give a commentary to the passengers on approaching the South Pole. This time it was Peter Mulgrew, a well-known New Zealand adventurer, who had walked from McMurdo (the DC-10's first objective) to the South Pole on an expedition led by Sir Edmund Hillary, famed for climbing Mount Everest.

On board the DC-10 were 237 passengers. Though most were New Zealanders, people of other nationalities, including 24 Japanese and 23 Americans, were present. There was also a sizable cabin crew to indulge them. In all, 257 excited souls were departing from Auckland that November 28, 1979 morning, never to return.

McMurdo is US exploration base 2,384 miles from Auckland. With a thousand visitors in summer and two hundred or so residents in winter, it had long been the main point of entry for those venturing into Antarctica. Late November would be summer, with daylight lasting twenty-four hours a day. Nearby is the much smaller New Zealand Scott base.

McMurdo has an ice runway able to take certain types of aircraft—weather conditions permitting. There was no question of the DC-10 touching down there; they only wanted to avail themselves of the base's radar facilities so they could be safely talked down to get a better view.

Next to McMurdo was Ross Island, with various peaks, the highest being Mount Erebus (12,450 feet) and Mount Terror (10,700 feet), named after the two ships, under the command of Englishman Captain Ross, that came to the area in 1841, and, fearful of being iced in, did not attempt to make land.

Mount Erebus is actually an active volcano. In satellite photos the red-hot magma at the peak looks scary against the snowy whiteness of the surrounding slopes. Despite appearances, Ross Island is a true island, being joined to mainland Antarctica only by a vast ice shelf.

The planned outward route from Auckland would take the DC-10 out over the sea, over New Zealand's South Island and on to the Antarctic. They would pass over a number of waypoints, including a couple of tiny islands, which would enable them to confirm their position and hence the accuracy of their INS navigation system. Interestingly, the nearer they got to the South Pole, the greater the discrepancy between their true heading and magnetic heading would become, attaining as much as 160 degrees at McMurdo. Indeed, at McMurdo anyone on the ground would find the compass needle pointing downward!

The last waypoint before McMurdo was Cape Hallet. According to the unrevised map the pilots had worked on prior to the trip, the McMurdo waypoint was on low-lying ground, with just sea in between the two waypoints. The nearest they would come to Mount Erebus should have been twenty to twenty-five nautical miles, and with relatively little distance covered after passing over Cape Hallet, there could hardly be an INS error of that magnitude.

In fact, it was a pity the INS was performing so well. With the McMurdo waypoint reprogrammed to be two degrees ten seconds of longitude farther east, the direct track to it from Cape Hallet passed not over the sea but right over the mountain!

While still some way off McMurdo, the DC-10 was able to contact the base on VHF, but as the base replied on HF, both parties opted for HF, as that is less affected by distance and obstacles. In addition, being so far from anywhere, there would not be the usual interference from other transmissions.

McMurdo informed them that there was still low overcast at two thousand feet with some snow, but that visibility remained about forty miles. The DC-10 asked permission to descend to Flight Level 160 (sixteen thousand feet), but was only authorized to descend to Flight Level 180, eighteen thousand feet.

McMurdo then said:

> Within range of forty miles of McMurdo we have radar that will, if you desire, let you down to 1,500 feet on radar vectors. Over.

This pleased the DC-10 captain and first officer, who were getting worried their trip would be a washout. Thinking that the visibility was good enough for them to make a descent, perhaps performing an orbit (circle) to keep out of the cloud ahead, the captain said:

> Well, actually it's clear out here if we get down.

To which the flight engineer said:

> It's not clear on the right-hand side here.

The first officer said:

> No.

Despite the comment by the flight engineer, the first officer called McMurdo at the captain's request, saying:

> We'd like further descent and we could orbit in our present position, which is approximately forty-three miles north, descending in VMC [visual meteorological conditions].

McMurdo gave permission and asked to be kept informed of their height, and the DC-10 acknowledged and confirmed they were vacating Flight Level 180.

With only three miles to go before they should be on the McMurdo radar, it is surprising that the captain should opt to go to the trouble of circling to descend in a clear part of sky where visibility on the right was not so clear. Was he trying to give his passengers a better view?

For a moment, the VHF radio seemed to be working. They then reverted to HF without considering that the inability to communicate by line-of-sight VHF could be because they were in the wrong location.

They then called McMurdo on HF:

901 ... still negative contact on VHF. We are VMC and we'd like to let down on a grid of one eight zero and proceed visually to McMurdo.

McMurdo replied:

New Zealand 901, maintain VMC. Keep us advised of your altitude as you approach McMurdo ...

Conversing in the cockpit, the captain said he had VMC "This way," and that he was "Go[ing?]." At the same moment, Mulgrew, the guide, frustrated at not being able to earn his keep, was saying to the others in the cockpit:

Ah well, you can't talk if you can't see anything.

Adding:

Here you go. There's some land ahead.

The first officer informed McMurdo that they were descending from six thousand feet to two thousand and that they were VMC. The captain said he had heard there was clear weather at a place called Wright Valley and seemed to be asking Mulgrew whether he could get them there. He thought he could.

Perhaps getting worried, the flight engineer said:

Where's Erebus in relation to us at the moment?

After saying, "Left, about twenty or twenty-five miles," and some discussion, Mulgrew admitted he did not really know. They then finally all agreed that conditions did not look good.

The flight engineer, expressing his concern more strongly, said:

I don't like this.

The captain, presumably relying on the INS, said they were twenty-nine nautical miles north and that they would have to climb out of it.

First officer:

You're clear to turn right. There's no high ground if you do a 180.

Thereupon the GPWS (ground proximity warning system), operating with a six-second delay to prevent spurious warnings, went:

Whoop! Whoop! Pull up! Whoop! Whoop!

The first officer announced they were at "five hundred feet."

GPWS:

Pull up!

The first officer declared:

Four hundred feet.

GPWS:

Whoop! Whoop! Pull up! Whoop! Whoop! Pull up!

Politely, the captain asked for go-around power; he did not push the throttles through the gates (to get maximum power at the sacrifice of engine life), so he could not have seen anything to indicate a mountain was ahead.

The last words on the CVR tape were from the GPWS:

Whoop! Whoop! Pull—

Five minutes later the McMurdo controller, wanting to find out why they were taking so long to confirm their descent to 2,000 feet, called them both on VHF and HF and repeatedly thereafter, to no avail.

The DFDR showed the aircraft had made two descending orbits either side of its track as it descended and leveled out at 1,500 feet. Continuing on at that height, it had impacted the thirteen-degree-upward slope of Mount Erebus at 240 knots, with a nose-up angle of ten degrees as the pilots attempted their last-minute go-around.

It was not until the next day and eleven hours later that a US Navy Hercules searching the northern part of Ross Island reported finding the wreckage and the absence of survivors.

With the landing gear up and only a thin layer of powdery snow covering the hard layer of ice coating the slope, there was nothing to cushion the shock, and the aircraft skidded up the slope for a distance of some 550 meters, breaking up as it went. Traces of fire showed on the ice.

Unlike crashes where there is some warning of impending disaster, passengers in CFIT disasters are not usually wearing seat belts, and certainly not in the brace position. In this case, many were thrown clear as the aircraft broke up. A large number of the bodies were surprisingly intact, but postmortems showed they died from the initial impact.

Had the pilots seen the slope ahead and raised the nose earlier and more sharply, the aircraft might have been climbing and losing speed, making the shock of the impact with the incline much less severe, and raising the possibility that there might have been survivors. However, one wonders, as with the JAL crash, how many traumatized survivors could last a whole night in the cold.

So great was the shock to that small nation, and so great was the scandal, that two inquiries were launched:

1. The usual New Zealand Office of Air Accidents inquiry, led by Chief Inspector Chippendale.

2. A Royal Commission of Inquiry, under the royal commissioner, Mr. Justice P. T. Mahon, who overturned the findings of the aviation experts of the Office of Air Accidents.

(1) The former attributed the accident to:
The decision of the captain to continue the flight at low level toward an area of poor surface and horizon definition when the crew were not certain of their position, and their subsequent inability to detect the rising terrain that intercepted the aircraft's flight path.

(2) The royal commissioner summed up his findings as follows:
"In my opinion the single dominant and effective cause of the disaster was the mistake made by those airline officials who programmed the aircraft to fly directly at Mount Erebus and omitted to tell the aircrew. That mistake is directly attributable not so much to the persons who made it, but to the incompetent administrative airline procedures, which made the mistake possible.
Neither Captain Collins nor the flight engineers made any error which contributed to the disaster and were not responsible for its occurrence."

The fact that aviation experts and legal experts could come to such different conclusions is interesting. In the present author's view, the truth lies somewhere between these two extremes and is tipped against the pilots.

The airline's mistake in not informing the pilots of the revision they had made is blatant, and it is more than understandable that the Royal Commission was revolted at the way the chief executive attempted to cover it up. They were also of course dismayed that the flag carrier, albeit of a tiny country, was run close to the CEO's chest like a private fiefdom. These sentiments, rather than technical matters, were bound to influence "professionals used to working in the judicial domain."

The actual key to the accident was a phenomenon called "whiteout," well known to pilots operating in areas with snow and mountains, whereby the snowy slopes merge with the clouds, giving the impression there is just thin cloud in the distance. The horizon disappears. A pilot without experience of the phenomenon or proper training in Arctic flying would be left with the impression it is safe to fly visually (VMC), as he *believes* he can see in the immediate vicinity. Under visual flight rules, this gave the pilots the regulatory justification for descending, as Captain Gordon Vette, author of Impact Erebus, pointed out at the tribunal.

If the key to the disaster was whiteout, then the airline should be faulted for treating the Antarctic flights as perks to be shared around and not choosing pilots with Antarctic experience. Failing that, briefing on whiteout and Arctic flying should have been better. On the other hand, it is difficult to understand why the pilots did not become concerned not only by their inability to communicate with McMurdo by VHF but also by their failure to lock on to either of the (VHF) radio beacons there. The two flight engineers were getting increasingly worried, and finally the captain agreed they should go around. However, this decision came too late.

The captain was known to be methodical and to show good judgment. One can only conclude that the wish to please the passengers got the better of him, with the lack of knowledge of how deceptive a whiteout can be as a mitigating factor. He evidently thought he had adequate visibility (VMC), and the fact that he performed two orbits to remain under VMC proves he wanted to stick to regulations and not fly blindly into cloud. As he was just three miles from the forty miles out line from where the base radar would have been able to talk him down, he must have been pretty sure of the weather conditions.

When they realized they could not see where they were, they should have acted immediately. Possibly the presence of the tour guide in the cockpit, and his interceding to give his opinion as to where they might be, slowed decision-making. In normal flying, pilots wisely try to avoid being disturbed at critical moments by anyone. In fact, it is a legal obligation under today's sterile cockpit rules.

The airline's monumental blunder was just one factor in an accident that would not have happened had due weight been given to command responsibility, cockpit resource management (CRM), and the need to explore all possibilities when something (in this case the VHF radio and radio beacons) is not working—a dramatic and sorry lesson for everyone.

Air Inter Airbus A320 Plows into Mountain (Sainte-Odile, 1992)

A ground proximity warning system (GPWS) would have warned the pilots

> Facing increasing competition from high-speed trains, France's domestic airline, Air Inter, with many short-distance flights, was shortening flight times, descending unusually quickly on landing. Since this would easily set off the ground proximity warning systems (GPWS) fitted to virtually every airliner in the world, Air Inter, with the "connivance" of France's aviation regulatory body, did not equip their airliners with them, perhaps knowing their pilots would disable them.
>
> *Air Inter Flight 148, January 20, 1992*

Another notable feature of this incident was that two experienced captains were flying the aircraft. From a cockpit resources management (CRM) point of view, that is undesirable, as the normal senior-versus-junior relationship breaks down, and corners are sometimes cut, as both think they know everything and do not think they need to be bossed. Despite their rank, the Air Inter captains making up the duo had little experience of flying the A320—only 160 and 60 hours respectively.

Air Inter was France's major domestic airline until it merged with Air France on April 1, 1997. Unlike British European Airways, which was the poor relation when it merged with long-haul BOAC to form British Airways, the working conditions and salaries at Air Inter were even more favorable than at Air France, for Air Inter had for a long time been carrying businesspeople and affluent passengers on protected routes.

The flight that led to this A320 disaster was one such flight, a short cross-country hop from France's second city, Lyon, to Strasbourg with ninety passengers, four cabin crew, and two pilots. It was winter, it was dark, and although ground visibility was six miles (ten kilometers), there were clouds above 600 feet, with the tops some 1,500 feet higher. The Vosges Mountains were nearby.

The captain flying the aircraft had approval from the tower to make a straight-in VOR DME approach and landing. Finding he was too high and going too fast, he asked the controller to be allowed to continue as if landing but overfly the runway, and on reaching the opposite outer marker, double back downwind using the ILS until able to turn back again to make a visual approach to the runway.

However, he was informed that if he did this he would probably be kept in a holding pattern until the three aircraft waiting to depart had done so. The A320 captain then decided that landing as they originally intended, using the VOR DME approach, would be preferable to hanging around in a holding pattern.

With the approval of the approach controller, who told him to maintain 5,000 feet and gave him the required vectors that should have brought him in the vicinity of Andlau and well-positioned for a run straight to the airport over relatively low ground, the captain executed the almost 360-degree circuit.

Although there were mountains in the vicinity, and even higher peaks northwest of Andlau, flying at a height of 5,000 feet and with no need to descend steeply, there should have been an ample safety margin, even if off course. According to the radar, after being at 5,000 feet for a minute, the aircraft descended 2,700 feet in the next minute, which was a ridiculous rate of descent in the circumstances.

With no warning, they hit one of Mount La Bloss's ridges at the 2,625-foot (800-meter) level. Had they been just a little higher or made a slightly wider turn to actually pass over Andlau, the controlled flight into terrain (CFIT) would not have occurred. Of course, they should have been very much higher at that point.

At the time of impact the aircraft was, unbelievably, descending at an angle of twelve degrees and banking at about eighteen degrees to the left. It cut a swath through the trees on the pine-tree-covered slope and caught fire. Eight passengers and a cabin attendant survived. Had the emergency locator beacons been capable of withstanding the fire, another six passengers could have been saved. As it was, survivors had to huddle near a fire at the tail to keep warm, soon joined by some journalists who had found rescue vehicles waiting on a road lower down—the journalists had simply gone up and followed the smell of fuel. On finding the wreckage they had phoned the rescue services, who still did not come.

The cockpit voice recorder (CVR) was recovered intact, while the flight data recorder (FDR), located only a few inches away, was too damaged to provide any useful information, despite being designed to withstand fires of 1,100°C for thirty minutes. One reason for the damage was that the FDR had been left untouched lying for a long time in the vicinity of flames. This was because accusations that the French investigators and authorities had tampered with the recorders in an earlier Air France Airbus A320 crash at an airshow in 1988 meant that only someone from the judiciary could take them.

However, the quick access recorder (QAR) used for maintenance and not designed to resist fire but situated in the nose was relatively undamaged and provided useful data though not nearly as much as the main one would have.

It took the investigators six hundred and seventy-six days to issue their final report.

They could not understand why the aircraft had descended so rapidly just immediately prior to the impact. They had noticed a coincidence that the aircraft should have had a glide slope of -3.3 degrees but that the pilot must have entered it into the flight control unit (FCU) as a vertical descent of −3,300 feet a minute, probably because it was set in the "Heading/Vertical Speed" mode instead of in the "Track/Flight Path" mode.

It was a mistake easily made and not easily noticed because the designers of the instrument panel only showed the figure entered into the FCU to two decimal places, thus -3,300 feet displayed as -33, which when quickly scanned visually could be taken for -3.3. However, when this was tried in the simulator, the aircraft still missed the mountain by a fair margin. Nonplussed, the investigators went to the manufacturer, Airbus, to see if they could explain why the aircraft had descended even faster.

They found out that to cope with emergency situations the descent is enhanced.

The pilots received no warning that they were sinking at an alarming rate or that they were about to hit the ground, because there was no ground proximity warning system (GPWS).

The French weekly magazine *Le Point* is said to have questioned Air Inter on this point and to have been told that Air Inter's pilots only flew in France and knew the terrain so well they did not need GPWS.

Apparently, only five weeks before the crash, M. Frantzen, director of the aeronautical training and technical control service, enjoined Air Inter by letter to reconnect these alarm systems but was sent packing. The sophisticated GPWS that has prevented many crashes would not only have warned the pilots of the proximity of the terrain but also of the abnormally high sink rate, an alert that the pilots in the Indian Airlines disaster described above ignored.

In the absence of FDR evidence, due to heat damage to the tape, attributing the cause of the crash to the easy confusion over the input mode seems quite plausible, but some people are not convinced the whole truth has been revealed.

On a website called EuroCockpit, there is a piece on the affair saying that instead of VOR Bendix + DME TRT, as used by Air France, Air Inter had a different type of system (VOR + DME-Collins). It also claims that Lufthansa, among others, had forbidden the use of that system for VOR DME approaches six months before the crash, and that Air Inter did likewise after the crash and an incident at Bordeaux.

According to EuroCockpit, it was a problem of hurried certification, and it says it is rumored that the woman elevated to head the Inspection Générale de l'Aviation Civile was previously responsible for VOR DME certification. Also, it is claimed that the records of receiving reports sent by Air Inter staff to Airbus and the BEA regarding problems with the Collins version have disappeared without trace, even though there are records of them being sent.

Allegedly, the CVR tape for the St Odile crash showed the pilots were somewhat casual, perhaps because it was two captains together. Macarthur Job suggested the nature of Air Inter's routes, with many automatic landings in bad weather, may have made them depend too much on the computers in controlling height.

That said, the disarming of the GPWS by Air Inter seems unforgivable. One reason for so many false alerts was to compete with France's ultrafast new trains Air Inter was bringing its flights into land faster than the norm with rapid descent rates. For a long time, pilots blocked the introduction of the TCAS collision alert system on similar grounds and said that it just meant an extra thing to think about.

This accident is also notable for the continuing litigation involved, with proceedings taking place even in 2009, some seventeen years later.

The Echo Association, founded by the victims' families, commented on the failure of previous efforts.

They said, "Six people have been "examined" in investigations, headed by three judges, but experts have failed to determine the exact cause of the Airbus A320 crash. It is not so much a fact that no reasons were found but that one court would come to one conclusion, for it to be rejected by a higher court."

One has to have some sympathy for the victims' families seeking justice for aviation in France is a very close-knit incestuous community.

757 Captain Selects Wrong Beacon, Loses Way (Cali, 1995)
Air brakes prevent pilots extricating themselves

> How could a state-of-the-art airliner with a sophisticated, almost automatic navigation system go off course?
>
> *American Airlines Flight 965, December 20, 1995*

So much of modern airliner piloting is routine, and on regularly scheduled flights the dispatchers, or the pilots, load a previously prepared flight plan into the flight management system with all the waypoints and so on. It prevents errors and, of course, saves considerable time. However, it means pilots, not having to enter the waypoints themselves, do not see where the designation for a waypoint beacon might be confusing.

American Airlines Flight 965 to Cali, Colombia, from Miami had taken off two hours late because of the delayed arrival of connecting passengers from the US's northeast, hit by a storm, and a very long wait before entering the runway due to congestion. The experienced pilots of the Boeing 757, with 163 passengers and 8 crew, were therefore keen to make up time. The fifty-seven-year-old captain had flown to Cali dozens of times, the last time only six days earlier, but this was the thirty-seven-year-old copilot's first trip there, which is one reason that he might have deferred to the captain's judgment more than usual.

Flying over Cuban airspace, they began their approach into Cali, situated on a plain at the end of a long valley straddled by high mountains, some of which are almost fourteen thousand feet high, after only about three hours. Unlike most airports, where the approach controller follows incoming aircraft on his or her radar, Cali could not do so, because FARC antigovernment fighters had blown up the installation. Instead, the controller had to rely on the pilots reporting their positions as they passed over various beacons, and a key part of this was a radio beacon far up the valley called Tuluá.

The captain told the controller his distance-measuring equipment (DME) showed they were sixty-three nautical miles from Cali, whereupon, while telling him to report on passing Tuluá, the controller confirmed he was "cleared directly" to Cali. For some reason, the captain thought this meant he could forget about Tuluá, although in his read back he mechanically ended with "report Tuluá." He reset the flight management system to take them directly to Cali, in the process erasing all the waypoints in between, including Tuluá.

Cali Airport only had one runway, roughly aligned with the valley, and with the prevailing wind usually coming from the valley, aircraft would normally skirt the airport and land from the opposite direction. This was what was programmed into the flight management system, meaning the pilots could land almost automatically with no hassle. However, noticing that the wind had dropped, the controller gave them the option of a straight-in landing.

This appealed to the captain, aware that the delay in departing from Miami meant the cabin attendants might not have sufficient rest time, and the return flight the following morning would be delayed. The pilots were happy to accept the offer, although it would require some feverish activity, losing height, and consulting paper charts to note waypoints and frequencies. They did not have time to do this in a calm and collected way and keep a proper eye on what the aircraft was doing.

The first officer deployed the air brakes, which are rectangular panels on top of the wings that flip up and not only slow it but push it downward without putting the nose down and making the passengers uncomfortable. This made the aircraft descend more rapidly.

Fumbling with the chart, they saw they must enter a waypoint called Rozo into the system. Meanwhile, the controller repeatedly asked them to confirm when passing Tuluá, but they had no idea where it was, having erased it from their computer and with no time to find a way to find it.

The captain asked the controller if it was okay to go to Rozo and was told what to do after that, with the controller again asking him to report passing Tuluá.

Realizing they must anyway have already passed it, the captain entered Rozo into the system by pressing "R" on the panel. Ten options came up, none of which were Rozo, but as the nearest one is usually shown first, and that could only be Rozo, the captain hastily and fatally confirmed that, without getting his copilot to verify. If the copilot had verified, he would have seen the system showing a track taking them more than ninety degrees to the left, which was obviously wrong.

The flight management system duly executed his instruction and, with the aircraft descending, gently but surely turned it left through an angle of some 120 degrees so it passed over the mountains and headed for the Romeo beacon, at Bogota Airport, 172 miles (300 kilometers) away.

There is a rule that, apart from when cruising, one pilot should constantly watch what the aircraft is doing, but in this case, they were both too busy concentrating on the charts and did not notice the aircraft turning. After a minute or so, they began to realize they were obviously heading for the wrong place. For a moment, they considered inputting Tuluá, but the captain then decided they should head straight for Cali, which they did know. What they did not realize was that while still high they had passed over the mountains into the next, almost parallel valley.

The controller with no radar called them to ascertain their height, which they confirmed was nine thousand feet.

Heading for the Cali CLO beacon took them over the high mountain ridge separating the valleys, and with them lower than before, the ground proximity warning system (GPWS) alerted them to terrain. Unsure of how high it was, they applied full power to climb, but still the warning system kept sounding, and as they got closer it said, "Pull up! Pull up!"

Had they not left the air brakes deployed, the powerful Boeing 757 would almost certainly have gained enough height to surmount the ridge, but as it was it crashed into the trees at the top, with part of the aircraft sliding over.

The crash site was six miles south of Tuluá VOR and twenty-eight miles north of the runway. As the controller had no operating radar, the location of the crash was not immediately obvious, and rescuers did not reach it until the following morning to look for survivors. They found five, one of whom died later in hospital of internal injuries.

The Inquiry

Many items from the wreckage were spirited away by thieves arriving by military helicopter and ended up on the black market in the US, with American Airlines publishing a list with the part numbers to try and dissuade purchasers. Fortunately, the cockpit voice recorder and flight data recorder were recovered intact, as was the flight management system, which was not much use, since its memory did not at the time retain deleted settings.

The inquiry was carried out very professionally by the Bolivians with the help of the NTSB and found amongst other things that:

1. The flight crew had insufficient time to prepare for the approach to Runway 19 before beginning the approach.

2. The flight crew failed to discontinue the approach despite their confusion regarding elements of the approach and numerous cues indicating the inadvisability of continuing.

3. Many important differences existed between the display of identical navigation data on approach charts and on FMS-generated displays, despite the fact that the same supplier provided AA with the navigational data.

4. The AA965 flight crew was not informed or aware of the fact that the "R" identifier that appeared on the approach (Rozo) did not correspond to the "R" identifier (Romeo) that they entered and executed as an FMS command.

5. One of the AA965 pilots selected a direct course to the Romeo NDB, believing that it was the Rozo NDB and, upon executing the selection in the FMS, permitted a turn of the airplane toward Romeo without having verified that it was the correct selection and without having first obtained approval of the other pilot, contrary to AA's procedures.

6. The incorrect FMS entry led to the airplane departing the inbound course to Cali and turning it toward Bogota. The subsequent turn to intercept the extended centerline of the runway led to the turn toward high terrain.
7. The descent was continuous from FL 230 until the crash.
8. Neither pilot recognized that the speed brakes were extended during the GPWS escape maneuver due to the lack of clues available to alert them about the extended condition.
9. Considering the remote, mountainous terrain, the search and rescue response was timely and effective.
10. Although five passengers initially survived, this was considered a nonsurvivable accident due to the destruction of the cabin.
11. The Cali approach controller followed applicable ICAO and Colombian air traffic control rules and did not contribute to the cause of the accident.
12. AA training policies did not include provision for keeping pilots' flight training records, which indicated any details of pilot performance.

Elsewhere in this book we have regretted how difficult it is to discover the finer reservations an airline might have had regarding a pilot's performance, other than that he or she finally, after failing, passed the required tests during training and route checking. Some airlines may avoid keeping discoverable records lest the information be used against them in litigation.

Aftermath

American Airlines were sued, and though they were obliged to pay in a number of cases, they in turn sued Jeppesen and Honeywell, who made the navigation computer database and did not include the coordinates of Rozo under the identifier "R." The jury trial in Miami found that Jeppesen was 30 percent at fault for the crash, Honeywell was 10 percent at fault, and American Airlines 60 percent.

Having unsophisticated juries decide these complex situations is questionable. One only has to remember the SilkAir "suicide" crash, where the jury decided Parker Hannifin were responsible without being allowed to see the contrary NTSB's findings.

757 Hits Sea as ATC Confirms Altitude 9,700 Feet (Lima, 1996)
"Two-dollar-per-hour aircraft polisher" brings down jet

> Air traffic control radar, unlike military radar, usually only indicates the altitude reported by the aircraft's transponder, which is the same as that shown by the aircraft's altimeter. When the worried pilots asked ATC to confirm their altitude, the controller was unwittingly only confirming the erroneous one they were seeing on their altimeter.
>
> *Aeroperú Flight 603, October 2, 1996*

The pilots had flown their sophisticated computerized Aeroperú 757 out over the sea for safety while they tried to sort out the cascade of problems they were having. Their altimeter showed them at 9,700 feet, and being a dark night, they were not surprised they could not see the water far below.

The ground proximity warning system (GPWS), which unlike the barometric altimeter works by bouncing radio waves off the ground, warned:

Too low! Terrain!

Too low! Terrain!

Too low! Terrain!

The computer had, they thought, been playing up, and it was just one of many silly warnings they were receiving. Anyway, air traffic control back in Lima had just confirmed their indicated height was truly 9,700 feet. Concerned nevertheless by the continuous GPWS alarms, the copilot again questioned ATC regarding their 9,700-foot altitude, which ATC reconfirmed. The captain was more concerned with their airspeed, which ATC confirmed was two hundred (knots), meaning they were in danger of stalling.

Before they could remedy that, the aircraft hit the sea, bounced 200 feet into the air, inverted, and fell back again, drowning any surviving occupants.

Nightmare

For the Aeroperú pilots, the flight had been a nightmare from the moment they lifted off on departure from Lima. The incorrect data that was making the altimeter and airspeed indicator give erroneous readings was making the aircraft's computers believe the aircraft was flying outside the normal envelope. In consequence, it was issuing a barrage of warnings and setting off a tirade off aural alarms.

Pilots are trained to react to alarms, and when these come in rapid succession for no rhyme or reason and seem unconnected, one can imagine the crew's confusion. For instance, reducing engine power and deploying the air brakes in response to an overspeed alert did not seem to have any effect—at least according to the airspeed indicator, which was not functioning properly.

Sometimes they would get a stick shake, indicating they were about to stall because of inadequate airspeed while being simultaneously warned they were going too fast.

Perhaps out of habit or because they were too busy coping with the crises and different alarms, or even regressing mentally, the Aeroperú pilots failed to note that, in contrast to the barometric altimeter, the changes in height shown by the radio altimeter corresponded with what they would expect, according to their engine settings and pitch—climbing, level, or descending.

With the incessant warnings and the need to cope with them, it is understandable that the pilots had little time to think. Hardly had they tried to deal with one when another sounded, and audio alarms are particularly stressful.

To give some idea of the confusion, here is part of the CVR (translated from Spanish), beginning about seven minutes before they struck the water. Times in parentheses are elapsed time in minutes and seconds from takeoff (00:00).

22:42 First officer to ATC:

We have terrain alarm; we have terrain alarm!

22:46 Lima ATC:

Roger, according to the monitor… it indicates Flight Level 120 [12,000 feet] over the sea, heading northwest course of three hundred.

22:55 First officer to ATC:

We have terrain alarm and we are supposed to be at 10,000 feet?

23:00 Lima ATC:

According to monitor you have one, zero, five [10,500 feet].

23:03 Mechanical voice alert:

Wind shear! Wind shear! Wind shear!

23:06 First officer to ATC:

We have all computers crazy here…

23:07 Mechanical voice alert:

Too low, terrain!

23:07 Captain:

Shit, we have everything!

23:14 Mechanical voice alert:

Wind shear! Wind shear! Wind shear!

23:17 Captain:

Shit, what the hell have these assholes [maintenance] done?

23:47 First officer to ATC:

We don't have… We have, like, 370 knots. Are we descending now?

23:56 Lima ATC:

It shows the same speed . . . you have approximately 200 speed.

23:56 First officer to ATC:

Speed of 200?

23:58 Lima ATC:

Speed of 220 over the ground, speed reducing slightly.

24:05 Captain:

Shit! We will stall now . . .

24:07 Mechanical voice alert:

Sink rate! Sink rate! Sink rate!

Sink rate!

They managed to avoid stalling and climbed, with the barometric altimeters showing over 9,000 feet. They thought they had leveled out at Flight Level 100 (10,000 feet), a nice, safe altitude. The flight data recorder (FDR) later showed that the radio altimeter indicated they had only climbed to 2,400 feet, then re-descended to 1,300 feet, and had finally gone no higher than 4,000 feet.

Meanwhile, by noting their different locations with his civilian radar, which could independently determine distance and direction (but not true height, which depended on that reported from the aircraft by the transponder), the air traffic controller had calculated their groundspeed to be 200 knots, when the pilots' instruments were showing their airspeed to be 370 knots. To have a difference of that magnitude between groundspeed and airspeed would require a wind over 200 mph. The first officer was surprised, but for some reason it did not immediately make him doubt the overspeed warnings. Perhaps his reaction to them was an understandable automatic reaction to alarms in general.

Unsure of their height, they had gone out over the sea to play it safe. Had they remained over land, they might very well have seen some lights that would have shown how extremely low they were, and because of the presence of hills and mountains would have given more weight to the accurate readings shown by their radio altimeter.

The controller had informed them that a Boeing 707 taking off from Lima was coming to assist them and would guide them in. Without waiting, they thought that already being relatively near the airport it would be a good idea to try to lock onto the airport's instrument landing system (ILS) glide path, which in normal circumstances would have enabled them to descend safely without relying on their instruments. However, unbeknown to them, they were far too low to pick up the beam.

Unfortunately for the occupants of the Aeroperú 757, civil aviation radar, which is not intended for tracking an enemy trying to sneak in, relies on the transponder in the aircraft to give it its altitude and ID. So, when the Aeroperú crew asked ATC to confirm their height, ATC was only repeating what the barometric altimeters in the aircraft showed.

Some suggest that this fact was overlooked because all were convinced it was a computer problem. The airline did try to contact the pilots and find out what was happening, but staff there were also led to understand it was a computer problem. Had it been daytime, with senior staff at hand both at the company and at ATC, the pilots might have been better advised.

On a sophisticated aircraft with so much run by computer, it is too easy to attribute a real problem to the computer, as happened when Captain Piché started transferring fuel to the side from which it was leaking and had to glide eighty miles to salvation.

Some accounts say the pilots were sensibly using their radio altimeter to go back to Lima, but the following cockpit voice recorder (CVR) transcript does not seem to bear this out.

29:49 Captain:

How can it be flying at this speed if we are going down with all the power cut off?

29:56 First officer to ATC:

Can you tell me the altitude please?

Because we have the climb that doesn't ... ?

29:59 Captain:

Nine.

30:00 GPWS:

Too low! Terrain! Too low, terrain! Too low, terrain!

[These warnings continue until aircraft inverts at 30:50.]

30:01 ATC Lima:

Yes, you keep nine seven hundred (9,700) according to presentation, sir.

30:04 First officer to ATC:

Nine seven hundred?

30:09 ATC Lima:

Yes, correct. What is the indicated altitude on board? Have you any visual reference?

30:12 To ATC:

Nine seven hundred, but it indicates "Too low, terrain!" ... Are you sure you have us on the radar at fifty miles?

30:21 Captain:

Hey, look ... With 370 we have ... have ...

[Presumably, the captain was referring to the airspeed indicated by his instruments.]

30:29 First officer:

Have what? 370 of what?

30:36 First officer:

Do we lower the gear?

30:38 Captain:

But what do we do with the gear?
Don't know . . . that.

30:40 [Sound of impact.]

30:41 First officer to ATC:

We are hitting water.

30:44 First officer:

Pull it up.

30:48 Captain:

I have it, I have it!

30:50 GPWS:

. . . Too low, terrain! . . .

[The ground proximity warning system (GPWS) voice alert ceases, because the bottom of the aircraft is no longer facing the sea below but sideways.]

30:55 Captain:

We are going to invert! [Go upside down.]

30:57

Whoop! Whoop! Pu— [Voice alert sounds.]

30:59 Sound of impact.

End of recording, with barometric altimeters indicating 9,700 feet.

All on board—sixty-one passengers and nine crew—lost their lives.

The FDR showed the 757 was traveling at 260 knots and descending at an angle of ten degrees when the left wingtip and left engine pod first snagged the water. Ingestion of water no doubt led to failure of that engine.

With the right engine already at a low power setting because the pilots thought they were flying too fast, their airspeed would have fallen quickly to stalling speed as the aircraft "bounced" and rose to two hundred feet, before inverting and falling back into the sea.

A lawyer made much ado of the suffering the passengers endured in the half hour preceding the impact with the sea. Though the constant changes in speed and height as the pilots responded to the continuous false alerts must have been scary, they can hardly have been expecting to fly straight into the sea.

A judge ordered Aeroperú and the two-dollar-per-hour aircraft polisher who had left the masking tape on the pitot probes and static inlets to pay $29 million to the families of the seventy dead. Another judge had already sentenced the hapless worker to two years in jail.

This incident ultimately led to the closure of the airline.

On Taking Over the Aircraft, Captain Forgets Timer (Busan, 2002)
Timer was set to tell pilots when to turn

> Air China, the flag carrier of the People's Republic of China, had an impeccable safety record—forty-seven years without a fatal accident—partly because it was able to avoid flying in adverse weather conditions.
>
> *Air China Flight 129, April 15, 2002*

The Boeing 767 on the flight from Beijing to Busan, in South Korea, was seventeen years old and not equipped with the enhanced ground proximity warning system (EGPWS), which, besides warning pilots when getting too near the ground, tells them if they are in danger of hitting ground rising ahead of them.

Weather at Busan was cloudy and misty, and as the 767 came in to land the controller advised them that a change of wind direction meant that instead of coming straight in, as they first planned, they would have to make a circular approach and land on the runway from the opposite direction—something quite routine and which the pilots should have been prepared for, and that was included in an approach briefing, in which all eventualities were reviewed, such as having to abort the landing at the last minute because another aircraft is still on the runway.

They approached the runway on the downwind leg twenty knots too fast and made a left turn a little late, and not the aggressive, sharp one of forty-five degrees they were supposed to make. With the extra speed, this meant they were farther along than they should have been and quickly reached the end of the runway, where they were to hold course for precisely twenty seconds before executing a 180-degree base turn to line up with the runway for touchdown.

With the timer set, the copilot remarked that the wind was making the aircraft difficult to handle. At this critical moment, the captain took over the controls without allotting tasks, notably keeping a watch on the timer. There was some confusion, with the captain flying straight on, although the timer was showing twenty-five or so seconds.

Flying into cloud, they had lost sight of the runway and should anyway have executed a go-around, since a circular approach is visual, and the runway must therefore always be in sight. Well past the point he would have been after the time was up, the captain began the 180-degree turn, allowing the aircraft to sink below the minimum safe height, which the copilot duly brought to his attention.

Copilot:

Pull up! Pull up!

But it was too late. The aircraft plowed into trees near the top of a mountain. The tail hit first, as the aircraft was pitching up as the pilots tried to save it— though in landing configuration with little thrust, there was little hope of that.

Perhaps because the tall trees cushioned the impact, 37 of the 166 people on board survived, including the captain. He was interviewed many times but never explained why he had not done a go-around on being unable to see the runway.

Apart from that gross error, a whole series of factors contributed to the crash. The pilots had never done a circular approach to that airport either for real or in the simulator. Of course, local knowledge would have helped, since they would have been aware that one should never go beyond the expressway, as mountains lay behind it.

Air China now class Busan's Gimhae Airport as one needing special care. As airliners always approached from the south, the charts did not show the area to the north with the mountains. Having only practiced circular approaches at Peking Airport, which is surrounded by low terrain, the pilots may not have been sufficiently aware of the lurking dangers, such as at Busan. Another factor was work overload due to unpreparedness.

It is notable that a captain taking over at a critical moment proved to be the trigger for the disaster. Had BA's Captain Burkill taken over in those final few seconds when his Boeing 777 lost thrust on coming into London's Heathrow airport, as described in Chapter 2, the outcome could have been very different. He was (wrongly) criticized by some for that at the time.

Chapter 14
UNIQUE CASES

Without an engineer's understanding of how systems work, pilots have nevertheless at times been tempted to experiment or try out their own "clever" ways to solve a problem. While obvious with hindsight, the outcomes can be unpredictable and sometimes fatal.

Swissair Flight 306, September 4, 1963
Air Florida Flight 90, January 13, 1982
Nigeria Airways Flight 220, July 11, 1991

Taxiing with Brakes on to Clear Fog, Etc. (Zurich, 1963)

Zurich Airport was shrouded in fog at six in the morning when the Swissair Caravelle was due to take off for nearby Geneva before going on to Rome. Despite the poor visibility, controllers let an airport vehicle guide it to the runway. Finding the fog too thick to take off, the pilots obtained permission to taxi halfway down and back, purportedly to check conditions farther down but in fact to cleverly clear the fog. They would do this by proceeding with the brakes half on and the engines at a higher than usual power to blow the fog away. Having been relatively successful, the pilots duly took off without realizing the Caravelle's brake discs had become incredibly hot.

This heat was progressively conducted to the wheel rims, also raising their temperature abnormally. However, it was not until the wheels had been retracted after takeoff that the temperature of the tires and hydraulic lines reached critical levels. At that point the overheated tires exploded, damaging hydraulic lines, causing the fluid to leak and catch fire due to the hot metal. All eighty people on board died in the subsequent crash. They included forty-three children.

It was taxiing at high thrust settings with the brakes half on that was the cause of the overheating and a tire bursting. However, it was having highly flammable hydraulic fluid that made disaster inevitable.

Air Florida Flight 90, Washington DC, 1982

Another example of "clever" unorthodox maneuvers that can be fatal was the Air Florida Flight 90 crash into the icy Potomac River. The captain of the Boeing 737 had "cleverly" used reverse thrust to help the aircraft move back from the stand and then, contrary to instructions for snowy conditions, "cleverly" taxied closely behind another aircraft on the assumption that heat from the engines of that aircraft would help deice his 737.

Using reverse thrust in the presence of snow had sucked it into the engines and following close behind another aircraft would have made the situation worse, since snow on the wings, instead of falling off, would become slush oozing down and freezing on the wings' leading edges and the inlet to the engines. Furthermore, the pilots failed to switch on engine deicing when they came to that point on the checklist, meaning the snow that had been sucked into the engines by reverse thrust froze solid and affected their performance.

Nigeria Airways Flight 2120—Tires on Fire in Wheel Well, July 1991

Maintenance workers a few days before had decided the Douglas DC-8 jetliner should have its tires replaced, as they were losing pressure. However, the owner, Canadian airline Nationair, which operated it on behalf of Nigeria Airways, stopped them doing that so their flights could run to schedule.

The aircraft made several takeoffs and landings with those low pressures, but when it took off from Jeddah, over the objections of the maintenance engineer, who had been unable to find nitrogen to inflate the tires, to take 257 pilgrims and 14 crew back to Nigeria, the aircraft had to taxi three miles (almost five kilometers) over hot taxiways to the beginning of the runway. The underinflation of the number two tire meant all the weight that axle was meant to support was borne by the number one tire, which flexed beyond normal limits as it rolled, resulting in overheating. During the takeoff run, the pilots heard the bang as the number one tire failed but continued the run, with the number two tire failing as well. For some reason—perhaps the extreme weight it was bearing—the number two tire ceased rotating. Friction with the runway wore down the rubber and allowed the metal rim to contact the runway. This in turn heated up and no doubt produced sparks, initiating a self-sustaining fire in the wheel well.

Fed by hydraulic fluid and fuel, the blowtorch effect of the slipstream even meant it got hot enough to start burning through the cabin floor. With bodies and seats dropping out, the aircraft crashed short of the runway, killing all on board.

Following this disaster, manufacturers decided to install fire detectors in wheel wells. This incident, caused by cutting corners on maintenance to be more competitive in a difficult market, contributed to the demise of the airline.

Teen Survives Two-Mile Free Fall in Just Her Seat (Peru, 1971)
The dense jungle that broke her fall was almost her undoing

> We describe this event in detail because it is not only remarkable but uplifting, with many valuable lessons regarding the will to survive and human fortitude.
>
> *Lansa Flight 508, December 24, 1971*

The Lockheed L-188 Electra, the only large turboprop airliner made in the US, had stubby wings and four powerful engines with huge propellers whose prop wash mostly passed over flaps that gave considerable extra wing area when deployed. It was exceptionally good for taking off from short high-altitude runways. In concept it was an excellent aircraft, and it served later as the basic airframe for the Lockheed P3-Orion anti-submarine and maritime surveillance aircraft, of which 650 were produced.

However, a phenomenon called "whirl mode flutter," which was not understood at the time, produced vibrations in the outer engine nacelles that when transmitted to the wings at frequencies coinciding with the harmonics of the outer wing panels would cause the wings to go up and down violently enough to break off.

After a couple of crashes due to loss of a wing, the FAA restricted its speed, until Lockheed eventually solved the problem and modified the wings at great expense. Because of the early crashes, US orders dried up and Lockheed suffered a considerable loss both from cessation of the program and from lawsuits. It had missed the window of opportunity, since faster, pure-jet aircraft, such as the 727 and 737, were in the offing.

Existing aircraft were in many cases sold abroad, including to South America, where they performed admirably on some routes for years. The rigid wings were not particularly resistant to exceptional stresses, though many were subsequently reinforced.

It is alleged the LANSA Lockheed L-188A Electra that took off at midday on Christmas Eve, 1971, from Lima, on Peru's coast, was an early production model and an amalgam of several aircraft. It was to make the short flight inland over the Andes and then the Peruvian rain forest to its first stop, Pucallpa, before going on to Iquitos, even farther inland. There were ninety-two people on board—eighty-six passengers and six crew. Flying at an altitude of twenty-one thousand feet in good weather, the pilots were musing about how concerned they were about getting back to Lima in time to celebrate Christmas with their families, especially as the flight was already seven hours late.

For twenty-five minutes everything was fine, but then it became increasingly cloudy, and a storm front could be seen ahead even by the passengers. No doubt because they were determined to avoid further delay, the pilots flew straight into it. The clouds became almost black, and passed by the windows, as the passenger who survived said, "like menacing living creatures." There was extreme turbulence, accompanied by lightning. It got worse, with Christmas presents falling from the overhead bins. Finally, a lightning bolt struck the right engine, and fuel in the wing caught fire. The aircraft dived vertically, with the pilots struggling for control, at which point the rigid wing broke off.

With passengers screaming, the fuselage fell almost vertically, with objects inside tumbling forward, and then broke.

The Search

Aircraft were unable to locate the wreckage, which was obscured by the canopy of tall trees. Also, the torrential rain meant there were no plumes of smoke. The authorities were inundated with erroneous reports of locations of the crash, perhaps in part because the storm had created sounds and flashes suggestive of a crash. There were so many that the police commander said anyone reporting further sites would be arrested, which obviously put a damper on the flow of information. One of the reports had been correct but was ignored.

There were some survivors, but they died before rescuers reached them more than ten days later, guided to the spot, unbelievably, by a passenger thrown from the aircraft who had miraculously survived the two-mile (three-kilometer) free fall.

Miraculous Free Fall

That person was seventeen-year-old Juliane Koepcke, a half-Peruvian, half-German high school senior studying in Lima, who had been sitting by a window at the rear with her mother next to her and a heavily built man in the aisle seat.

According to Juliane, when the fuselage broke apart everything went quiet, and she found herself alone, with only the sound of wind in her ears. Her mother and the man in the aisle seat gone, she was falling to earth alone, strapped in at the end of her row of seats. Losing and regaining consciousness, she saw the forest below moving in circles like "green cauliflower, like broccoli" as she fell toward it, pinned painfully in place by her seat belt biting into her stomach.

She only really came to at nine o'clock the following morning, meaning she had remained comatose and in shock the whole afternoon and night. There was no sign of anyone else. No debris. She had a broken collar bone, a gash in her calf, a deep puncture wound, abrasions of varying severity, concussion, and an eye injury brought about by the change in air pressure.

Having lost her glasses, she found it difficult to see distant objects anyway. She had also lost a shoe and would have to put the foot with the remaining shoe forward every time there was something in the water potentially dangerous. She was wearing a short skirt that did little to protect her from the cold at night and from mosquito and insect bites, and which a Peruvian newspaper later drew alluringly in a comic strip depicting her ordeal.

She was totally alone, injured, in a dense jungle, and danger lurked everywhere. However, Juliane had several things in her favor:

1. Her parents had inculcated the notion that solutions could always be found, provided one thought calmly and logically;
2. From an early age, she had spent several periods at her father's research station in the rain forest, even going on trips with her parents deep into it with no handholding. She had learnt about piranhas (only dangerous in still water), stingrays, snakes, caimans, poisonous stonefish, and other dangers, including certain plants.
3. Most importantly, her father had taught her that if lost she should look for a stream and follow that, as it would lead her to a river and eventually people.

She shouted to no avail and wandered around in the hope of finding her mother, until finally finding a stream and following it to a creek. She only had boiled sweets to eat, which soon ran out. Thinking back later, she believed the muddy matter she took in when drinking from the stream and river must have provided some sustenance, as well as hepatitis. On the fourth day, she heard the flapping of large wings that could be only those of vultures that might be preying on the dead from the aircraft. Around the next bend she saw a row of seats. The three occupants, two men and a woman, were upside down with their feet sticking up grotesquely in the air, their heads buried deep in the ground. She looked around but could not find any sign of people.

On the fifth or sixth day, she heard the call of hoatzin (stink birds), and, having been told by her ornithologist mother that they are usually associated with open expanses of water, she felt more confident. However, just ahead, where the creek should feed into the river, the way was blocked by impenetrable vegetation. With tremendous effort she forced her way through the thickets to the side in the direction of the sound of the birds and finally reached the river. Weak and hungry, she half swam, half floated down the moving water in the middle, encountering caimans and also logs that could have broken her bones.

As evening approached on the tenth day, she flopped down on a gravel bank and dozed off. When she opened her eyes, she thought she must be dreaming, for a boat was moored nearby. She swam over to it. It was real; moreover, a path with foot marks led from it to a primitive shelter, which in her weakened state took her hours to crawl to. In it was the outboard motor for the boat and a can of gasoline, some of which she poured onto her maggot-infested wounds, another trick her father had taught her. She later told London's *Daily Mail* that thirty-five worms came out of her arms alone.

Only the next day at dusk did three forest workers return to the shelter by pure chance, as the weather looked bad. Fortunately, she was able to explain in Spanish who she was, as, shocked by her completely bloodshot eyes, they thought she was the water goddess. They tended her as best they could, giving her food, she could hardly get down. Two more men arrived out of the darkness, and after a night with all of them together on the hard shelter floor—she did not dare say she was more comfortable on the ground outside—they took her back to civilization in the boat.

She was able to explain where the crash site was, as the creek she had followed, Stingray Creek, was the only one blocked at the junction with the river by vegetation. When rescue crews following her clues finally located the aircraft, they discovered that as many as fourteen others had initially survived the crash but, unable to seek help, had died awaiting rescue.

Juliane Attributed Her Surviving the Fall to a Combination of:

1. A powerful updraft produced by the storm;
2. Her sitting very tightly buckled in at the end of the seat bench, which meant it was spinning around the heavier end like a maple leaf. Though not conscious for the entire fall, she remembered the jungle below seeming to come toward her, moving in circles;
3. The seat was underneath her, with her facing upwards, so when it hit, it went through the foliage, protecting her like a boat in water;
4. Luckily coming down between high trees linked by lianas, which are long, woody vines that use trees to climb up to the forest canopy to access sunlight. In her book, she said that a man working on the recovery of the bodies told her the only well-preserved seat bench he found was in a spot between giant trees connected by a dense network of lianas.

The investigation concluded the crash was due to pilot error. They had flown into the storm, and their maneuvers following the lightning strike, which must have weakened the wing, very likely overstressed it.

Rogue Controller Deletes Data (JFK, 1979 or 1980)
As Soviet airliner lands with Soviet ambassador

One should not forget that air traffic controllers, like bank staff, are vulnerable or might have their own agendas. In addition, ATC equipment on the ground could be subject to sabotage. In Japan, some activists managed to find out where the cables linking ground radar facilities passed and brought down Tokyo's ATC system by cutting them. In the following case, a US air traffic controller allegedly deleted data.

A report we found on the Internet, dated Thursday, January 31, 1980, that we have been unable to retrieve again to attribute properly and headlined "FAA Acts to Remove a Controller" said:

The Federal Aviation Administration moved yesterday to dismiss an air traffic controller for allegedly tampering with radar data and contributing to the "potential endangerment" of a Soviet airliner being guided to a landing at Kennedy International Airport last January.

The announcement of the FAA action said that "important flight data" on the Soviet plane, a four-jet Ilyushin 62 operated by the airline Aeroflot, had been "deliberately erased" as the aircraft approached Kennedy. Among those on board the plane was the Soviet ambassador to the United States, Anatoly F. Dobrynin.

There are ways a rogue controller—or worse, controllers acting in unison—could put aircraft in great danger, but no more will be said, as the author does not want this to become a terrorist's handbook.

1981 Industrial Action by US Air Traffic Controllers; Reagan's Response

Prof. Joseph McCartin, who has written a book, *Collision Course*, on the origins of the strike and the aftermath, which led to President Reagan firing eleven thousand air traffic controllers, says that in the lead-up to the strike, tension was brewing. New York controllers threatened not to handle aircraft from the Soviet Union or Iran, citing as their reasons the Soviet invasion of Afghanistan and the US hostages being held by Iranian revolutionaries.

Also in Miami, the FAA concluded that controller Ron Palmer directed a Braniff International Airways jet into a thunderstorm in retaliation for the airline's refusal to honor controllers' FAM flight requests. (FAM stands for "familiarization," the idea being for controllers to get the feel of what it was like to be in the cockpit. These were very popular with controllers but were curtailed following 9/11 for security reasons.)

There were no major accidents while new controllers were engaged and trained following the sacking, though some pro-union people say the Air Florida Flight 90 crash into the icy Potomac River in January 1982 can be partly attributed to the use of a less qualified controller.

Standing up to air traffic controllers is difficult. The French ones were and are notorious, with the country's location meaning they can cause wide and costly disruption. In 1973 an exasperated French government had military personnel take over, but the attempt to defeat the controllers was short-lived, for that very day there was a midair collision at Nantes, near the Atlantic coast, between aircraft of two Spanish airlines.

The union mocked the government, saying it proved that they, as they had always claimed, were indispensable.

747's Terrifying 30,000-Foot Plunge (Pacific, 1985)
Pilot whose mistakes caused initial upset then saved them twice over

> An airliner is somewhat like a two-legged human walking assuredly and
> stably. When either trips up, the fall can be precipitous and recovery not
> easy, especially when unable to see the ground. However, airliners often
> have much more height to play with.
>
> *China Airlines Flight 006, February 19, 1985*

China Airlines, with the call sign "Dynasty," is not the flagship airline of the People's Republic of China but that of the offshore island Taiwan.

Their flight from Taiwan to Los Angeles was the regularly scheduled one in a Boeing 747SP (special performance), designed for what at the time were considered ultra-long-haul flights. For such a long flight there were two crews, but the captain and first officer from the primary crew were back in their seats as it approached the US coast, with the captain having returned early from his rest, saying he had found it difficult to sleep.

Everything was going smoothly, with the performance management system and autopilot maintaining the aircraft on its course at Mach 0.85. They were flying in what is referred to as "coffin corner," namely at a considerable height, where there is relatively little margin between going too fast, in which case air over some parts of the aircraft would exceed the speed of sound, (Mach 1) producing shock waves, and going too slowly in the thin air, causing the aircraft to stall. Normally, with the autothrust constantly making minor adjustments, this posed no problem, though the pilots would always be aware that maintaining the correct airspeed was essential.

As they approached reporting point Redo, about 300 nautical miles (550 kilometers) northwest of San Francisco, they encountered some clear air turbulence, which fortunately, as proved later, prompted the captain to tell the passengers to fasten their seat belts. As the airspeed fluctuated between Mach 0.84 (251 KIAS) and 0.88M (264 KIAS), the autothrottle moved the throttles forward and aft to maintain the commanded cruise Mach number (0.85 M).

When the airspeed again rose to Mach 0.88, the autothrottle reduced power, and then when the airspeed fell to Mach 0.84 immediately applied more. Engines one, two, and three responded, while engine number four did not. The flight engineer moved the throttle for that engine back and forth, with no perceptible response, until it flamed out.

The captain asked for the performance charts to ascertain the three-engine en route cruise altitude and told the first officer to request permission for them to fly at a lower altitude. The flight engineer began going through the checklist in preparation for a restart at lower altitude but was cut short by the captain ordering him to restart it immediately, even though doing so was virtually impossible at that altitude and contrary to SOP (standard operating procedure), which said a restart should only be attempted below thirty thousand feet.

The Boeing 747 can fly perfectly well on three engines and easily stay up with two, so this was by no means a crisis and indeed something that happened from time to time. However, with one engine, and particularly an outboard one, not working, thrust was asymmetric, with the aircraft so much wanting to yaw to the right that application of the rudder was required to compensate. The autopilot on the 747, unlike the Airbus A330, had no control over the rudder, as moving the ailerons was largely sufficient for making normal turns and compensating for any yaw. The autopilot was therefore struggling to keep the aircraft on course, and even with the control column controlling the ailerons hard over to the left, the aircraft was tipping and turning to the right.

Instead of immediately taking over manual control, the captain concerned himself with the restarting the number four engine—something impossible at that great height. With one less engine, the aircraft was losing speed and descending into the clouds.

When the captain belatedly assumed manual control, the right bank was already reaching sixty degrees, and since he failed to apply left rudder and was ill prepared, the aircraft lurched even farther to the right and tipped downward. So focused was he on his airspeed that he did not monitor the attitude indicator, and when he turned his attention to it he found it unbelievable—something he had never seen, as in a 747 the maximum might be forty-five degrees, and they were way beyond that—and exclaimed it had failed. Faced with all the attitude indicators showing "crazy" readings, the crew believed they had failed, something highly improbable.

In cloud and without recourse to their attitude indicators, the pilots were at a loss as the aircraft plunged downward.

The terrifying drop was in two parts, with the second part worse than the first. The first part, a 10,310-foot descent to 30,132 feet in thirty-three seconds, where the aircraft seemed to stabilize somewhat, and the second part, from 30,132 feet to 9,577 feet in one minute fifty-two seconds of greater hell.

The 747 had broken out of the clouds at 11,000 feet (3,400 meters), allowing the captain to see the horizon and, by surprising airmanship, bring the plane under control, leveling out at 9,600 feet (2,900 meters). However, this subjected the passengers to tremendous g-forces. Engines one, two, and three, which the crew had mistakenly assumed had flamed out, appeared to be working again, the problematic engine number four restarted.

With all four engines working normally, and the aircraft seeming to be in good order, they sought permission to climb and proceed to Los Angeles. However, the inboard main landing was down and could not be retracted because of failure in one of the hydraulic systems. The extra drag meant they would not have enough fuel (and reserves) to reach Los Angeles, so they requested permission to land at San Francisco and were allotted a straight-in landing, which was fortunate, as the elevators in the tail had been compromised, probably by part of the main wheel bay, which was blown off when the aircraft exceeded its maximum airspeed during the drop.

By another dint of good airmanship, the captain brought the 747 in to a perfect touchdown, adjusting engine power to control the rate and angle of descent.

All 251 passengers and 23 crew had survived, with 22 minor injuries and two serious casualties, one of whom was a male flight attendant, who spent just over two days in hospital with acute back strain, while the other was a passenger with lacerations and bone fractures on his right foot—serious injuries are officially defined as "any injury which (1) requires hospitalization for more than forty-eight hours, commencing within seven days from the date the injury was received; (2) results in a fracture of any bone (except simple fractures of fingers, toes, or nose)."

Passengers Terrified

This must have been one of the most frightening passenger experiences ever. With the aircraft diving vertically and at various angles, and unattached passengers and flight attendants floating to the ceiling at times or pressed down or up with forces reaching 5 g at others, it was so traumatic that some could not remember the details when interviewed; others cite the odor of urine and food, as noodles and dishes flew around, since smells link one to events.

While mentally passengers were certain they were going to die, physically they were like astronauts under training being given the experience of weightlessness with no hard impacts. Compared with passengers hitting the ceiling panels on other airliners caught up in clear turbulence, it was relatively benign, though the fact that so many had their seat belts fastened helped considerably.

The Aircraft

Damage to the aircraft caused by the overspeed and parts breaking off and striking others was much greater than the pilots realized.

For instance, the elevators in the tail had been struck by the broken-off cover of the main wheel bay. With a large segment missing, the captain had had to land the aircraft by varying engine power to adjust the pitch on coming in to land.

The wings were bent permanently two to three inches upward at the wingtips, but within the manufacturer's allowable tolerances. This may seem a lot, but in view of the length of the wings did not represent a great degree of distortion. Indeed, after repairs, the aircraft remained in service with the company for more than eleven years before being transferred to other owners.

The Investigation

To protect pilot privacy, the cockpit voice recording was overwritten every thirty minutes, so there was nothing of relevance left when power was switched off on arrival in San Francisco. Investigators had to fall back on interviews with the aircrew. While the cockpit data recorder proved most helpful, unraveling it took considerable time, as there were gaps where it had undergone the exceptional g-forces during the precipitous descent.

The near calamity had been so unnecessary, for if the captain had followed standard operating procedure and descended to thirty thousand feet before trying to restart the number four engine, there would have been no problem.

He had never experienced an in-flight engine failure, though had coped with one on takeoff in the simulator. Although not mentioned in the official report, it is alleged elsewhere that he was not aware that the autopilot had no effect on the rudder. However, the Boeing training manual for the 747 had very precise instructions regarding the actions to be taken in the event of an engine failure, with particular regard to the use of the rudder and the setting of the rudder trim so the aircraft would fly straight without recourse to use of the ailerons. Since the captain had so many flying hours (15,494, of which 3,748 were in Boeing 747s), those training manual instructions were possibly a distant memory.

The investigators concluded rather that the captain's initial mistakes were due to fatigue, which he himself would not admit, as it was his natural time for sleeping and in the days prior to the flight he had flown several sectors in different time zones.

Though the precipitating factor was the hanging of the number four engine, the probable cause of this incident was the captain's preoccupation with an inflight malfunction and his failure to monitor properly the airplane's flight instruments, which resulted in his losing control of the airplane. The contributing factor was the captain relying for too long on the autopilot instead of getting the feel of the aircraft early on before things got out of hand.

747 Cargo Hold Door Opens Midflight (Honolulu, 1989)
Dogged persistence of parents of victim helped make NTSB reconsider

> The incident described below, involving the eighty-ninth production-line Boeing 747, manufactured in late 1970, shows that short circuits not only present a fire risk but also the risk of triggering something more unexpected—in this case, the unlocking of a cargo hold door.
> Highlighted the danger of investigators concluding in any investigation the most probable cause is the cause.
>
> *United Airlines Flight 811, February 24, 1989*

The aircraft used for United Airlines Flight UA811, from Honolulu to Auckland, New Zealand, on February 24, 1989, with 337 passengers on board, was an old Boeing 747 that had seen considerable service. Honolulu was essentially a refueling stop—the airline also had a nonstop flight from Los Angeles to Auckland, for which they used a newer long-range 747.

The aircraft received clearance to take off at 1:52 a.m. local time and climbed out as usual, except for deviating somewhat from the normal course to avoid some thunderstorm cells, their presence ultimately proving fortunate, as they made the captain keep the "Fasten Seat Belt!" signs switched on.

Sixteen minutes into the flight, with the aircraft having passed twenty-two thousand feet on its climb to its thirty-thousand-plus cruising height for the long transpacific leg to Auckland, the aircraft juddered, and the aircrew heard a thud that appeared to have come from behind and below them. There followed a tremendous explosion, which for a moment tilted the aircraft leftward.

The roar of the wind became unbelievably loud, and the misting of the air confirmed that the aircraft had suffered a rapid decompression. Failure of the lights added to the fears of many that this was to be another Lockerbie. However, the emergency lighting soon kicked in, and people began to realize the aircraft was not (at least for the time being) in the process of breaking up. The cabin pressure warning horn prompted the crew to put on their oxygen masks, but to their consternation they found there appeared to be no oxygen.

In putting the aircraft into an emergency dive to get to a level where they could breathe more easily, the captain might, as in the Aloha case, have been putting the integrity of the aircraft at risk. At the time, there was no way he could have known the extent of the damage. In fact, even without an emergency dive the flight crew should have been able to stay operational until reaching a reasonably comfortable height.

The first officer entered 7700 (the emergency code) into the transponder and called Honolulu Approach Control to formalize the emergency. While able to communicate with far-off ATC, the pilots could not communicate with the cabin crew just behind them. Meanwhile, the instruments were showing the number three engine was failing and would have to be shut down. After that was completed, the flight engineer went back to inspect the damage.

In the early 747s, the upper cabin behind the flight deck was quite small. The engineer was astounded to see a three-meter-wide section of the skin below the windows had separated from the stringers and formers on the right-hand side and open sky visible between the gaps. This accounted for the noise of the wind in the cockpit. Worse was to come.

On descending to the larger lower-deck business-class cabin below, the flight engineer saw that there the stringers and formers had disappeared as well, leaving a three-meter-wide gaping hole extending from floor level to the ceiling. What was more, the floor had failed just above the cargo hold door below, and the five rows of twin seats that had been attached to it there had disappeared, presumably together with a number of passengers, as the aircraft had been quite full.

The remaining business-class passengers had tried to move back to the main cabin, where the scene was almost unimaginable. People were screaming, some were hysterical, and others were already in the brace position, waiting for the aircraft to crash. Because the volume of air in the main passenger cabins had been so much greater, the effect of the mass of air rushing forward when the sudden decompression occurred had been even greater too. The escaping air had hurled meal trays and glasses with such force that some were even embedded in the bulkheads. It had also dislodged wall and ceiling panels, and there was dust and debris everywhere. The projectiles had injured a number of passengers, and even those who were unscathed seemed distressed because of the wind and the stress.

The cabin crew valiantly tried to get the situation in hand, but the noise of the wind made communication with the passengers virtually impossible. One positive was that having descended to a lower altitude, the passengers, whose oxygen masks were not working, were coping.

In assessing the damage, the flight engineer was particularly concerned to see long streaks of flame emanating sporadically from the number four engine (the one on the extreme right). The violence of the decompression had ejected debris, or possibly a passenger, not only into the number three engine, which had been shut down, but even as far as the outboard engine. Hurrying back to the flight deck, the flight engineer found his colleagues waiting for him to assist with the shutdown of that engine, which they duly carried out.

With engine power on only one side, the 747 limped back toward Honolulu, with the reduced speed having somewhat lessened the disturbing noise of the wind. Twenty minutes later the distraught and terrified passengers had their spirits raised by the sight of Honolulu, which they had left shortly before.

When the pilots lowered the flaps to land, a flap asymmetry appeared, and the captain had to limit the amount of flap to ten degrees. He decided to come in at 190 to 200 knots not only because of the limited amount of flap, but also because the aircraft might be difficult to control at the normal lower speed due to damage to some control surfaces, having both engines on one side shut down, and the gaping hole. He made a perfect landing and managed to bring the aircraft safely to a halt using the wheel brakes and idle reverse thrust on the two good engines.

The captain ordered an emergency evacuation, in the course of which some passengers—and surprisingly, it is said, all the cabin crew—sustained some form of injury. A number of the passengers had been injured when the decompression occurred.

Nine business-class passengers had been ejected altogether—eight in the seats they were sitting in and the ninth, who had been sitting just across the aisle from that block of seats, without his seat, making one wonder whether he was properly strapped in.

Things could have been much worse, and the fact that the aircraft remained intact is perhaps a credit to Boeing, especially as it was an early-model 747. Both the flight crew and the cabin crew performed well in very difficult circumstances. Valuable lessons were learned, such as ensuring the pilots and cabin crew are guaranteed to have oxygen available; the oxygen requirement is much greater for someone engaged in mental activity as well as physical.

The Investigation

It was immediately obvious that the untoward opening of the cargo hold door on the front right-hand side was responsible for the explosive decompression and associated structural damage. Cargo hold doors, as in the case of those on the DC-10, have often been problematic on high-flying jets as in order to leave more room for cargo they often open outwards, unlike the plug doors in the passenger cabins that seal more tightly the higher the aircraft.

An extremely complicated mechanism, consisting of pushrods, cranks, cams, and latches to mention only a few of the items, is required to cope with the width. In addition, on large aircraft an electric motor is usually incorporated to do the basic locking and unlocking, with manual locks completing the process. There should be no danger of the unlocking function operating in flight, as the motor uses electrical power only available on the ground either from the APU (auxiliary power unit) at the rear of the aircraft, or from the airport electricity supply.

Without the actual door, the NTSB felt they had to surmise what happened and produced a report concluding that since in their view a normal door could not open in flight, the locking mechanism must have been damaged by poor ground handling as alleged in the case of a cargo hold door on a 747 that took off from London's Heathrow in 1987.

Parents of Victim Pursue Their Own Investigation

It is rare for relatives of victims to have any voice in accident investigations, let alone carry out their own investigation. Yet this is what happened here, and led to the NTSB revising its conclusions.

Kevin and Susan Campbell, parents of Lee Campbell who was one of the passengers sucked out of the business class cabin—they expressed the hope he was sucked into the inboard engine and did not have to suffer the long fall to earth—followed the investigation from the outset. Having money and time, they went to Honolulu and travelled the country to get technical information, and badgered the NTSB. When the latter held a public meeting to explain their conclusion that the door had been damaged by poor ground handling the Campbell's were so incensed that, when told they could take some documentation, they in doing so scooped up a considerable amount of material intended for the NTSB's internal use. This provided many useful leads, enabling them to finally persuade the Board to reconsider and prove the matter one way or the other by retrieving the door.

The incident had occurred in February 1989, and in September the following year by an incredible feat of technology the US Navy recovered the cargo door (in two pieces) from a depth of 14,200 feet. The recovered door showed it had been unlocked electrically and the manual locking pins had insufficient strength to prevent the door opening. This contradicted the April 1990 NTSB report, which concluded the electrical unlocking of the door prior to takeoff could only be considered a remote possibility.

This shows the danger, in the absence of definitive evidence, of assuming without any proof that what was the most probable cause by far was actually the cause. The NTSB published a correction.

Improvements, too late for those nine business-class passengers, were subsequently made, and there has not been an accident since due to the opening of a Boeing 747 cargo door.

China Airlines and El Al 747F Engine Mounts (1991 and 1992)
Boeing redesigns engine mounts at great expense

> It is not so widely known that Boeing spent a considerable amount of money redesigning the engine-supporting pylons on its early 747s and refitting many aircraft following two crashes of the cargo versions.
>
> *China Airlines 747 Freighter, December 29, 1991*
> *El Al 747 Freighter, October 4, 1992*

The supporting pylons for engines mounted under the wings of airliners are usually designed to allow the engine to possibly fall off in certain circumstances, such as when exceptional torque is produced by an engine seizing up; when an engine is seriously on fire; and when ditching in water, so the wing is not torn off, or so the aircraft does not spin around disastrously and cartwheel.

On December 29, 1991, the inboard number three engine came off an eleven-year-old China Airlines Boeing 747-2R7F freighter as it was climbing out under high power from Taiwan's Chiang Kai-Shek International Airport. Because it was under high power it initially shot forward before falling back and glancing off the outboard number four engine set farther back, which in turn also fell off. The aircraft subsequently crashed, killing all five occupants.

Though the number three engine was retrieved from the sea some distance from the main debris, the pylon was only discovered seven months later.

Nine months after that disaster, exactly the same thing happened to an El Al Cargo 747 climbing out of Amsterdam's Schiphol Airport. Besides knocking off the outboard engine the number three engine took away a large section of the leading edge of the wing. The captain just about managed to keep the aircraft under control with limited aileron function and were able to fly on. However, when they slowed the aircraft to return to the airport to make an emergency landing, the damaged wing stalled, and the aircraft rolled over very much like the DC-10 did on slowing after losing an engine at Chicago. The aircraft plowed into an eleven-story apartment complex in Amsterdam's down-market district of Bijlmer.

The last words from the crew were disturbingly matter-of-fact:

Going down . . . eh . . . 1862, going down, going down. Copied, going down?

In addition to the four occupants of the aircraft, forty-seven people were killed on the ground, with the total perhaps more, as the building housed a number of unregistered illegal immigrants. The disaster immediately received considerable publicity, because the dramatic pictures probably made people consider just how much damage an aircraft plunging into a building can cause.

The mystery surrounding the nature of the aircraft's cargo was to give the story legs. Amsterdam's Schiphol Airport was one of several European airports allowing El Al to transship cargo unsupervised, and as a result the details of the manifest were not public knowledge. After claiming the freighter had been carrying innocuous flowers and perfume, the Israeli government eventually admitted that a precursor for making nerve gas (sarin) had been on board, but claimed it was for testing filters, and the amount was small. It was a very sensitive matter, especially as the aircraft had come from the United States and it would look bad if the Americans were seen to be providing Israel with materials for chemical warfare. Depleted uranium was also present.

Residents reported seeing men in white suits sifting through the wreckage.

People, and notably rescuers in the area of the crash, where the debris burned for an hour and smoldered for several more, reported all sorts of ailments and depression, and there were accusations of a cover-up. As in the case of 9/11, people would anyway have been affected by the untold number of constituents in the cloud of smoke and dust when an aircraft, laden with fuel just after takeoff, hits a building. The nature of the cargo is not very relevant here, except to remark that some very nasty things probably pass over our heads from time to time in aircraft cargo holds and diplomatic bags.

A fourteen-inch crack was found in one of the fuse pins used for attaching the engines on the El Al freighter and designed to shear as a safety measure under the special circumstances mentioned above, but not in normal service. Such cracks can develop when metal is under repetitive flexing conditions, especially if there is an initial pit on the surface due to corrosion, in a phenomenon called *stress-assisted corrosion*, where the stress facilitates the penetration of the corrosion. The initial pit would probably have been due to the surface being left vulnerable to corrosion (passive) due to lack of primer or faulty coating during manufacture. Incidentally, shortly before, Boeing had issued warnings concerning the pins used for attaching the Pratt & Whitney engines in question.

Following these two incidents, Boeing notified users about the necessary checks and then redesigned the entire mounting system. Modifying existing aircraft cost the company a considerable amount of money, as it involved a lengthy task, said by one commentator to be forty days. In addition, a considerable number of susceptible 747s had been delivered.

This accident with so few people on board made the Boeing 747 safer; unfortunately, there were ten times more victims on the ground. Siting or expanding airports where aircraft have to overfly cities is obviously not ideal.

In-Flight Deployment of Thrust Reversers (Lauda Air and TAM)
Only meant for slowing aircraft on the ground

> In-flight deployment of a thrust reverser, especially when the engines are at high-power settings, can be disastrous, for you have one engine pushing the aircraft forward and the other pulling it quite powerfully backward.
>
> *Lauda Air 004, May 26, 1991*
> *TAM Flight 402, October 31, 1996*

Lauda Air 004

Start-up airlines on long-distance routes with, say, less than half a dozen sizable aircraft are loath to take any out of service for nonessential checking, as, with no replacement, this can mean canceling flights or even services outright. For instance, when a Lauda Air 767 crashed in Thailand on its way back to Vienna, the airline lost a quarter of its capacity and had to temporarily cancel some services to Australia.

Though not mentioned in the TV documentary *Mayday*, featuring and indeed eulogizing the airline's founder, Niki Lauda, sixty-one error messages concerning that 767's thrust reverser were found in its computer. Even though there was a Boeing representative at Lauda Air's Vienna home base, he was not consulted.

The 767, with 213 passengers and 10 crew, had taken off for Vienna from Bangkok's Don Mueang Airport at 23:02 local time. Four minutes later the pilots received a visual warning concerning the thrust reverser on the left-hand engine, but their documentation said it was just advisory, perhaps meaning there could be problems with it on landing, and they continued regardless. Eleven minutes later, with the copilot saying the reverser had deployed, the aircraft tipped to the left and plunged downward so fast that it broke up at about four thousand feet and came down in an isolated mountainous area six hundred meters above sea level. Debris was spread over a square kilometer of forest. There could be no survivors.

The Thais sent in their own investigators, and they were joined by a team from the NTSB led by Bob MacIntosh, who found that local people were removing items that could be essential for the inquiry. The black boxes were found, but the cockpit data recorder was too damaged by fire to be of any use; however, the voice recorder could be read.

The left engine was found with the thrust reverser obviously deployed, so with confirmation from the voice recorder that it had been, there was no doubt that was the cause of the crash. But without the data recorder evidence, and with the wiring either burned or stolen, there was no way to determine why it had deployed. To obtain certification from the FAA, Boeing had tried deploying the thrust reverser in flight and found recovery possible, but it was later found this had been with a pilot expecting it and at the relatively low airspeed of 250 knots. Acknowledging such an occurrence could be fatal at higher speed or with no expectation. Boeing sent telexes to all users and incorporated extra safety devices, such as locks, not only on the 767 but also on other aircraft.

TAM Flight 402

The Brazilian TAM air crash occurred five years later, when manufacturers were much more aware of the danger of in-flight deployment of a thrust reverser. The makers of the popular Fokker 100 twin-jet came up with an ingenious way to ensure that deployment of a thrust reverser, especially on takeoff, would not cripple their aircraft. They simply had a cable link the thrust reverser to the thrust lever so that if it did accidentally deploy it would pull the lever back to cut power to that engine. As twin-jets are designed to be able to easily take off with one engine, there would be no problem.

As the TAM 402 Fokker 100 took off from Sao Paulo, the captain's attention was drawn to a warning that the autothrust was not working properly, in itself not serious enough for him to reject the takeoff, as the levers could be worked manually. Perhaps because of an intermittently faulty micro-switch and a short circuit, the right engine thrust reverser deployed, causing the aircraft to tip and veer to the right, but just when it looked bad the captain regained control thanks to the safety mechanism pulling the throttle back and cutting power to that engine.

However, the copilot, thinking the lever had come back on its own, pushed it forward, and again the aircraft misbehaved before the safety system operated. The captain, with the autothrottle warnings in his mind, ordered for it to be switched off. The copilot then forcefully held the right-hand throttle lever forward to prevent it returning, putting so much strain on the cable that a link in the middle (there so that it could be disconnected for servicing) parted.

The aircraft then plunged, with the right wing down, into a housing area just beyond the airport, killing all ninety-six occupants and, surprisingly, only three people on the ground, despite raging fires from burning fuel. The aircraft manufacturer had considered such an event so unlikely it did not insist on training for such a mode.

Captain's Son Disengages A310 Autopilot (Russia, 1994)
Failure to notice partial disengagement of autopilot

> The captain on this flight allowed his teenage son to sit in his seat for a
> moment.
> Full details only became generally available thanks to diligent research
> by Macarthur Job and his assistants. The following distillation is based
> (with permission) on his account in *Air Disaster* (Vol. 3).
>
> *Aeroflot Flight 593, March 23, 1994*

The Aeroflot A310 was on a regular scheduled flight from Moscow to Hong Kong.
The three-man flight crew consisted of Captain Danilov, the commander, Captain
Kudrinsky, who was the reserve commander, needed for such a long trip, and
Second Pilot Piskarev. In addition, there were sixty-five passengers and nine
flight attendants. Among the passengers was an off-duty captain.

The presence in the passenger cabin of Captain Kudrinsky's family, and
notably his son, El'dar, fifteen, and daughter, Yana, thirteen, would make the trip
more memorable than his usual flights. He would be able to relax with them
before being called to the cockpit to take over from Captain Danilov, who was
flying the most difficult sectors out of Moscow—near sensitive areas and where
there was always considerable traffic. In accordance with regulations, medical
staff at Moscow checked the pilots' blood alcohol levels and blood pressure
before the flight, and they recorded "nothing untoward." The presence of alcohol
was not given as a possible contributory factor, though there were suggestions
that the captain was drinking with his family. One story said he struggled to get
back to the cockpit from the cabin.

When Captain Kudrinsky finally took over from Captain Danilov, with
Piskarev still acting as copilot, it seems that the latter, a short man, at some point
pushed his seat right back, either to relax after his hard work or to get up for a
moment, and left it pushed back.

Just as they were about to pass near the city of Novosibirsk, in the middle of
Siberia, Makarov, the off-duty captain, brought the two kids to the cockpit, no
doubt knowing it would be the last straightforward sector before crossing the
frontier into Mongolia.

Vacating his seat, Captain Kudrinsky allowed his daughter to take his place,
adjusting the seat's height so she could see out of the windscreen. He told his
daughter to hold the controls and adjusted the heading selector on the autopilot
slightly, first in one direction and then in the opposite direction. In the girl's
gentle hands, the control column duly moved as the autopilot banked slightly to
adjust the course to the selected heading. The idea was to let her get the feel of

flying the aircraft. Evidently not too enthralled, she soon gave up her seat to her eager elder brother.

El'dar took her place, with Makarov filming the event with a video camera. Instead of being passive like his sister, El'dar, having taken in what his father had said about how the controls worked, asked whether he could move the control wheel. His father said he could, whereupon El'dar sharply turned it three to four degrees to the left before his father could anticipate his move by turning the autopilot heading selector, as he had done for the young girl. Kudrinsky then returned the selector to NAV (automatic navigation) mode, and the aircraft banked the other way to resume its preset track.

By exerting a very positive force on the control column, the boy caused the autopilot to disengage in the banking axis. (As already mentioned in the context of the TriStar Everglades crash, manufacturers program autopilots to disengage if a force exceeding a certain limit is applied to the control column.) This meant that a strong boy with one minute's flying experience and no training, and Piskarev, who was short and had his seat so far back he was almost out of reach of the controls, were the only people controlling the banking (rolling) of the aircraft.

At the time the A310 did not have any system, aural or visual, to warn pilots of such an autopilot disconnect, probably because the manufacturer thought it could only happen if a pilot moved the control column sharply, in which case he would know what he had done and would be aware of what was happening. With none of the professional pilots in the cockpit realizing what had happened, El'dar kept pushing the control column (yoke) to the right, perhaps because he was right-handed. In consequence, the aircraft increased its right-hand bank from fifteen to twenty degrees in seven seconds. However, with Kudrinsky chatting to his daughter, and Piskarev assuming the autopilot was doing everything, no one noticed the gradual increase in bank, which due to the turn would at night only have been evident from the artificial horizon.

With the bank progressively increasing and attaining forty-five degrees forty seconds after Kudrinsky, expecting the aircraft would look after itself, had reset the autopilot to NAV, it was El'dar who realized something odd was happening. He interrupted his father to ask why the aircraft was turning, something he also could only have determined from the instruments.

Asked by his father whether it was turning by itself, El'dar said it was, perhaps not realizing the effect of his hand on the control column. The three proper pilots then launched into a discussion as to why the autopilot was making the aircraft turn on its own and concluded that it must be a holding pattern. Soon the bank had attained fifty degrees, still without the three pilots realizing anything was amiss.

Pilots, learning to fly on tiny aircraft, know that when an aircraft banks steeply, the amount of lift decreases. In consequence, they increase engine power and raise the nose when doing so to avoid sinking. Likewise, the A310's autopilot increased the engine power and raised the nose in order to maintain its programmed altitude and speed.

It was Makarov, the off-duty captain, no doubt feeling the g-forces on his feet, who noticed something was wrong. No sooner had he uttered a warning than the aircraft, in order to maintain its height, pitched upward. It then began buffeting, a sure sign it was about to stall.

Kudrinsky called out:

Hold the control column. Hold it!

Piskarev immediately turned his control column yoke to the left to counter the increasingly precipitous rightward bank and would no doubt have succeeded in righting the aircraft had El'dar not also followed his father's instructions, probably intended for Piskarev.

Not knowing what action to take, the boy gamely clutched the yoke, maintaining it firmly in an almost neutral position. This resulted in an unequal battle between the boy ensconced in the perfect position and the ill-placed Piskarev. The latter was only five foot three inches (160 centimeters) tall, and with his seat right back, his shoulder harness would be keeping him at arm's length from the control column and unable to reach the rudder pedals. As a result, Piskarev's efforts were insufficient to halt the dangerously increasing bank, which, by causing the aircraft to lose lift, was making the autopilot raise the nose farther to maintain height. Finally, Piskarev must have knowingly overridden the autopilot by pushing the control column sharply forward, thus causing it to disengage completely.

This time there was an aural warning that the autopilot had disengaged. This was followed first by the nonurgent warning that the aircraft had departed from its designated height and then by the more pressing warning of an imminent stall. With these warnings coming in quick succession and increasing stress levels, all three pilots started shouting conflicting instructions to El'dar. If the boy had had even minimal flying experience, he might have known instinctively what to do to put the aircraft on a level keel. However, with the general confusion and the situation becoming increasingly scary, there was not much hope. He was probably transfixed.

Meanwhile, the Alpha floor safety function (mentioned in connection with the Habsheim air show Airbus crash) sensed that the aircraft was going into a stall and pushed the nose down by adjusting the tail trim just as Piskarev was trying to do the same by pushing the control column forward. This put the aircraft into a forty-degree dive that was to last thirteen seconds. With the engines still supplying normal cruising thrust, the airspeed continued to build up alarmingly even after the overspeed alarm had sounded.

Piskarev yanked back the control column to pull the aircraft out of its precipitous dive just as Kudrinsky, trying to get back in his seat, was shouting to the boy, "*Get out! Get out!*" The powerful g-forces resulting from the aircraft beginning to level out multiplied the weight of the boy, making it even more difficult for him to clamber out.

With the aircraft leveling out and going into a climb, Piskarev, no doubt concerned by the overspeed warning, called for the engine thrust to be reduced regardless of the fact that he was holding the control column fully back. Why he kept the control column fully back is difficult to comprehend. There have even been suggestions that he was using it to pull himself and his seat forward. Meanwhile, the reduction in g-forces had allowed El'dar to vacate his seat and let his father slip in, albeit with it in the back position, which was not to help matters later.

With full-up elevator and airspeed of only ninety-seven knots, the aircraft slewed to the right and then slipped into a downward spin to the left. Belatedly Piskarev asked for full throttle but having the elevators still in the full-up position, this was not destined to help recovery. As the aircraft reached two hundred knots, it went into an uncontrolled spin that was to last for a minute.

With the aircraft in a twenty-degree dive and the fully back control column(s) keeping the aircraft in a semi-stall, Kudrinsky, now back in his seat, was finally managing to get out of the spin using the rudder.

So Near Salvation

They leveled out with only a little over 1,000 feet (350 m) of altitude remaining, which would have been fine had they not at the time been sinking at a rate of 13,870 feet a minute (70 meters a second). With insufficient time to arrest their descent, they plunged onto the trees on a snow-covered slope of a 2,000-foot (600-meter) hill some three hundred nautical miles southwest of the Siberian capital, Novosibirsk. Rescuers found the fire and explosion impact had destroyed much of the wreckage—not that anyone could have survived the impact anyway.

Had it not been for a cockpit video recording taken by the off-duty captain, it would not have been possible to determine exactly what happened, and particularly who did what.

Autopilot Takes Unconscious Pilots to Destination (Greece, 2005)
Flight director takes hapless pilots up and onward

> Not realizing cabin pressurization was set to "Manual" when they took off, the 737 pilots lost consciousness due to lack of oxygen as the flight director and autopilot took them up to cruising height and onward to their destination, Athens.
>
> *Helios Airways Flight 522, August 14, 2005*

At 9:00 a.m. local time, the Helios Airways Boeing 737-300 took off from Larnaca, Cyprus, for Athens, where it was to make a stopover on its way to Prague. On board were 121 people, including the 2 pilots and 4 cabin crew.

As the aircraft climbed through ten thousand feet, the cabin altitude (air pressure) warning horn sounded. As the same warning sound is also used to indicate a configuration mistake, such as failure to extend flaps at takeoff, the pilots assumed it was merely a false alert. They even discussed the problem with their maintenance people at Larnaca, but with some difficulty, as the captain was German, the first officer Cypriot, and the maintenance engineer English.

On the advice of maintenance about how to silence the irritating horn, the captain left his seat to trip the relevant circuit breaker behind his seat, but due to the extra exertion and lack of oxygen, he collapsed before he could do so. The last communication received by ATC was eleven minutes after takeoff as the aircraft was passing through twenty-two thousand feet, the height at which lack of oxygen renders one fully comatose.

The aircraft continued to climb to its cruising height of thirty-four thousand feet and continued toward Athens, as programmed in the flight management control system (FMCS). As it approached their airspace, Greek air traffic controllers were surprised at not being able to communicate but were initially not too concerned, as they had heard about the air-conditioning problem the pilots had mentioned only four minutes after takeoff from Larnaca. Becoming concerned that some sort of hijacking might be involved, they alerted the Greek air force but told them to hold off, thinking the Helios flight was sorting out its technical problems.

Finally, at 10:55 a.m. two air force F-16s were scrambled. On intercepting the aircraft, they reported that the first officer was slumped over his seat and that the captain's seat was unoccupied. They then said two people had entered the cockpit wearing oxygen masks. The pilots also noticed that the passengers' oxygen masks had deployed.

The cockpit voice recorder later showed someone with a very labored male voice had made Mayday calls, which the controllers had not picked up, no doubt because the frequency was still set for the point of departure.

The first thing a professional airline pilot taking over control would have done would have been to bring the aircraft down to a height where breathing would have been easier. However, the pilot in question was flight attendant Prodromou, a private pilot with only a few hours' experience on a Cessna. Being unfamiliar with the 737 controls, he might well have wanted to seek advice over the radio before performing a maneuver from which he might not be able to recover, and there is evidence he hesitated for a long time.

Greek investigators later said Prodromou, together with a female colleague, used an emergency oxygen kit, opened the cockpit door using a code, and managed to fly the plane for ten to twelve minutes before it crashed from lack of fuel.

Postmortem examinations carried out after the accident showed that many of the passengers were alive when they hit the ground. Of course, unlike the pilots, they had the benefit of the oxygen masks that deploy automatically, and while these only intended to supply oxygen long enough to allow the pilots to bring an aircraft suffering decompression down to a level where breathing is possible without masks, they would have helped.

The Time Line

The details of what happened on that flight are shown by the following time line, where the time in hours and minutes is the elapsed time from departure:

00 hours 00 minutes (09:07 Cyprus time)

Departure from Larnaca for Athens with an estimated flight time of one hour twenty minutes.

00:04

> Crew reports problems—cabin altitude warning horn sounds at ten thousand feet, but because the same horn also denotes Takeoff Configuration Warning, crew assume it is a false alert. At fourteen thousand feet, passenger cabin oxygen masks deployment lamp (presumably) illuminates in cockpit.

00:07

> "Master Caution" alarm sounds at seventeen thousand feet. Instruments show an apparent problem with the avionics cooling system, caused by the thinness of the circulating air.

> The German-born captain and Cypriot first officer spend five or more minutes discussing the problem of the horn with the

English maintenance engineer at their base on Cyprus, who says the horn can be silenced by tripping the circuit breaker on a panel behind the captain's seat. The captain leaves his seat, but because of the exertion he collapses before being able to trip the breaker.

The last contact with the aircraft was as it climbed through 28,900 feet.

00:11

Autopilot takes aircraft up to its 34,000-foot cruising height, where it levels out and flies on for *one hour* toward Kea Island.

01:13

Aircraft overflies Kea VOR at 34,000 feet and continues at that height, ready to turn into the Athens Airport Approach. It duly turns, but as there is no pilot to initiate descent, it stays at 34,000 feet, and the flight director, interpreting it as a missed approach, flies on before executing a roughly 290-degree turn to bring itself back to the Kea beacon.

01:31

Aircraft joins Kea holding pattern, making repeated circuits.

02:17

As the aircraft executes its sixth circuit in the Kea holding pattern, it is intercepted by two Greek F-16 aircraft. They report, "Captain not in his seat; first officer slumped over controls; passenger oxygen masks dangling."

02:42

Pilots of the F-16s see a male entering the aircraft's cockpit, sitting in the captain's seat, and trying to regain control of the plane.

02:43

Left engine flames out due to fuel depletion; aircraft loses height.

02:47

Cockpit voice recorder CVR records two Mayday messages transmitted with a labored voice. They are not received on the ground, as the frequency is still set to that for the departure point, Larnaca.

02:53

The right engine flames out at an altitude of 7,100 feet.

02:56 (12:03 local time)

The aircraft crashes into hilly terrain near the village of Grammatikos, about twenty miles northwest of Athens International Airport.

According to early reports, even though the fuel had essentially run out, the flammable brush and grass meant fires smoldered for several hours, making many of the bodies difficult to identify. However, the body of a female flight attendant was found in the cockpit area, and the blood of a male flight attendant with some general aviation (private pilot) experience was found slumped on the controls.

The medical report suggested that some passengers were alive, but no doubt unconscious, when the aircraft hit the ground. This is somewhat surprising in view of the fact that the passengers' emergency oxygen masks only supply oxygen for a limited time.

The final report by the Greek investigators made the following conclusions.

Direct Causes

1. Nonrecognition that the cabin pressurization mode selector was in the MAN (manual) position during performance of the preflight procedure, "Before Start" checklist, and "After Takeoff" checklist.

2. Nonidentification of the warnings and the reasons for the activation of the warnings (cabin altitude warning horn, passenger oxygen masks deployment indication, indication, "Master Caution") and continuation of the climb.

3. Incapacitation of the flight crew due to hypoxia, resulting in continuation of the flight under the control of the flight management computer and the autopilot, until depletion of the fuel and engine flameout, and subsequent impact of the aircraft with the ground.

Latent Causes

1. The operator's deficiencies in organization, quality management, and safety culture, documented diachronically as findings in numerous audits.

2. The regulatory authority's diachronic inadequate execution of its oversight responsibilities to ensure the safety of operations of the airlines under its supervision. Inadequate responses to findings of deficiencies documented in numerous audits.

3. Inadequate application of crew resource management (CRM) principles by the flight crew.

4. Ineffectiveness and inadequacy of measures taken by the manufacturer in response to the previous pressurization incidents in this particular type of aircraft, both with regard to modifications to aircraft systems as well as guidance to the crews.

Contributing Factors to the Accident

1. Omission of returning the pressurization mode selector to "Auto" after unscheduled maintenance on the aircraft.
2. Lack of specific procedures on an international basis for cabin crew to address the situation of loss of pressurization, passenger oxygen masks deployment, and continuation of the aircraft's ascent (climb).
3. Ineffectiveness of international aviation authorities in enforcing implementation of corrective action plans after relevant audits.

The investigators essentially said that while the maintenance people contributed to the accident by not setting the cabin pressurization mode to automatic, the pilots should have picked this up in three separate checks, and should have subsequently realized the nature of the problem, if not from the warning horn, then from the deployment of the passenger cabin oxygen masks and the Master Caution alert.

Flight International has said that Helios's lawyers claimed the pilots could not have made such mistakes, despite the nonvolatile memory in the pressurization mode unit having showed the controls were set to "Manual" during the flight.

The attempts by the male flight attendant with some piloting experience, in the presence of a female flight attendant, to save the aircraft were not really part of the inquiry. It seems they were only able to gain entry into the cockpit using a code very late on, by which time the oxygen they had was possibly running out, and the fuel certainly was. Had they been able to gain access earlier, and had they been able to communicate by radio for advice on how to fly the aircraft, it might have been a somewhat different story. (It has been argued by some that measures taken since 9/11 to make it impossible to gain access to the flight deck without authorization from the pilots could be counterproductive in such circumstances.)

The report criticized the regulatory authorities, saying this was not the first time such a problem had occurred and that those authorities should have insisted the manufacturer did more to prevent such an error remaining undetected.

Founded in 1999, Helios Airways changed its name to A Jet Aviation Limited in April 2006, while continuing operations with its two remaining Boeing 737-800 aircraft.

In *The Crash Detectives*, Christine Negroni says that a Helios mechanic and three executives facing criminal charges in the accident hired experts to help defend them. According to them, examination of scoring on the back of the selector knob showed it might have been pushed to the off position in the crash. That would have meant the crash was due to mechanical failure.

Her point is that one can never be sure, though one must wonder about conclusions drawn by those with an interest in the outcome. Even then they only said "might."

Chapter 15
MILITARY ACTION

Korean Airlines Flight 902 Off Course (Kola, Russia, 1978)
Russian military shocked at its vulnerability

> Some passengers had noticed that the sun was on the wrong side of the
> aircraft but apparently thought the pilots must know what they were
> doing. Little did they realize they were about to be shot down over the
> Soviet Union.
>
> *Korean Airlines Flight 902, April 20, 1978*

The Korean Airlines 707, en route from Amsterdam to Anchorage, Alaska, had for some unknown reason deviated from its course soon after reaching Iceland. The sweeping deviation to the right was so gradual that it could not have been made manually, and the likely cause was a drift in the inertial navigation system, or the inputting of erroneous waypoint coordinates.

The Korean pilots' failure to notice they were on the wrong heading was surprising, when even a passenger had noted the sun was in the wrong position and led to suggestions that they were preoccupied with something extraneous, such as playing cards. Soon after penetrating Soviet airspace, they found a Russian fighter flying alongside on their right, signaling them to land.

Not that it made any difference. As the American space journalist and historian Jim Oberg pointed out, the ICAO (International Civil Aviation Organization) says the aircraft ordering another to land should fly on the left, perhaps because that is the side on which the captain always sits and he can thus more easily observe the indications of the intercepting fighter. Oberg goes on to say the Korean Airlines captain maintains he lowered his landing gear and flashed his navigation lights to indicate he was going to comply and follow.

Apparently, from the initial size of the blip on their radar, the Russians first thought it would prove to be a harmless Boeing 747, but when the fighter pilot announced it was a civilian 707, they had their doubts, as the same design under the designation RC-135 was used by the Americans for intelligence purposes.

Despite the strongest possible objections by the Russian fighter pilot, intercepted by the American listening posts, the general in charge ordered that the aircraft be brought down, no doubt on the assumption that the civilian markings were just a trick, for why would a civilian aircraft fly so far into Russian territory? (In his defense, it has to be said that the use of civilian aircraft or civilian-looking aircraft to provoke responses for analysis was not unknown.)

With great reservations, the Russian pilot fired off two missiles, the first of which failed to detonate, while the second exploded in the vicinity of the 707's left wing, damaging the fuselage and killing two passengers. The Korean pilot initiated a rapid descent to a height where the passengers could breathe without their oxygen masks, as is standard practice. In addition, the passenger cabin would have been getting cold quickly due to the damage to the fuselage.

In so doing, he entered low cloud and incidentally became invisible to the fighters, which had overshot him. On turning back, they failed to locate him, and, unaccompanied, the pilot flew on for almost an hour, looking for somewhere to land, before finally successfully bringing the aircraft down on a frozen lake.

The fact that a possibly hostile aircraft could fly low and undetected over its territory for such a long time came as a great shock to the Russian military. Though it realized there had been no intrigue, it feared the Americans might have learned much from the event and might exploit that newfound loophole for military purposes.

The passengers were quickly returned, and the crew somewhat later.

Korean Airlines Flight 007 Off Course (Sakhalin Island, 1983)

Korean Airlines again provokes paranoid Soviet defense forces

> The event previously described led to a big shake-up in the Soviet air defense system and recognition that inferior technology was making it vulnerable. Heads rolled, and the climate was such that all involved would do their utmost to ensure another rogue—perhaps a more sinister one—did not get through. This nervousness, and even the inferior technology, possibly led to a replay, with a much less happy conclusion.
>
> *Korean Airlines Flight 007, September 1, 1983*

On September 1, 1983, Korean Airlines Flight KAL007, en route from Alaska to Seoul, had a similar course deviation as that incurred by KAL902 five years earlier. However, this time the Boeing 747 had already passed over militarily sensitive areas of Eastern Russia and was heading possibly for the even more sensitive area around Vladivostok. The aircraft was eventually shot down by a Russian fighter just off the northern coast of Japan. All 269 passengers and crew lost their lives.

The United States, and notably President Reagan, made much of the fact that the "Evil Empire" had shot down an innocent airliner. Some have expressed doubts about what happened, and books, seemingly enthusiastically researched, have raised all sorts of possibilities, with most of them making the States at the very least look devious.

Two TV films were made about the incident, but they appeared before more information became available after the opening up of the Soviet Union. The first, *Shootdown* (1988), starring Angela Lansbury, was essentially about a mother (Lansbury) who had lost a son on the flight trying to get the truth out of the Russians, and not least the US government, via a kind of deep throat contact. ("Deep throat" was a term used to describe the official leaking information to journalists Bob Woodward and Carl Bernstein during the Watergate affair. A highly placed informant.) The second, *Tailspin* (1989), was more of a documentary, using the facts then available.

One positive outcome of the incident was the decision by President Reagan to allow the civilian use of GPS.

President Reagan greatly embarrassed the Soviet leaders by claiming the shooting down of the "innocent" airliner was an uncivilized act. However, we have now learned that the Soviet leadership was quite paranoid and thought it quite possible the United States might at some point carry out a preemptive nuclear strike, and it would thus be especially wary of the United States getting more information with a spy plane. Indeed, shortly after this incident a situation developed in which erroneous Soviet satellite detections of US missile launches against the background of an actually harmless NATO communication-in-nuclear-war exercise could have triggered a nuclear exchange by mistake.

In addition to the two incidents described in this chapter, family-owned Korean Airlines aircraft were involved in a succession of accidents and incidents over the years, partly due to senior Korean Air Force officers being fast-tracked to captains, their arrogance, and junior pilots who, according to Korean culture and military training, did not dare question them. As said in the piece on the Korean Air Cargo crash at London's Gatwick, the airline had a rethink with help from Delta, especially as regards CRM, and, having regained respectability, seems to be one of the safest.

Were it not sad, it is amusing to remember how the Nutgate scandal on December 5, 2014 showed how pernicious the family culture at the airline had become. It involved Heather Cho, the eldest daughter of the owner, promoted to senior vice-president for cabin services and safety.

On being served the nuts in the first-class cabin on a Korean Air flight about to take off from New York's JFK airport, Cho shamed and insulted flight attendants by shouting, throwing a service manual, and forcing them to kneel before her. The flight was delayed while there was a change of staff. With airline staff pressured to hush it up, it finally became public and worldwide knowledge, as it highlighted aspects of Korean society, where those in power have sway over the little people. A chastened Cho was given a prison sentence for endangering aircraft operations.

US Warship Downs Iranian Airliner (Persian Gulf, 1988)

Engaging ragtag gunboats

A fabulously expensive US warship designed to fight World War III, with a "gung ho" captain, found itself larking around with Iranian gunboats, and in the process shot down an Iranian airliner on a scheduled flight. The loss of the Pan Am 747 over Lockerbie, Scotland, was just possibly a consequence of this.

Iran Air Flight 655 and USS Vincennes, July 3, 1988

The "Sea of Lies" article in *Newsweek* (July 12, 1992), still available online today, is one of the best in-depth accounts of what happened, covering the events leading up to the incident, the incident itself, and the cover-up that followed.

Rather like many of the narratives in this book, it benefits from having been produced long afterward, when much more information was available. It further benefits from being written by a whole group of writers and researchers in collaboration with ABC News *Nightline*.

Here we just give the basic facts. Anyone really interested in the details should consult that excellent *Newsweek* article, which was used as the basis for much of this account.

The Captain

Much attention has been focused on the gung-ho tendencies of Captain William C. Rogers III, commander of the USS *Vincennes* guided missile cruiser that shot down the Iranian airliner. According to a fellow officer on another ship nearby, he had shown excessive aggressiveness in an earlier action in the area, while others maintained he would take gratuitous risks to gain promotion.

Rogers had served on the staff of the Chief of Naval Operations and supposedly had friends in high places. Someone pointed out "actual combat action during his dream posting in command of the ultra-sophisticated Aegis cruiser would have been his ticket to flag rank." One officer responsible for his training noted his failure to stick to battle plans.

However, Captain Rogers was not operating in isolation, but directly under the orders of fleet headquarters in Bahrain, and the *Vincennes* operated on their time, whereas Iranian time was a confusing thirty minutes ahead of that. This half-hour difference may have contributed to the confusion when an officer on the *Vincennes* checked the time the Iranian aircraft took off against the published timetable.

Key Characters Not on the *Vincennes*

1. Captain McKenna was the surface warfare commander in Bahrain.
2. Captain Watkins, on Admiral Lee's staff in Bahrain, who at a crucial moment questioned Rogers's location relative to the gunboats and their bearing.

3. Rear Admiral Leighton (Snuffy) Smith, commander of Carrier Battle Group 6, was on the aircraft carrier USS *Forrestal* some two hundred miles to the southeast. Though he was not in command of Captain Rogers, there were situations in which he could intercede, having dispatched two F-14 fighters and two A-7 attack planes to hold off fifty miles away from the "engagement" between the *Vincennes*'s helicopter and some Iranian gunboats.

4. Apparently, the rear admiral did not intervene at a critical moment because of the well-established navy principle that the commander on the spot has to make the final decision.

5. Mark Collier, twenty-five, piloting the *Vincennes*'s helicopter, Ocean Lord, under the command of Lieutenant Roger Huff, sitting in the copilot's seat, who followed the Iranian gunboats. He was said to have been unable to resist dropping down to have a closer look at what they were doing, and in consequence had a few shots fired in his direction. (Captain Rogers was then able to use this as justification for engaging the gunboats, which in fact did not even come within five thousand yards of the *Vincennes*.)

6. Captain David Carlson, on the frigate USS *Sides*, nineteen miles to the east, and the captain and officers of the USS *Montgomery*, a somewhat less sophisticated warship compared with the *Vincennes*, which had been allegedly pursued by the gunboats, giving the *Vincennes* the original pretext to join in.

Then there were the officers and crew of the *Vincennes* itself, who from all accounts were not all up to the high standard one would expect on such a sophisticated and costly ship—not that that prevented them from getting decorated. (This under par performance on the part of some may have been partly due to their lack of experience of such situations.)

Key People on the *Vincennes*

1. Commander Guillory, the tactical officer for surface warfare, who had previously been a personnel officer. Rogers had little faith in him, and he was said by colleagues at the time in personnel not to be a computer man, preferring to use pieces of paper stuck on his screen! Although Guillory gave actual orders for the ship's five-inch gun to fire when ready—first at a launch eight thousand yards away—Rogers had really assumed Guillory's role himself.

2. Petty Officer Andrew Anderson, whose task in the ship's Command Information Center (CIC) was to identify any air traffic within range of the ship.

3. Petty Officer John Leach, sitting next to Anderson and asked by Anderson for an opinion—cut short by Lieutenant Clay Zocher.

4. Lieutenant Clay Zocher, the petty officer's immediate boss, who, according to Newsweek, had stood on this watch only twice before during general quarters and had never mastered the computer routines for his console.

5. Lieutenant Commander Scott Lustig, the vessel's tactical commander for air warfare, who at the crucial moment of deciding whether to fire or not asked Captain Rogers, "What do we do?"

   Lustig was ultimately awarded the navy's Commendation Medal for "heroic achievement," because of his "ability to maintain his poise and confidence under fire enabled him to quickly and precisely complete the firing procedure."

The Ship

Conceived in the 1960s at the height of the Cold War, the Aegis-class missile cruisers were designed to prevent US naval fleets, including aircraft carriers, getting overwhelmed by an onslaught of missiles. Allegedly, they could simultaneously track as many as two hundred missiles or bogies (enemy aircraft) within three hundred miles at the same time, as well as follow what might be threatening the fleet from under the water.

Essentially, they were narrow-hulled boats powered by two powerful aircraft engines, with a phalanx of computers and monitors below linked to massive arrays of antennas on the superstructure. They could accelerate quickly to high speeds thanks to the powerful engines.

The *Vincennes* was the first Aegis-class cruiser to be deployed in a combat situation. However, deploying it in the narrow waters of the Persian Gulf and letting it stray from international waters to engage with Iranian gunboats was like pitting a Ferrari against bumper cars on a fairground and at the same time requiring its occupants to identify foe from friend in the crowd.

The crews of other navy vessels in the area regarded the crew of the *Vincennes* as arrogant.

A number of US naval vessels were operating there, ostensibly to protect shipping but covertly to ensure Iraq was not defeated by Iran in the Iran–Iraq War (in which the US was supposedly neutral). The initial reason for including an asset that was as sophisticated and costly as the *Vincennes* was that the Chinese had supposedly supplied the Iranians with a missile that other line ships could not counter.

The Ship's Combat Information Center (CIC)

This was the nerve center of the ship and somewhat like a computer games arcade, but much better. Here the various personnel watched their monitors, with the information they were seeing and annotating being collated by computer to be shown as symbols on two large screens at the front of the room in front of the captain and his two battle commanders.

Officers and personnel in the CIC wore headphones and communicated over several channels, with left and right ears usually listening to different circuits. Rogers and his key officers in the CIC were all on the same circuit, but, unbelievably, according to *Newsweek*, so was half of the ship!

Crew members had discovered they could tap into the "command net" to hear the action over their Sony Walkmans. However, in so doing, they drained power and the volume faded. Whenever it got too low, Lustig had to yell, "Switch" so everyone could turn to an alternate command circuit. Then the hackers would switch to that channel, too. Besides giving rise to the possibility that vital information could be missed or words misheard, this does not give the impression it was a responsibly run ship. If they had wanted to enable the crew to listen in without compromising their own communications, they surely had the technological wherewithal to do so.

The Iranian Airliner

Iran Air Flight 655 (IR655, an Airbus 300 with 290 people on board), took off from Bandar Abbas Airport in Iran at 09:50 on July 3, 1988 to make the simple twenty-nine-minute hop across the Persian Gulf to Bahrain. Allowing twelve or so minutes to start the engines, taxi to the end of the runway, and perform the usual checks, it would normally have become airborne at about 09:02.

However, a delay of about fifteen minutes due to a passenger having problems at immigration meant it only became airborne at 10:17 Iran time. Senior US officers later disingenuously claimed it took off twenty-seven minutes behind schedule, when in fact it was only fifteen minutes, as airliners do not normally actually lift off at their advertised departure times, which refer to the moment the doors are closed, and they are ready to depart from the gate.

Such a short flight would only involve the aircraft climbing to fourteen thousand feet and remaining there for a little while before descending into Bahrain. Furthermore, the route was not only a straight line but also in a twenty-mile-wide commercial airline corridor designated Amber 59.

A pilot flying along the centerline would have had ten miles' leeway on either side and would not be expecting trouble. On occasions, but not on that day, the control tower would warn civilian aircraft about military action taking place on the sea en route.

The Ghost Ship

One reason subsequently given to the Senate Armed Services Committee for the *Vincennes*'s presence was that "it had been racing that morning to rescue a Liberian tanker, the *Stoval*." No such ship appears in any ship registry. In fact, the *"Stoval"* was a fictitious ship, created by fake radio messages designed to entice the Iranian gunboats out of their lair, and was a trial run for an American sting operation.

What Happened?

The more realistic reason for the *Vincennes* being in the area was Rogers's desire to get at the Iranian gunboats that had been trailing a lesser US warship, the USS *Montgomery*. In so doing, he had sent up a helicopter to watch them.

08:40

> Captain McKenna, in Bahrain, returned to his command to find Rogers was forty miles to the north of where he thought he had told him to stay. Questioned about this, Rogers said he was supporting his helicopter and having difficulties communicating with it.

When told by McKenna to go back south, Rogers replied with incredulity, "You want me to what?" According to *Newsweek*, McKenna could hear the guffaws in the *Vincennes*'s combat information center. Furious, McKenna gave a direct order for the *Vincennes* and the *Montgomery* to come south.

Obliged to comply, Rogers duly obeyed but left the helicopter watching over the gunboats.

Mark Collier, the helicopter pilot, followed the gunboats as they retreated northward and finally decided to lose height to get a closer look, saying later that he wanted to see how many men were on board and what armament they carried. Evidently not pleased by this, one of them fired a burst of antiaircraft fire.

Collier later described it as "eight to ten bursts of light" and "sparks ... just like a big spark" one hundred yards from the helicopter.

The occupants decided to "get out of there," with the helicopter commander, Lieutenant Roger Huff, sitting in the copilot's seat radioing the *Vincennes*, "Trinity Sword. This is Ocean Lord 25. We're taking fire. Executing evasion."

(Under the navy's rules of engagement, this allowed Rogers to engage the enemy, even though it had just been one burst of fire and might have been little more than a warning shot.)

Ordering "General Quarters" and "Full Power," Rogers reversed course and proceeded northwards at thirty knots.

09:28

Two F-14 fighters and two A-7 attack planes took off from the aircraft carrier USS *Forestall* with orders to hold off eighty miles away from the action—Rear Admiral Leighton Smith on the *Forestall* did not want them shot down by mistake.

09:39

Rogers radioed fleet headquarters in Bahrain, declaring his intention to open fire. Not happy with the situation, they queried his position and the bearing of the gunboats.

Captain Watkins in Bahrain then asked:

Are the contacts clearing the area?

In fact, the gunboats were milling around randomly a long way away, probably thinking they were safe in their own territorial waters and it was not a question of them "clearing the area." Rogers continued to argue in favor of an attack, and, although lookouts on the bridge said the boats were milling around, used the fact that at one point in time they said a couple were heading in the direction of the *Vincennes* as the clincher. He reported to Bahrain that the gunboats were "gathering speed and showing hostile intent," and that he intended to open fire.

09:41

> Admiral Lee finally agreed to Rogers's planned attack, just as the *Vincennes* was passing into Iranian waters.

> To suggest that the *Vincennes* was under attack, when the nearest launch to be shot at was more than eight thousand yards away, was ridiculous, and it seemed no one on the *Vincennes* thought that was the case.

09:45:30

> The Iranian Airbus A300 received takeoff clearance and lifted off a minute later from Bandar Abbas Airport, in Iran, some fifty-five miles to the northeast. Its route southwest to Dubai would take it over the *Vincennes*, which really should have been forty miles to the south.

> Captain Rogers, virtually pushing aside Guillory, his tactical officer for surface warfare, set the big consolidated data screen to sixteen miles in order to concentrate on the gunboats. Note:

1. The blip of the Iran Air Flight 655 Airbus A300 climbing out of Bandar Abbas Airport had appeared on the Vincennes's radar, and as a precautionary measure had been tagged "Assumed Hostile," as the airport handled both civilian and military traffic.

2. Petty Officer Anderson, responsible for identifying any aircraft within range of the Vincennes, got the Aegis radar to interrogate the aircraft's transponder using the IFF system, and got a Mode 3 return, indicating it was a commair (commercial airliner). To make sure, he scanned the flight schedule for the airport but somehow missed Flight 655. He was said to have been confused by the half-hour time difference, the four different time zones in the Gulf, and not helped by the fact that the lights in the Combat Information Center flickered each time the ship's five-inch gun fired a round at the Iranian gunboats.

3. Just as he was wondering aloud to the petty officer next to him whether the blip might possibly be an Iranian warplane, he was overheard by his superior, Lieutenant Zocher, who was worrying about an Iranian P-3 patrol proceeding down the coast. Fearing that the two aircraft might be coordinating an attack, he decided to pass the information to his superior, Lieutenant Commander Scot Lustig, the ship's tactical commander for air warfare.

4. Instead of getting the two F-16s from the Forestall to check it out in the little time available, Lustig ordered Zocher to warn the aircraft it was approaching a US warship.

09:47

> Captain Rogers was not only engrossed in his battle with the gunboats but making it difficult for others to work, with the ship shuddering and keeling

over, as he shouted to the gun crew to load faster and ordered hard right rudder so that his stern gun could bear as well.

As already mentioned, the audio link tying in the key officers was subject to fading due to the demands of crew members listening in with their Walkmans and had to be switched repeatedly.

09:50

The audio link picked up the words "possible Astro" (code word for an F-14 uttered by a person never identified).

Petty Officer Anderson re-interrogated what he thought was the aircraft's transponder, but, in what was to prove fatal, forgot to reset the range and received a response from a military aircraft still on the runway at Bandar Abbas Airport. Naturally, it was Mode 2: military aircraft.

09:51

Anderson exclaimed "possible Astro" just as Rogers had swung the ship right round for his forward gun to bear again. When it jammed after getting off eleven rounds, he ordered hard rudder, with the result that the ship swung round, with books and papers falling off the consoles as the ship keeled over.

With the blip thirty-two miles away, Lustig said to Rogers:

What do we do?

Though it is possible to blame Rogers for creating the predicament by chasing after the gunboats and placing the *Vincennes* where it was, one cannot say he callously shot it down without some cause.

First, he must have had in mind the case a year earlier of the USS *Stark*, which was struck and almost sunk by two anti-ship missiles launched from a lone Iraqi fighter while its captain had been in the toilet.

The *Vincennes* issued three further warnings:

Iranian fighter . . . you are steering into danger and are subject to United States naval defensive measures.

Rogers could not be certain he was dealing with an enemy warplane. For one thing, it was, at 7,000 feet, somewhat high for an attacking aircraft. In addition, another officer, Lieutenant William Mountford, warned "possible commair."

Subsequent playback of the tapes of the Combat Information Center data revealed that Petty Officer Anderson's screen must have shown the aircraft almost at 12,000 feet at 380 knots and still climbing. Yet he and Petty Officer Leach were chanting that it was descending and picking up speed. Anderson was shouting out that the speed was 455 knots, the altitude 7,800 feet and descending. If Rogers was going to fire, he would have to do so before the aircraft narrowed the distance to ten miles.

09:54:05

> With the unsuspecting aircraft eleven miles away, Rogers used the firing key to free up the SM-2 missiles for firing. The missiles did not launch immediately, as Lieutenant Zocher was so unsettled that he pressed the wrong keys, it is said, twenty-three times. In the end, an experienced petty officer leant over and did it for him.

The crew of the Airbus had not heard any of the warnings issued by the *Vincennes*, as all of the four radio bandwidths had been set to ATC (air traffic control) frequencies. Also, on the three occasions they used a civilian frequency, the way they expressed themselves was confusing in that they mentioned groundspeed and not airspeed for the aircraft, which was different. Had they interrogated the transponder, the Iran Air pilots would have known it concerned them.

The first of the missiles from the *Vincennes* blew the left wing off the Airbus, which fell out of the sky in full view of the shocked crew of the *Montgomery*. Somewhat farther away, the radar operative on the USS *Sides* was telling its captain he reckoned it had been a civilian airliner.

Conclusion

The Iranian airliner had had 274 passengers and 16 crew on board, and all perished. The US paid $61.8 million in compensation to discontinue a case brought by Iran against the US in the International Court of Justice in 1989, all the while not admitting responsibility.

The US Navy faced a conundrum, for if they determined the captain was at fault captains in future might hesitate to defend their ships. Actually, it was not so much Captain Roger's final decision that was questionable but his setting up and fixation on the brawl he had initiated with the gunboats in Iranian territorial waters—unable to communicate with him because of this the captain of the aircraft carrier Forrestal did not allow his two F-16 fighters to check the aircraft's identity lest they be shot down by "trigger happy" Rogers—the radar operator not resetting his distance and hence picking up the transponder of a fighter on the runway at the airport, and the operator erroneously finally confirming to the captain the aircraft was descending when it was climbing, that might be construed as wish fulfillment.

One thing is for sure, it was crazy having an ultra-sophisticated warship costing a fortune engaging speedboats posing no real threat, a task for which she was furthermore neither suited, nor designed. To gloss over the affair, the US Navy gave medals to Captain Rogers (and other members of the crew) but, tellingly, not another ship.

Australian "Four Corners" TV Program

The 40-minute "Four Corners" program https://youtu.be/u17zMS2hSfo that includes footage filmed by the BBC on board the Vincennes at the time makes it all much more real. It makes our account look more than fair and mentions new points. For instance, while briefly on station the Vincennes had entered Iranian waters ten times. The US possibly retrieved the airliner's cockpit voice recorder but did not reveal the contents. Ends with footage of Captain Rogers leaving the ship on return to the US to a hero's welcome.

MH17 777 Shot Down by SAM Missile (Ukraine, 2014)

Malaysian Airlines was not the only airline following that route to save money

> Luck did not seem to be on Malaysia Airlines side in 2014, for they had
> already that year "lost" a Boeing 777 in mysterious circumstances.
> It was a matter of being in the wrong place at the wrong time, but should
> they and other more revered airlines have risked being there?
>
> *Malaysia Airlines Flight 17, July 17, 2014*

Background

Taking the shorter route eastwards from Western Europe, as did more than fifty other airlines per day, the Malaysian Airways 777 was flying over a war zone on the Ukrainian side of the border with Russia.

And, to make money from fees for those overflights, the Ukrainian government did not want to ban them outright, so instead of declaring the zone closed to civilian aircraft they declared they could fly over so long as they were over thirty-two thousand feet. Above that they would be presumed safe.

Ben Sandilands, on his blog, *Plane Talking*, has pointed out that, strictly speaking, airlines should not have allowed their twin-engine aircraft to fly that route, as they would not be able to maintain that altitude should an engine fail. Anyway, the BUK missile that shot down the Malaysian 777 was effective up to an altitude of eighty thousand feet, more than double the height it was at.

Three days before, on July 14, a Ukrainian Antonov 26 transport aircraft had been shot down at twenty-one thousand feet, and US intelligence, with its satellites and intercepts, and Ukraine intelligence must have been aware that ground-to-air missile launchers that could only be the BUK were being operated in the area. They should have seen that overflights by civilian airliners were stopped, though to be fair, getting decisions on such matters takes time.

There had been warnings three months earlier from the FAA, but they were for areas outside the one where MH17 was shot down. Aviation safety authorities in the United States and Europe generally warned about potential risks pilots faced flying in or near Ukraine airspace around that time.

The Incident and Investigation

The flight, with mostly Dutch passengers, was a regular scheduled one from Amsterdam to Kuala Lumpur with 283 passengers and 15 crew. The aircraft, a Boeing 777-200ER, disappeared from radar as it flew over Ukraine towards Russia about thirty-one miles (fifty kilometers) from the border, with most wreckage found about six miles (ten kilometers) farther on.

A Singapore Airlines 777 was only some thirty miles (fifty kilometers) away when the incident happened, and two Air India airliners not much farther away than that.

There were no survivors, and the wreckage was in territory controlled by insurgents and difficult to access. With only limited resources of their own and almost two-thirds of the passengers Dutch, the Ukrainians asked the Dutch to take charge of the investigation, which they carried out very thoroughly.

They collected a considerable amount of debris and applied it to a frame representing the aircraft and were able to show that a missile had exploded top left of the cockpit.

Aftermath

Immediately after the loss of MH17, the airspace was closed and airlines that had been using it changed to alternative routes.

The Russians vetoed an attempt to have the case investigated by the UN, and great efforts were made to try and prove who was responsible, with some claiming Russian personnel were present. President Putin was also blamed by the Americans and others.

All one can really say is that it was surely a mistake and that allowing ground-to-surface missiles to get into the hands of difficult-to-control units is dangerous. Surprising few airliners have been shot down considering the number of shoulder-launched SAMs supplied by the US to insurgents in Afghanistan to fight the Russians, which ended up on the open market and became easy to obtain at one time.

It seems only El Al, the Israeli airline, have airliners with expensive countermeasures, though some special aircraft for VIPs, such as Air Force One for the United States' president, obviously do.

Chapter 16
TERRORISM

Air India 747—Deadliest Bomb on a Plane (North Atlantic, 1985)
A second bomb almost simultaneously blew up an Air India flight from Tokyo

> This was the first instance of a jumbo jet being downed by an onboard
> explosive device.
>
> *Air India Flight 182, June 23, 1985*

At 06:19 UTC on June 23, 1985, two Japanese baggage handlers were killed, and four others injured at Tokyo's Narita Airport as an item of luggage was being transferred to Air India Flight 301. It had come from Montreal on a CP Air flight. Whoever had set the timer had apparently not realized Japan does not have summer time, in which case the bomb would have gone off when the aircraft was airborne.

One hour later, Air India Flight 182, bound for London from Montreal 329 people on board, exploded in midair off the west coast of Ireland, and fell from 31,000 feet (9,400 meters) into the sea with no survivors.

Backdrop
In 1983, armed Sikh separatists took over Sikhism's holiest shrine, the Golden Temple complex at Amritsar, India. From there the separatist leader agitated for a Sikh homeland called Khalistan in the northern Indian state of Punjab. In early June 1984, Indian prime minister Indira Gandhi ordered an attack by the Indian army on the temple in which the separatist leader and his supporters were killed, together with hundreds or more innocent pilgrims. This not only led to demonstrations around the world but to the assassination of the Indian prime minister by two of her Sikh bodyguards, in turn leading to the 1984 anti-Sikh riots, in which thousands of Sikhs in India were killed.

Earlier, a number of militants wanted by the Indian government had fled abroad, with countries refusing to return them. Many ended up in Canada, and in Vancouver there were demonstrations and firebrand speeches by activist leaders seeking revenge. There was also intimidation of moderates—one of them, a prominent Sikh lawyer, was beaten up by someone with a metal bar in February 1985, and he wrote to the prime minister that April saying he should act before something serious happened.

The Canadian Security Intelligence Service (CSIS), created only a year before, were in fact following a Sikh leader called Parmar, even tailing him on June 4, 1985 to Duncan, on Vancouver Island, where he met Inderjit Singh Reyat, a local Sikh marine mechanic, and another man. They drove to a wood, where the agents lost sight of them. They did hear what sounded like an explosion but could not be sure.

The Investigation

The CVR and FDR for Air India Flight 182 were recovered after two weeks, but all they could show was that the flight had been completely normal. Despite the great depth of the water, some large pieces of wreckage were recovered before severe weather set in and showed that an explosion had occurred in the front luggage bay just aft of the avionics bay.

However, tiny pieces that might have provided clues as to who might have made the explosive device were scattered over a wide area deep under water, unlike at Lockerbie, where by unbelievable luck a key one was found. However, things were different in Tokyo, where the bomb had gone off in a confined space with it even possible to recover fragments embedded in the walls.

On the fragments, the Japanese investigators found the serial number of the Sanyo tuner that held the explosive. It was one in a batch of two thousand that had only been sold in British Columbia. Canada's Mounted Police were eventually able to find the shop in Duncan that had sold it to a Sikh called Reyat!

How Was It Allowed to Happen?

A browbeaten airline check-in clerk, to her everlasting regret, allowed luggage to be booked through without a confirmed booking, the two aircraft were allowed (in those days) to depart with luggage of no-shows, and a case that raised suspicions was not properly checked. Also, the Canadian government had ignored repeated warnings.

Prosecutors hoped bomb maker Reyat would testify against his coconspirators, but he led them on a song and dance, saying at the trials he could not remember, resulting in him being later convicted of perjury. As it was the then policy of the newly created CSIS to erase tapes of intercepts following their translation, it meant evidence against others did not stand up in court.

Witnesses were murdered, threatened, and the suspects walked free.

The official inquiry in 2010 said in part "a cascading series of errors contributed to the failure of our police and security forces to prevent this atrocity. The level of error, incompetence, and inattention which took place before the flight was sadly mirrored in many ways for many years, in how authorities, governments, and institutions dealt with the aftermath."

Failed Precursor to 9/11 (Marseille, 1994)

Terrorists intended to crash A300 Airbus on Eiffel Tower or Elysée Palace

> In 1994, French special forces stormed a hijacked Air France airliner that terrorists intended to use for a 9/11-style outrage in Paris. The authorities claimed it was one of the most successful operations of its kind ever.
>
> *Air France Flight 8969, December 24 to 26, 1994*

At 11:15 a.m. on December 24, 1994, Air France flight AF8969, an A300 Airbus, was about to depart for Paris from Algiers Airport, with 220 passengers and 12 crew. As Algeria was in a state of civil war, passengers did not think it untoward that four armed men dressed in sharp blue uniforms with Air Algérie insignia claiming to be security agents came on board to make a final passport check before departure. That is until there were shouts of "Allahu Akbar!" ("God is great!") from the four "security men," who, having abandoned their passport checks and stayed inside, shut and locked the doors.

Three of them quickly proceeded to the cockpit, while the fourth kept the passengers at bay by waving a gun at them. For those at the back of the aircraft, it was quite difficult to work out what was going on, and later some even remained unaware that selected passengers had been used as bargaining chips and shot.

Algerian special forces, already on high alert at the airport, as the country was in a state of civil war, came to see what was going on and quickly confirmed the individuals were terrorists. They actually belonged to the Algerian Armed Islamic Group (AIG).

It being Christmas Eve, many top French officials were away on vacation and, as usual, Prime Minister Balladur was at his Alpine chalet, in Chamonix, where he stayed glued to the phone line for the first day. The French foreign minister and interior minister were in Paris handling the situation, their roles being to deal with the Algerian authorities and set up a team ready to go into action and overcome the hijackers should that prove necessary.

The hijackers' immediate demand had been the release from house arrest of two leaders of the Islamic Salvation Front (FIS), an Algerian organization banned in Algeria. The hijackers negotiated via the pilots with the Algerian interior minister, who had rushed to the airport and was using the radio equipment in the control tower.

After an hour or so the hijackers seemed more intent on going to Paris than seeking the release of the two political leaders. To try and move things along, they shot an Algerian policeman, whom they had identified during the passport checks. He was led away asking not to be killed, as he had a wife and child. His body, still quivering, was dumped outside the aircraft. With the Algerian authorities still adamant, the hijackers then killed the commercial attaché at the Vietnamese embassy in Algeria, who was the only person on board who was neither Algerian nor French; the poor man, thinking he was being released, went back for his coat.

The Algerian authorities insisted on the release of the elderly, invalids, children, and so on. By the end of the second day, Christmas Day, the hijackers had released sixty-three in all. Their departure probably made it easier for the terrorists to handle things in the aircraft, where they established a strict and oppressive regime without being rude.

First, they made women and men sit separately and use different toilets. They also ordered the women, including the flight attendants, to cover their faces with whatever material was available, which included cut-up blankets. Then, applying psychological intimidation, the hijackers instituted a regime whereby they would treat the passengers harshly for twenty minutes, allow them to relax for twenty minutes, and then repeat the process. The intention of this "torture," as some described it, was probably to keep the passengers off balance.

The authorities managed to identify the leader of the hijackers and even had his mother brought to the airport to try to persuade him to let the passengers go. This only seemed to provoke him further.

Meanwhile, the French had not been wasting time. Right at the start they put their Special Deployment Unit (National Gendarmes Intervention Group, or GIGN) on standby. This was an elite police unit trained along military lines for intervening in hijacks and hostage situations. Although they had not been able to get the Algerian authorities to agree to let them help, a team of forty of these elite commandos set off in an A300 Airbus for Palma de Mallorca, in Spanish territory, some two hundred miles north of Algiers.

Balladur, the French prime minister, had returned to Paris to take overall command, and as the evening of the second day approached, the hijackers, worn down and tired, realized they were getting nowhere slowly. After threatening to blow up the aircraft, they issued an ultimatum. If the boarding ramp was not pulled back and the plane not allowed to take off before 9:30 p.m., they would kill a hostage every half hour.

Announcing that the first to be shot would be Yannick Beugnet, a cook at the French ambassador's residence in Algiers, they brought him to the flight deck and allowed him to plead for his life over the radio. The cook was then shot and his body thrown onto the tarmac. Balladur was livid at the news, telling the Algerian prime minister they would hold him responsible. He then called the Algerian president, saying he expected to see the imminent arrival of the aircraft on French soil.

The aircraft took off in the early hours of the next day, December 26, and at 3:30 a.m. landed at Marseille, where the hijackers had been told a refueling stop was necessary. There was an element of truth in this in that much of the fuel initially loaded in Algiers for the flight to Paris had been consumed by the auxiliary power unit (APU), which had been kept running to keep the lights and air-conditioning working. The Special Deployment Unit had taken off from Majorca in its A300 and followed them to Marseille, touching down at an adjacent military base.

In Majorca, the commandos had had time to fully familiarize themselves with the layout of an A300 and prepare. On the ground in Marseille, they went through their routines again. Meanwhile, the Marseille chief of police was conducting negotiations with the hijackers, the aim being to wear them down further and, if possible, delay any action until dusk, which would be quite early, as it was the middle of winter.

The French had received information that the hijackers intended to blow up the aircraft over Paris, and very likely on a landmark there. Passengers released in Algeria had described how explosives were set up in a way that demolition experts said indicated they would cause the aircraft to explode. This belief was further reinforced by the terrorists' demand for twenty-seven tonnes of fuel, when only ten tonnes would largely suffice. Anyway, the French were determined not to let the aircraft leave Marseille.

At dawn, the hijackers issued an ultimatum, but the negotiators were able not only to extend this but also exploit it by saying some servicing was needed. They sent in commandos in disguise, ostensibly to provide food and water, clean and service the toilets, but in reality, to plant snooping devices, identify who was who and where. They noted that the doors were neither blocked nor booby-trapped. As the afternoon wore on, with no fuel delivered to the aircraft, the negotiators drew things out by saying that holding a press conference in Marseille would be preferable, as the press was there. They even exploited that by having the hijackers move the passengers toward the back of the aircraft, ostensibly to make more room for the press conference, but in fact to make more room for the commandos and to enable them to separate any hijacker in the cockpit—and possibly in charge—from the others.

As 5:00 p.m. approached, the exhausted hijackers knew things were getting critical for them, and their leader came into the cabin to select a hostage to kill—the youngest member of the cabin crew. However, it seems he was hesitating to put his threat into execution, perhaps out of fear of provoking the French. Instead, in their frustration the hijackers let off some rounds in the direction of the control tower, where they knew the scheming negotiators would be. The commandos would eventually have moved in anyway, but their firing of shots at the control tower resulted in Prime Minister Balladur giving the head of the special forces carte blanche.

Their plan was to have three teams, who would use boarding ramps to access the doors. Two would creep up from the back, trying to keep out of sight of the hijackers, and open the doors at the back to provide a good escape route for the hostages. The third, led by Major Favier, would have the most difficult role of penetrating the aircraft via the front right door to separate hijackers on the flight deck from their companions.

Despite the commandos having planned and rehearsed the operation so meticulously, the ramp they were standing on proved to be too high to allow them to open fully the right front door they had prized ajar. Then, on quickly pulling it back to allow enough space, they almost left a commando dangling between the fuselage and the ramp. Fortunately, he had a good grip on the door and was able to hang on there until it was opened successfully. They were met with a hail of bullets, with one lucky shot from the hijackers hitting a gendarme's gun, setting off the rounds it contained and knocking him backward down the steps with quite severe injuries.

As hostages evacuated from the rear doors, the commandos progressively overcame the hijackers, though one held on, firing round after round, for almost twenty minutes. Sharpshooters firing from the control tower were at the beginning unable to get shots in because of the presence of the aircrew. The first officer then slid out of a broken cockpit window, only to injure himself quite seriously on hitting the ground after dropping from such a height. No hostage was killed—the commandos had flak jackets and the aircrew said the bodies of the hijackers had at times shielded them.

Another reason given for the survival of all the crew was that they had built up a rapport with the hijackers, who could so easily have killed them out of spite. Favier, the commando leader, suggested it was a matter of mutual respect. Was it the Stockholm syndrome in reverse? (The term "Stockholm syndrome" comes from a famous case in Stockholm, where hostages kept for several days in a bank ended up empathizing with the hostage takers.)

Bomb under Seat Was Trial for Bojinka Plot (Okinawa, 1994)
Bombs were to be placed on eleven US airliners

> Even when closely following people planning a terrorist act, it can be
> difficult or impossible for the security services to be sure of their
> intentions, let alone have material that would stand up in court. Here, a
> chance fire led them to a laptop computer revealing all.
>
> *Philippines Airlines Flight 434, December 11, 1994*

The massive Boeing 747 bound for Tokyo had stopped at the Cebu Airport, having come from Manila. There, some passengers, including one traveling under a false name who had spent some considerable time in the toilet and then sat for a while in seat 26K, which was not his, had descended.

Most of the 273 passengers boarding at Cebu were Japanese. There was a crew of twenty, including the captain, fifty-seven-year-old Eduardo "Ed" Reyes, Second Officer Jamie Herrera, and Flight Engineer Dexter Commendador. All three were former air force pilots.

Having taken off thirty-eight minutes late due to congestion, the 747 was flying at thirty-three thousand feet and not yet beyond Okinawa when there was an explosion. The autopilot immediately corrected the bank to the right, and since there was no cabin depressurization, there was no need for an emergency descent.

Back in the cabin, eleven passengers were injured, and the one in 26K seemed to be in extremis. Passengers tried to move away from where the explosion occurred, but flight attendants ordered them to remain seated. Lead economy-class flight attendant Fernando Bayot did move the injured passenger in 27K, who had injuries to his legs, before trying to help the Japanese man in 26K, below whom there was a hole in the cabin floor. He lifted him up, only to find that much of his lower body was missing and had fallen into the cargo hold below. Though the man then expired, Bayot instructed the flight attendants to administer oxygen and make it look as if he were being treated, thus avoiding panic. Eleven passengers were injured, some seriously. After Bayot told him the situation, the captain sent the flight engineer to assess the damage.

Though the autopilot was keeping the aircraft level, input of turns had no effect. The captain was loath to disengage it for fear that the aircraft would enter an irrecoverable bank, but finally they had to risk it, only to find the yoke had no effect either. With Tokyo two hours away, they decided to make an emergency landing at Naha, on Okinawa, but only managed to make themselves understood after an American took over from the Japanese controllers. Using differential engine thrust, they managed to make the wide turn toward Naha.

Following the first officer's suggestion, they dumped thirty-six tonnes of fuel to reduce landing speed and for there to be less fuel to feed a fire, should one break out. After a control check—flying the aircraft as if landing to see whether they would have the minimal control necessary—they touched down safely.

Apart from the professionalism of the aircrew, the passengers owed their survival to the fact that the 747 in question was an ex–Scandinavian Airlines version, where row 26 was not above the central fuel tank.

Link to WTC Bombing, Bojinka Plot to Down Eleven Airliners

The man who had planted the explosive device under seat 26K was Ramzi Yousef, a self-confident operative who had planned the bombing of the World Trade Center in 1993, leaving underlings to be arrested thanks to discovery of a part with a serial number from the van used.

In Manila, he was planning to kill the pope on his forthcoming visit, down eleven airliners using onboard bombs—Flight 434 was a test for that—and have a man called Murad fly a small plane into the CIA's headquarters.

A couple of weeks after placing the bomb on the Philippine Airlines flight, Yousef, Murad, and another conspirator were preparing bombs in an apartment that was near the papal nuncio's residence and on the route the pope would take. They were relaxed and enjoying soft drinks when Murad went to the sink to wash his hands. Bomb-making chemicals they had been mixing there had not been flushed away, and when water Murad was using splashed on them they coalesced, starting a fire with acrid smoke that could not be extinguished. Yousef and Murad fled into the street and watched from the other side as the fire services dealt with the fire. Yousef, protecting himself as usual, told Murad to go back and retrieve the laptop with all their plans.

The local police chief thought a fire on the papal route and so near the papal nuncio's residence, where he would be staying, worth investigating. She arrested Murad and declined to release him despite the offer of what was an incredible bribe for someone on her small salary.

Once the encryption was broken, the laptop revealed details of the so-called Bojinka Plot, including flights to be targeted, with even timings for the fuses. That alone could have resulted in some four thousand deaths—more than 9/11. There were even details regarding the planned crash into the CIA headquarters at Langley, Virginia. After painful interrogation, Murad revealed more details.

US Diplomatic Security Service (DSS) Corner Yousef

A partial fingerprint of Yousef's was also found. He had once again got away, but a month later he was arrested in Pakistan thanks to shrewd handling by Diplomatic Security Service (DSS) agents there of a walk-in at the US embassy, who by not going through prescribed channels avoided him being tipped off.

9/11—World Trade Center and Pentagon (USA, 2001)
Protocols for dealing with hijacks turned on their head

> Since there were relatively few passengers on board the early-morning flights, it was the 2,700 people on the ground killed by those aircraft that made the tragedy so terrible.
>
> *American Airlines Flight 11, September 11, 2001*
> *United Airlines Flight 175, September 11, 2001*
> *American Airlines Flight 77, September 11, 2001*
> *United Airlines Flight 93, September 11, 2001*

Nineteen terrorists hijacked four aircraft almost fully laden with fuel for long transcontinental flights. They intended to fly them into symbolic US buildings. Hijackers on three of the four aircraft succeeded in doing this.

With so many excellent accounts, documentaries, and movies covering the events of 9/11, it would be difficult to add usefully to them here. However, the time line below, presented in tabular form, is perhaps a useful tool for grasping how the overall situation developed. The operation depended largely on timing and doing the deeds before people, and in particular those on the aircraft, realized what was happening. The outcome could have been even worse had airport congestion not delayed the takeoff of the fourth aircraft, enabling the passengers to learn the hijackers' real intentions.

For unknown reasons, the four hijackers on that last flight, UA93, were rather slow in going into action, despite already being behind time due to the thirty-five-minute delay in taking off. When they did finally hijack the aircraft, passengers and cabin crew inevitably started phoning out, only to learn that their aircraft would probably soon become a flying bomb; if they did nothing they were goners anyway.

Outline of Events

The FAA's Herndon command center supervises air traffic control for the whole United States. The movie *United 93*, about the passengers in the last of the four aircraft to be hijacked, who rebelled after learning by phone what had happened to the other aircraft, is based to a large extent on what happened there. Its operations director, Ben Sliney, had an overall view of the unfolding situation, which he faced on his first day in the job, having returned to the agency after a period in private law practice. He has very cogently explained how difficult it was to assess what the hijackers were doing and which aircraft had been hijacked.

In an interview with Jeremy Vine on a program broadcast by the UK's BBC Radio 2 promoting the above movie, in which he played himself, Sliney said he heard about the hijacking of an airliner just before attending a meeting. He had not been unduly concerned, as there were well-tried protocols for dealing with hijacks, and hijacks took time to play out and usually ended with no loss of life or serious consequences. Interestingly, the FAA thought September was an odd time of the year for a hijack.

In its role of overseeing air traffic throughout the United States, the FAA's Air Traffic Control System Command Center at Herndon does not communicate with aircraft directly. However, it does receive all the necessary data relating to aircraft movements, weather patterns, delays and problems and special situations, such as military activity and the need to clear airspace for Air Force One, the US president's aircraft. It then gives instructions to ATC centers and other ATC facilities all over the States. It also communicates with North American Aerospace Defense Command (NORAD), responsible for defending North American aerospace.

While the center does not handle individual aircraft, its monitors do show the situation of air traffic throughout the country. If pilots enter the squawk code for a hijack into their transponder, this instantly shows up on its screens, as well as on those of the handling ATC. It is also shown on those at NORAD (North American Aerospace Defense Command), which meant that in the days when the controls for transponders were more primitive, pilots had to be careful not to mistakenly squawk the hijack code and attract the attention of a couple of fighters.

However, none of the four aircraft hijacked that day squawked that code, and as a result the controllers in direct contact were the only ones able to progressively deduct that a given aircraft had been hijacked, from the fact that it was not replying or obeying instructions, was changing course or height unduly, or had its transponder switched off. In some cases, controllers heard snippets of conversation in a foreign language, shouts, or in the case of the first aircraft to be hijacked, an announcement destined for the passengers, but which was transmitted because either one of the pilots keyed the radio mic, or a hijacker did so by mistake, saying, "We have some planes; just stay quiet and you will be okay." In every case the transponders were ultimately switched off, making it difficult to even be sure which aircraft was involved, as the controllers could just see the radar echo, without the flight number and height, on their primary radar.

On the next page we show the route taken by the four aircraft, from where and when they took off, and where and when they hit their objective, though the last one crashed before reaching it, whatever it was.

These diagrams are based on the official 9/11 report and are not proportional.

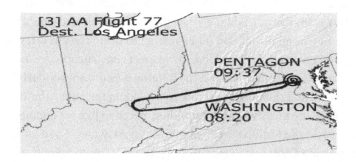

Thus, the FAA Herndon command center and operations manager, Sliney, only belatedly became aware of the gravity of the situation. Sliney said it was only when he saw CNN coverage of the burning North Tower of the World Trade Center, with most people assuming it had been struck by a light aircraft, that he realized from the size of the hole and the amount of fire that it must have been a big jet, and possibly Flight AA11.

Indeed, it was only when the second aircraft hit the South Tower while everyone was watching the burning North Tower on CNN that the FAA command center could grasp the likely implications of any other hijackings that day. They exchanged information with NORAD and at times were frustrated that the military liaison man could not seem to put them in the loop. However, in the BBC 2 interview, Sliney expressed sympathy for the military in view of the constraints under which it must have been operating. Who could order the shooting down of a civilian airliner, and an American one at that?

The first two aircraft struck the twin towers of the World Trade Center in almost perfect fashion from the hijackers' point of view. However, the third aircraft, having taken off from Washington's Dulles International Airport, not far from the Pentagon, was later seen heading for it, with its wings level, as if the pilot knew what he was doing.

Some have said it was a great feat of airmanship for a novice to hit the Pentagon at almost ground level, with the maneuver simulating a perfect landing. However, damage and the effect of the inevitable fire would surely have been much more significant had he been able to bring the aircraft down in the hollow at the center of the five-sided, low-aspect construction, and one wonders whether that was the intention and the piloting was not so wonderful after all.

It is thought that the fourth aircraft was heading for either the White House or, more likely, the massive Capitol Building, perched like a wedding cake at the end of the two-mile National Mall, making it an easy target and as symbolic as the World Trade Center. However, the thirty-five-minute takeoff delay, due to congestion at Newark Airport, coupled with the long time the hijackers for some reason took to take over the aircraft, meant that passengers and cabin crew would learn their likely fate by phone and not be so amenable.

From the phone and other evidence, and as shown in the film *Flight 93*, the passengers decided to take on the hijackers. They did so and very likely managed to penetrate the cockpit, breaking down the door with a food cart from the galley. Confusion apparently reigned, and the aircraft flew into the ground.

Some say the hijackers crashed it on purpose. Others have maintained a fighter shot it down just when the heroic passengers had gained control, a story that would not make for pleasant reading. Much material is available on the Internet for anyone interested in exploring the matter further, although as mentioned in Acknowledgments, one can waste a lot of time going down the conspiratorial alley without coming to any definite conclusions.

Hijackers' Modus Operandi

Each of the four hijacker groups included one or more members who had learned how to fly at various flying schools for light aircraft. Their instructors had been surprised that learning how to land did not seem to interest them. Using manuals for the airliners concerned, namely the Boeing 757 and 767, and no doubt PC flight simulation programs, they had acquired the necessary flying skills for their task.

Although some of their number were stopped and searched at the airport security checks, they were allowed to board. It seems bluff played a large part in that they would have a member go into the toilet and come out wearing something harmless made to look like a bomb that could be detonated at any moment if the passengers and cabin crew did not comply with instructions. Should that not suffice, they had other items, including pepper spray, that could be used to subdue or kill. The fact that there were relatively few passengers on board the aircraft made them easier to control, as it was possible to herd them into the rear of the aircraft, away from the events and the murdered aircrew at the front.

Timeline

Airline Flight Number [sequence]	American Airlines AA 11 [1]	United Airlines UA 175 [2]	American Airlines AA 77 [3]	United Airlines UA 93 [4]	Total Death Toll
Aircraft type	Boeing 767 ER	Boeing 767	Boeing 757	Boeing 757	
Departing	Boston	Boston	Dulles Washington	Newark New York	
Destination	Los Angeles	Los Angeles	Los Angeles	San Francisco	
Passengers	76	51	53	33	213
Crew	11	9	6	7	33
Hijackers	5	5	5	4	19
Total deaths [- hijackers]	92	65	64	44	265 [246]
Actual takeoff	07:59	08:14	08:20	08:42	
Delay in taking off		Approx. 10 min.	Approx. 10 min.	Approx. **35 min.**	

				09:00 a.m. UA issues system-wide warning of *'cockpit intrusion.'* Flight 93 replies 'Confirmed.'
Pilots alerted of danger				
Hijacked	08:14	08:42	08:52	09:10-09:20
Hijack suspected	08:20	Immediately	09:05	9:28
Sounds heard by ATC due to radio mike switch being left on	'We have some planes. Just stay quiet and you will be OK.'			Sounds of scuffling/ Arabic voices
NORAD informed by FAA	08:40	08:43	09:24?	No available data from NORAD (ongoing exchanges)
ATC certain	08:25	08:52	08:56	09:32
Time of impact	08:46	09:03	09:37	10:03 or 10:06 (disputed)
Level	95th-103rd floors	80th floor	Ground Level	Crash into terrain
	-----------	-----------	-------------	---------------
Location	N. Tower World Trade Center	S. Tower World Trade Center	Pentagon	Open country
	-----------	-----------	-------------	---------------
Effect	Catches fire; tower collapses on itself at 10:29	Catches fire; tower collapses on itself at 09:59	5-storey section collapses	(Passengers confronted hijackers)

Ground fatalities	**2,793** including woman who died months afterwards due to inhalation of cement, glass, lead and asbestos dust while fleeing the scene. *	125	0	
	Other ATC-related Events			
Between 09:03 and 09:07	**FAA's New York & Boston Regional Centers stop takeoffs and landings**		**NY Port Authority closes Newark Airport**	
09:25	**FAA National Command Center STOPS ALL US TAKEOFFS**			
09:45	**FAA orders all 4,500+** aircraft still airborne over the US to land. Orders those that have not yet reached US territory to turn back or land at non-US alternates.** ** Under instrument flight rules			

* Other than the woman cited, a difficult to determine number of people subsequently became ill or died from the aftereffects of the dust cloud arising from the collapse of the twin towers.

Hijack Protocols

Aircrew, cabin crew, and even passengers had learned that the way to deal with a hijacking was to at least initially comply with the hijackers' demands, and indeed in most cases this had worked well, with the authorities able to wear the hijackers down with endless negotiations on the ground. Normally, special forces would only quickly storm the aircraft if the hijackers commenced systematically murdering the occupants.

The pilots of the last aircraft to be hijacked did receive a warning about cockpit incursion, but unless they were aware that the protocols had changed and the likely consequences, they would have let the hijackers in if they threatened to kill a flight attendant unless they opened the flight deck door.

Nowadays the protocols are different. Cockpit doors are reinforced and are not opened, regardless of any threats made and include a system to that can be set to prevent their being opened even by someone knowing the code. Passengers are screened more carefully, with flights from abroad required to give details of passengers, and the aircraft turned back or required to land at an isolated airfield if anyone suspicious is on board. Also, with passengers and pilots aware that compliance with the hijackers' instructions is unlikely to save them, it will be virtually impossible for hijackers to take over aircraft as easily.

The Planners

The events of 9/11 did not represent original thinking in that terrorists had already tried to use an airliner as a flying bomb against the Eiffel Tower. The training of at least one member of each hijacking team to actually fly the aircraft well enough to hit an albeit easy-to-find target was possibly novel.

Probably the biggest issue for the 9/11 planners was how many teams could be prepared before the sheer number risked jeopardizing the entire mission through someone being caught due to bad luck, an error, or loose talk. The choice of early-morning flights, which are less subject to delay and have relatively few people on board, greatly increased the chances of success. Indeed, it was the terrorists in the delayed aircraft that failed in their mission.

The Result

Whether the great loss of life was intended or not, it gave the US administration a blank check to use on the international stage, making the instigators initially the losers. That said, it is difficult for people outside the States to realize just how much this event traumatized the country, not only in terms of loss of life but also by seeming to be a lance pointed at its commercial and political heart.

Intelligence Services, CIA, FBI, and NSA

The subsequent inquiry said there were missed opportunities to prevent the outrage which led to soul-searching at the CIA and FBI. However, it did not mention the fact that the National Security Agency was monitoring phone calls two of the hijackers were making from the US to a known Al-Qaeda operations headquarters in the Yemen as they installed themselves, took flying lessons, and purchased flight simulation software. The NSA did not inform the FBI of this perhaps because of laws, since modified, not allowing them to spy on Americans in the US. Also, and this is the perennial problem for all such agencies, instigating physical surveillance might make the suspects and in turn possibly very much bigger fish in their network aware that they are being monitored.

Unintended Consequences of Reinforced Cockpit Doors

In 2015, reinforced cockpit doors that could not be unlocked even with the code prevented the captain of a Germanwings flight from Barcelona to Dusseldorf regaining entry to the cockpit to stop his suicidal copilot crashing the aircraft and killing all one hundred and fifty people on board.

Liquid Bomb Plot Changed Flying Forever (UK to US, 2006)

Precautions now taken, though a nuisance, have made flying safer

> The British were closely monitoring a group on the point of blowing up
> transatlantic airliners using innocuous-looking liquids in drinks bottles.
> They wanted to delay the arrests until they had enough evidence to
> stand up in court. The Americans feared delay would result in tragedy
> and lay them open to accusations of not protecting their citizens.
>
> *UK arrests, August 9, 2006*

On December 5, 2005, a "person of interest" to the British security services was arriving back at London's Heathrow airport from Pakistan. They could not search him without alerting him, but they could get away with a look inside his checked-in suitcase if done quickly enough. To their disappointment, the only items of any note were a box of batteries and a packet of Tang, a sugar-based powdered soft drink with two orange dyes to make it look like orange juice when mixed with water, made famous through NASA space flights.

The owner of the case was twenty-five-year-old Ahmed Ali, a UK resident with dual British-Pakistani citizenship, who had graduated with a BSc in computer science from City University, London. Unemployed, he was living in a council flat in Walthamstow, East London. From a very young age, he had been meeting people going out to Pakistan and Afghanistan to give aid to those who, with US support, were fighting the Russians and who eventually became the Taliban. According to a 2008 article in London's *Daily Telegraph*, Ali was already, at age fourteen or fifteen, praising the Taliban and calling for shari'a to be introduced alongside British laws.

According to Pakistani surveillance, Ali had met up with a Rashid Rauf, who had fled the UK after possibly murdering his uncle. He not only had high-level links to Al-Qaeda but also bomb-making training. Therefore, anyone associating with such a man needed to be watched, and with Ali back in England, the need was to see in turn with whom he associated.

The closest associate proved to be twenty-six-year-old Tanvir Hussain, an unemployed single man with a child who had also been to university. Then there was Umar Islam, two years older, at twenty-eight. Surveillance continued with not much of note.

Then on July 15, Ali met an unknown individual, and the two of them went to a park, where they lay on the grass facing each other to talk without an observer being able to lip-read, proving they were highly trained professionals. After they separated, the unknown man was followed to High Wycombe, thirty miles outside London, and found to be Assad Sarwar. A university dropout who had once worked as a postman, Sarwar was living with his parents. He was buying a chemical, hydrogen peroxide, which though quite common and used by hairdressers can also be used to make bombs, such as those used in the London bombings.

Where plots are involved, the security services can spend ages devoting considerable resources—it can take thirty people to follow one individual if three shifts are needed—with nothing to show to a court of justice. Then things can move too quickly for comfort.

The group were found to be visiting a flat of which the security services were unaware. One day someone came out with a bag of rubbish, which they dumped in a bin quite some distance away so it would not be linked to that address. It proved to contain hollowed-out batteries.

Taking one of the greatest risks of the whole operation, the security services planted a bug in the apartment and just managed to get away before anyone returned. This audiovisual bug proved a godsend.

Thanks to it, they were able to see Ali and his associates emptying soft drinks bottles through a hole made in the base and replacing it with a liquid. Listening in, they heard talk about places in the US, such as Miami, Los Angeles, and so on. They also heard Wills being recorded. Ali was followed to an Internet café, where it was found he had downloaded a list of flights to the US leaving around the same time. The intentions were clear.

Meanwhile the Americans were getting nervous and were not sure the British had everyone in their sights—might someone make a dry run, leading to the loss of an airliner and two or three hundred deaths? The British wanted to wait to have more evidence for the courts, but their hand was forced when the Pakistanis arrested Rauf at the instigation, in one way or another, of the US.

From that night on, passengers were no longer allowed to bring liquids or gels with them on board flights, and mothers had to taste milk prepared for their babies. The regulations were later eased, but they still remain, as we are all too aware.

The fastidious work by the security services in this case certainly prevented many deaths, though estimates as to the number of aircraft that would have actually been downed vary widely. It is chilling to think that at least some of the perpetrators intended to bring their wives and children with them. Could it have been to make it easier to bring the materials through security, or could they have wanted their wives to share their imagined glory?

Considering the perpetrators were supposedly arrested, the panic ban on liquid in carry-on baggage, causing many flights from London to be canceled, seemed somewhat excessive.

"Shoe Bomber" and "Underpants Bomber"

Shows danger represented by individuals acting alone or almost alone

> Besides the grandiose plots, such as the Bojinka Plot and the Liquid
> Bomb Plot, a couple of failed ones by individuals are of note.
>
> *American Airlines Flight 63, December 22, 2001*
> *Northwest Airlines Flight 253, December 25, 2009*

Shoe Bomber

Richard Reid was born in England in 1973 with a Jamaican grandfather on his father's side. He was in and out of prison for petty crime, and on the last occasion his father, a habitué of prisons, recommended he convert to Islam for a better life and better food. Upon his release, he attended mosques, including one where the anti-American cleric Abu Hamza al-Masri was holding forth.

He then made trips to Pakistan and underwent training at terrorist camps. On his last trip, he and another British man, Saajid Badat, went on to Afghanistan, where they were given shoe bombs, clunky footwear containing explosive material. On December 5, 2001, he traveled to Brussels, declared his passport lost, and obtained a clean new one from the British embassy.

On December 16, 2001, he traveled to Paris, and the next day he bought a return ticket from there to Antigua via Miami, meaning he did not need documentation for the US. On December 21, 2001, he tried to board the flight, but his disheveled appearance, and the fact that he bought his ticket with cash and had no checked-in baggage, made a security agent suspicious. He missed his flight, with the French police giving him the all-clear to fly the next day, as he did not seem to have anything suspicious on him.

After spending an uncomfortable night in damp conditions, he boarded American Airlines Flight 63 to Miami. Midflight, flight attendant Hermis Moutardier, thinking she smelled a burnt match, saw Reid attempting to light one. He promised to desist, but a few minutes later, as he tried to grasp her, he revealed a shoe in his lap with a fuse leading to it, which he was trying to light.

Several people were needed to overpower Reid, who was six foot three. A disaster had been narrowly averted since the fuse was damp because of the extended wait overnight in humid conditions, and very likely also because of Reid sweating. The other man, Badat, did not board his intended flight but failed to warn the authorities about Reid. Reid never expressed regret and has said he thought that God had not wanted it to happen just then.

Underpants Bomber

The Airbus A330 flight from Amsterdam to Detroit on Christmas Day 2009 had Delta Airlines livery but a Northwest Airlines 253 flight number, as Northwest was a subsidiary. On board were 279 passengers, 8 flight attendants, and 3 pilots. Having taken off at about 08:45 local time, it was due to arrive in Detroit at 11:40 local time.

As the aircraft neared its destination, a young Nigerian man called Abdulmutallab went to the toilet, spending some twenty minutes there before returning to his seat. Complaining of an upset stomach, he snuggled down under a blanket.

Twenty minutes or so before landing, passengers heard popping noises resembling firecrackers, smelled an odor, and saw the Nigerian man's trousers on fire. Jasper Schuringa, from the Netherlands, sitting on the other end of the same row, saw Abdulmutallab shaking with his trousers open and clasping a burning object. Schuringa pulled it from him and tried to extinguish the flames with his hands before jettisoning it. Flight attendants extinguished the fire with a fire extinguisher and blankets, and another passenger removed a partially melted, smoking syringe from the man's hand.

Schuringa grabbed Abdulmutallab and dragged him to the business class area at the front of the plane, where he removed his clothes to check for weapons or other devices. Abdulmutallab was restrained with plastic handcuffs. The virtually undamaged aircraft landed shortly after, and Abdulmutallab was taken into custody and given treatment for his burns.

Like the shoe bomber, he had created a device by mixing pentaerythritol tetranitrate, or PETN, a major ingredient of Semtex belonging to the same chemical family as nitroglycerin, with triacetone triperoxide, or TATP, and other ingredients. Liquid acid injected via a syringe was to be the trigger. It failed because he had been wearing the underpants for so long they were soiled.

A Litany of Security Failures

There has hardly ever been a case of so many security failures. The man's father had alerted the US that he posed a danger; the British had flagged him up to the Americans but omitted his surname; the CIA knew about him but did not void his visa because they hoped he would lead them to bigger fish, which reminds one of 9/11, where the NSA were listening in to calls made by one of the hijackers to an Al-Qaeda HQ while undergoing pilot training in the US. Finally, even without the alerts, the Israeli company, normally so adept at profiling passengers, should have stopped Abdulmutallab boarding—nonsensical routing, ticket bought for cash, and only hand luggage.

Chapter 17
"COMPUTERS WORK
IN MYSTERIOUS WAYS"

("THEIR WONDERS TO PERFORM")

Preening A320 Crashes at Local Air Show (Habsheim, 1988)
For years and years, the sacked captain disputed the official findings

> The crash of the innovative Airbus 320 at a tiny air show with
> passengers on board shown around the world seemed a bad portent for
> what was to be—and still is—one of the most successful aircraft ever.
> Was the pilot responsible? Or was it the computer?
> *Air France Flight 296, June 26, 1988*

Coming from Paris, the brand-new A320 would stop briefly at Basel-Mulhouse Airport to pick up sightseers and then impress spectators at a tiny airshow at Habsheim, close to the border with Germany and Switzerland, before going on to show the young passengers France's highest mountain, Mont Blanc, twenty-five minutes' flight time to the south.

Provided the pilots did not sink below the minimum permitted height of 100 feet (30 meters), given in their instructions, and their track kept to that of the 3,300-foot (1,000-meter) paved runway, with an obstacle-free overrun space extending some 650 feet (200 meters) beyond the end, the display location was reasonably acceptable. However, unbeknown to the pilots, the air show's organizers had placed the spectators facing a grass landing strip used for ultralight aircraft and gliders, set at a forty-degree angle to the paved runway. Besides being only 2,264 feet (690 meters) in length, this grass landing strip ended a mere 200 feet (60 meters) from the edge of a dense forest of young oak and birch trees, with many reaching to a height of 40 feet (12 meters).

A quick glance from above and afar might well have given the impression the closely packed trees represented a different-color field or mere bushes. However, it would seem quite probable that the pilots, with no time to assess the situation, simply presumed the spectators would be located along a landing strip suitable for flying demonstrations, with no immediate obstacles beyond.

Air France management had officially approved the application from the organizers of the air show, but final arrangements were hurried and seemed somewhat haphazard. The company only gave the file with the details of the operation to the pilots just before they left Paris, with no verbal briefing about conditions. The map of the airfield did not have a key to the symbols used, which might have been obvious had it not been a black-and-white photocopy of a colored map. Air France had previously done flyovers at Habsheim air shows, and it may have seemed to all a routine affair.

Since the two pilots chosen for the jaunt, Captain Michel Asseline, forty-four, and copilot Captain Pierre Mazières, forty-five, were both management captains—always a dangerous proposition—Air France staff perhaps too easily assumed they would be able to responsibly assess the situation for themselves and be able to perform the task without oversight. However, neither of them had experience of giving demonstrations at air shows.

As head of Air France's A320 training division and as one of the people responsible for liaising with the manufacturer over the technical aspects of the introduction of the A320 into Air France's fleet, Asseline had experience of executing the novel dramatic maneuver he was planning. He would use the ability of the computer to fly closer to the stall limit than a human could normally manage. He told his doubting copilot he had done it twenty times. While true, this would certainly have been over airfields or rather airports with plenty of space and in an A320 probably carrying much less weight in terms of fuel and passengers, if any, and importantly plenty of time to stabilize his aircraft.

In the mere *five minutes,* it would take them to fly from Basel-Mulhouse airport to Habsheim, they would have to lose height and speed, and lower the flaps and landing gear, before performing a low-level pass at a hundred feet (thirty meters), followed by a dramatic climb-out to impress the spectators.

Asseline would reduce the airspeed to just above the minimum needed to maintain horizontal flight and let the computer do the rest—namely, raise the nose very high to keep the aircraft flying parallel to the ground. Of course, to do this he would have to disengage the Alpha floor safety mode, designed to rescue the aircraft from dangerously high angle-of-attack situations, such as might develop in wind shear and the operation in question. The Alpha floor safety mode would otherwise apply full takeoff/go-around power via the autothrottle and spoil the party. Though the computer could be relied on to not let aircraft stall at the minimum airspeed, it would be a risky procedure if there were any obstacles in the aircraft's path, as it would be difficult to climb away at that minimal airspeed, and especially so if the engines were still idling and would therefore take more time to spool up.

It is a tenet of airlines and even the military that pilots making a landing, which this maneuver resembled, conduct a stabilized approach. "Stabilized" means the aircraft should have steadied and be properly lined up, and, as far as possible, be at the right speed well before the runway. Hurried, last-minute, possibly ill-judged maneuvers are to be avoided. If not stabilized in good time one must abort the landing and go around. The short distance between Basel-Mulhouse, where they were to pick up the passengers, and Habsheim would give the Asseline scant time to descend and stabilize in the best of circumstances. In addition, their unfamiliarity with the area would mean they would be looking for landmarks.

With the sightseers including many children picked up at Bâle-Mulhouse Airport on board, and Asseline announcing, "I am going to demonstrate the progress made[1] by French aviation," the jaunt was taking on the trappings of a trip to the circus, which is in fact what it turned out to be: a circus act shown on TV worldwide. The subsequent inquiry suggested the presence of the sightseers created a holiday feeling in Captain Asseline.

The A320 turned toward Habsheim with maximum bank and climbed to the thousand-foot altitude authorized by the Basel-Mulhouse Airport control tower. Still at almost a thousand feet, they headed for Habsheim having to refer to a map showing it to the right of the motorway as the "airport" was too small to figure in their navigation system. As a result, they were late in identifying it, leaving them insufficient time to drop down to a hundred feet (thirty meters) and reduce speed to that required for the demonstration. Asseline suddenly realized that the spectators were placed to view performances along the axis of that grass landing strip, set at a forty-degree angle to that of the paved runway, and some distance to the left. If he maintained his course he would fly over them, so banked to the right to fly in front of them. Just when he should have been stabilizing the aircraft at a hundred feet, he lost further height perhaps because aircraft lose some lift when banking. Adding to its vulnerability was the fact that the engines were throttled back to idle so Asseline could get the airspeed down to that minimum flyable airspeed in time for the computer to show off by pushing the nose up at the sharpest possible angle without stalling.

The aircraft did just as it was expected to do, ending up with its nose pointing upward at a maximum of fifteen degrees and with its trajectory virtually horizontal over the grass landing strip, which would have been fine except that the pilots had started the maneuver too low. The dangling main wheels were just thirty feet (ten meters), instead of a hundred feet (thirty meters), above the grass strip.

Some have suggested that having the nose so high up and the tail so low made the pilots feel they were higher than they were. The small scale of the airfield might also have deceived a captain used to runways four times as long and control towers three times as high. Others contest this.

Moments earlier Mazières had warned Asseline about electricity pylons carrying power lines almost two miles (three kilometers) ahead, and the captain told him not to worry, as he too had seen them. What neither of them realized was that the dark-green patch immediately ahead was a forest with trees so densely packed together that they did not stand out when viewed from above from a distance. As mentioned, some were forty feet (thirteen meters) high.

With the trees dead ahead and as high, if not higher, than they were (in the nose-up cockpit), Mazières was the first to see the danger. He shouted, "Go around!" They applied maximum power, expecting to climb away. However, the aircraft was performing at the corner of its flight envelope, and furthermore having had to lose height with the engines throttled back, it had neither quickly available engine power nor any airspeed margin.

1. Engine Thrust

Unlike internal combustion engines on motor vehicles, Jet engines do not instantly give extra power, and this is particularly true when idling, as indeed these were. This meant the engines would take six, if not eight, seconds just to spool up and begin truly accelerating the aircraft. If the pilots had in a timely manner properly stabilized their approach, the engines would have been at an intermediate power setting and spooled up much more quickly.

2. Airspeed

If an aircraft is flying normally well above its stalling speed, simply pulling back on the sidestick or control column will make it rise almost immediately. However, here the computer had been "told" to fly at a speed so close to stalling speed its nose was already pitched as high up as it could be just to stay level. Even were the computers to allow the pilots to push the nose up much further, which is unlikely, the aircraft would have doubtless stalled. In addition, that nose-up attitude with the wheels down would already have been producing considerable drag and preventing it from gaining forward airspeed.

One video shows the preening Air France A320 approach and pass in front of the camera. Then, just above the ground, it continues, preening even more to try to maintain its height. Heading straight for the trees, it must surely heave itself up, but its resources seem to fail it, even though the audio on the video shows the engines were finally spooling up.

As if making a perfect landing, the A320's rear underside brushes the trees, with it still seeming possible it might lift off. It then sinks into the foliage like a drowning swimmer engulfed by waves. A few moments later a dense cloud of smoke billows up from a point slightly farther on. Some spectators wail. A woman cries, "My God!" ("Mon Dieu!")

The two airfield fire appliances, and an ambulance with a doctor who was attending the show, rushed to the scene. Appliances from local fire departments tried to join them, but the larger vehicles were unable to get through the closely spaced trees.

Most Occupants Survive

As the last of the mobile occupants, some with smoldering clothes, made it through the forward exit, the flight attendant there called out to check whether there was anyone left. She got no reply, since the three remaining had already been overcome by smoke inhalation. Exiting at the rear had gone more smoothly, apparently in part because a flight attendant there had experienced an evacuation.

Amazingly, 133 of the 136 people on board managed to escape before fire consumed the entire aircraft, except for the tail and part of a wing. The passengers owed their relative good fortune to their lack of hand luggage (a point to remember) and the fact that many were young, physically able first-time flyers who had noted the location of the exits during the preflight briefing. Unfortunately, this meant some would not know how to release their seat belt buckle (another point to remember, as they differ from those in cars).

The three that perished were a handicapped boy, a young girl, who had been unable to undo her seat belt, perhaps because she had not been able to do it immediately and people behind had pushed her seatback forward, trapping her against the belt, and a courageous woman who came back to help her. A female flight attendant helping people exit from the front apparently went back into the cabin to check there was no one left. Not being able to see because of the smoke, she called out and got no reply; perhaps the three were already unconscious due to smoke inhalation.

Thirty-four passengers required hospital treatment for burns and injuries. In addition to suffering from shock and smoke inhalation, the two pilots were briefly hospitalized for minor injuries to the head and collarbone.

TV news bulletins around the world replayed and replayed the dramatic footage of the crash, much to the chagrin of the aircraft's manufacturer (Airbus Industries) and Air France, who had only taken delivery of the aircraft some three days earlier. The media, and particularly competitors, made much ado about the crash of this innovative aircraft, representing the new so-called fly-by-wire technology, which, thanks to computers, was supposed to enable aircraft to fly more safely. Some people, particularly in America, mocked Airbus, saying it seemed its computer made a perfect landing—on trees.

With this very public loss of their new-generation aircraft, the manufacturer and Air France went into a panic. Many prospective orders were at stake, as was the reputation of Airbus Industries, a consortium of French, German, British, Spanish, and West German companies. For the French it was considered, no doubt justifiably, an affair of state, as it threatened France's budding aircraft industry at a delicate moment.

The Aftermath

For a crash where "only" three people died and "merely" thirty-four were injured enough to require hospitalization, a surprising amount of controversy and bitterness ensued for more than a decade, even though investigators recovered the black boxes intact, had the benefit of the pilots being able to explain their actions, and there being several video recordings (some with audio). Nobody trusted anybody.

The various parties—Captain Asseline, the survivors and the injured, Air France, Airbus Industries, and even the French government—fought legal battles and exchanged insults. An exasperated Air France official was reported as saying, with more than a grain of truth[2], something like, "There would have been less trouble had the people died."

There were accusations that officials had tampered with the black boxes. Air France sidelined as unfit for duty a pilot who came out in support of Captain Asseline. The man in question thereafter remained an implacable foe of the airline and the aviation authorities. He has had no end of trouble ever since on account of his comments and the critical books he wrote, even accusing a cabinet minister of being involved in the alleged tampering of the digital flight data recorder (DFDR) and cockpit voice recorder (CVR) tapes. Reduced almost to the state of a beggar, he pursued the fight on the Internet to keep life and soul together.

As is the case in so many dramatic events—even President Kennedy's assassination—securing the "crime scene" and evidence is not usually uppermost in people's minds at the time. In the rush to analyze the CVR and DFDR tapes, those involved were casual in their handling of those valuable items of evidence. Then, when a judge later ordered their surrender to him, they were not immediately available, leading to increased suspicion that the authorities had been in the process of doctoring them.

Captain Asseline first maintained that the accident had occurred because the engines failed to spool up. Then on seeing video showing how extremely low he was, he suggested the barometric altimeter might have been giving the wrong reading, because of a bug in the software. He also said he pulled the throttles back for a moment when the engines did not respond, as from his experience he knew that was the way to reset the system, but there was no indication of that.

The subsequent inquiry laid the blame for the disaster largely on Captain Asseline but also criticized Air France for its casual approach to the whole operation. His pilot's license was revoked, and Air France dismissed him while retaining the services of Mazières.

Asseline then wrote a book called *Le Pilote—Est-il coupable?* (*The Pilot—Is He Guilty?*), and for years contested the official report.

An Affair of State

It was an affair of state whatever way one looked at it. Most of those claiming there was a cover-up admit that if there was one, the government officials involved did so not for personal gain but with the best of motives.

Regardless of whether any credence can be given to the conspiracy theories, Asseline's envisaged maneuver, which Mazières questioned at the time, was a reckless act over a tiny airfield. Furthermore, Asseline was literally playing to the gallery, even making an additional announcement in perfect German for those of his sightseeing passengers unable to understand French. His statement that he was going to demonstrate the evolution of French aviation was almost the language of a stunt or test pilot, and not that of an airline pilot. In addition, at Basel-Mulhouse Airport he had been taking part in a press conference rather than concentrating on the operation in hand. One can understand how difficult it would have been in that frame of mind for him to perform his stunt away from the spectators. Apart from the rashness of undertaking such a maneuver at an unfamiliar site with an almost full load of passengers, another fault would seem to be that Asseline took the Air France regulations regarding the hundred-foot fly-past height in a landing configuration—in other words, with landing gear and flaps lowered—too literally.

The hundred-foot minimum height was surely set for large, flat, open areas, such as major airfields or airports. In a confined area, such as Habsheim airfield, the height of nearby obstacles, such as the trees and the electricity pylons in the distance, obviously should have been factored in.

Asseline would no doubt have easily missed the trees had he aimed to fly at that hundred feet plus thirty-five feet or say fifty feet to allow for obstacles in the general area. This would no doubt have been so, regardless of whether, as he later claimed, his altimeter misled him, his engines had not spooled up as fast[3] as claimed by the investigators, and the aircraft did not lift when he thought it should have done. In any case, as it was not a large airfield, he should have maintained a greater safety margin and have been prepared for, say, a bird strike affecting an engine. Anyway, there was just not enough time to get the aircraft stabilized.

Many find his fight over the years to prove he was right, harming Airbus and Air France in the process, difficult to comprehend. However, as any good driver involved in even a minor road accident knows, one tends to replay and replay events, seeking things in one's favor. This must be especially true when so much is at stake and one was so proud of one's status before.

While Air France management surely had much for which to answer, they were presumably expecting their senior pilot to perform a simple fly past, not a stunt.

Anyway, the captaincy of a ship or aircraft is associated with the concept of command responsibility. That is to say, regardless of the poor decisions made by others, the captain should exercise good judgment, not get carried away, and make appropriate decisions notwithstanding the pressures to which he or she may be subject.

Afterthought—and a Lesson

As said, the passengers were mostly youngish adults or nimble youngsters, with no luggage. Even so, it was it was a very close-run thing:

Had the passenger profile been typical of one of today's commercial flights, with more elderly passengers and many collecting their luggage, the tragedy would have been far greater.

Airbus's Response to TV Documentaries, Etc.

Some TV programs and documentaries using their own "independent" experts have added their twist to the events. Airbus later issued a detailed and illuminating response rebutting their views.

These arguments were really a great waste of time, for even if the so-called outside experts and Asseline had on occasion been right—a big if—the rushed last-minute maneuvering in a constricted area with no time to stabilize the aircraft was a recipe for disaster.

One final point: Boeing were laughingly telling people how the vaunted A320 computer simply took over and landed the aircraft on trees, thinking they were a runway; Airbus and others point out that it stopped the pilots stalling, in which case the aircraft would have plunged nose-first into those trees, with awful consequences.

---

[1] He used the French word "continuité," which is difficult to translate.

[2] Proved right in the case of the Concorde crash, where everyone died, and claims were settled easily.

[3] The manufacturer of the engines subsequently improved their response time.

Indian Airlines A320 in Wrong Descent Mode (Bangalore, 1990)
Another A320 crash; aircraft too sophisticated?

> Just as Airbus seemed to be leaving the bad publicity of the Habsheim
> fiasco behind it, there was another Airbus A320 crash.
>
> *Indian Airlines Flight 605, February 14, 1990*

The Indian Airlines A320 Airbus had taken off from Bombay (Mumbai) with 146 people on board, including two captains in the cockpit. Both captains were experienced but had comparatively few flying hours on the A320, which with its sophisticated computer systems differed considerably from the aircraft they had previously flown, notably turboprops and 737s.

On the final part of the flight into Bangalore, in Southern India, the route-checking captain in the right-hand seat was the pilot flying (PF). This was a strange state of affairs, as the check captain had only flown about sixty-eight hours in the A320 and was not certified by Airbus Industries to be a check captain on the A320. Furthermore, the Indian civil aviation authorities had instructed the airline that the captain in question be positively monitored with regard to matters such as the sophisticated flight management guidance system (FMGS) on the A320. Not only that, the people instructing him on the A320 had observed "numerous small errors and omissions" regarding the use of the FMGS and operation of the power controls. At one point the captain undergoing the check asked whether the check captain flying the aircraft realized they were in the OIDM mode and suggested they switch to the heading vertical speed mode. The suggestion was apparently ignored.

Making its final approach to Bangalore with its landing gear already lowered, the A320 was descending, but as it descended it started sinking rapidly and was soon below the intended glide path, with its speed less than it should have been for an approach. Although it was obvious they were not going to make the runway, the pilots took no action, despite two sink rate warnings and four altitude alerts.

Only when they were at 140 feet, still sinking rapidly, and with the computer pushing the nose up to a high angle of attack to compensate for the minimal airspeed, did the pilot nonflying (PNF) shout, "Hey, we're going down!" and push the throttle levers for a TOGA (takeoff/go-around). In fact, the Alpha floor safety mode (disabled at the Habsheim air show) ensured the computer was already spooling up the engines. Even with the engines speeding up faster than certified, and perhaps faster due to improvements made in the light of the Habsheim disaster, the high sink rate combined with excessively low airspeed meant saving the situation would prove impossible in the hundred or so feet available.

With the nose pointing upward, but less than it would had the computer not prevented the aircraft from going into a stall when the pilots pulled their sidesticks back fully, the aircraft came down hard on a golf course over a half kilometer from the runway. After bouncing and forcing its way through some small trees, it skidded over a green. Slowing more but not enough to prevent the embankment around the course shearing off the engines, the aircraft finally ended its headlong charge just before the airport's perimeter wall and caught fire.

Despite the proximity of the airport, the fire service took more than twenty minutes to reach the scene. This was due to the locking of a gate for security reasons, a bad airport road, and lack of a radio link with the control tower. By the time the fire service arrived, it could do nothing except help those who had already escaped. At Habsheim almost everyone on board escaped, even though the A320 had caught fire immediately. This was largely due to them being able to save themselves thanks to the soft landing. At Bangalore 54 people out of 146 survived. Many were seriously injured. Both pilots perished.

The inquiry found that, contrary to instructions in the operating manual, the aircraft had been descending in open/idle descent mode (OIDM), with too low an airspeed for an approach. It should have been in the heading/vertical speed mode so that the flight director kept the aircraft pointing in the desired direction and, most importantly, controlled the rate of descent, not letting dangerously high rates of descent develop. Thus, once again it seemed an A320 had crashed because a senior pilot had not used the sophisticated systems wisely and appropriately. Some say the aircraft was too sophisticated for Indian Airlines at the time.

Recriminations

After the crash, the Indian government claimed Airbus was undermining the reputation of Indian Airlines' pilots and said the check captain was an excellent pilot, even though he was not certified by Airbus to be a check captain testing other pilots.

A French newspaper countered by claiming that C. A. Fernandez, the pilot being checked out, had performed badly in stress situations, had been referred for extended basic training in India, and had not been certified as passing. As mentioned, the check captain flying the aircraft had been found wanting by Indian government officials during tests.

Despite this early scare and spat, which led to the temporary grounding of the airline's A320s, the company has since purchased or leased many A320s.

In-flight Upset of Qantas QF72 A330 (off North Australia, 2008)

Passengers with seat belts unfastened hit ceiling panels

> Almost ten years after this event the *Sydney Morning Herald* published a
> sober article based on an interview with the just retired Qantas captain
> who had experienced this traumatic event.
>
> It was picked up by the *Daily Mail* in much more dramatic terms, "'It
> went psycho': Qantas pilot reveals horror moment computers on flight
> QF72 BROKE DOWN - sending the aircraft nosediving towards the Indian
> Ocean with 303 passengers on board."
>
> Details had been public knowledge for many years; however, seeing it
> through the pilot's eyes is very illuminating, and adds a valuable
> perspective for the second edition of this book.
>
> *Qantas Flight 72, October 7, 2008*

One can walk nonchalantly down a street for years until one day one is mugged by a hooded man with a gun. Although you may come out of it unscathed you have to take time off work to recover, and whenever you walk down that street you are anxious and hyper alert. This can happen to pilots who for some reason find their computerized aircraft behaving wildly and seeming to do mad things.

As in the case of the Qantas Bangkok overrun, the Australian Air Transport Safety Bureau (ATSB) has produced very comprehensive reports, namely a fact-giving interim report and a final report on the incident we now describe.

As the nub of the reports is technical, namely the performance of the air data inertial reference unit (ADIRU), and the way the computers were programmed rather than human actions, it is really for techies wanting to know how fly-by-wire aircraft work or those who think it might have some relevance to the Air France A330 crash described next.

In the Passenger Cabins

The Qantas Airbus A330, with 303 passengers, 9 cabin crew and 3 flight crew on board, had taken off from Singapore for Perth on October 7, 2008, at 09:32 local time. It was a regular scheduled flight, with flight number QF72.

Having completed almost two-thirds of their journey and with the aircraft cruising at thirty-seven thousand feet, the passengers were relaxing after their meal, with some on their feet to visit the toilet, when the aircraft suddenly pitched slightly downward before pitching nose-down even more. After the aircraft leveled out and climbed to regain height, the same thing happened again, except that the upset was less violent.

The first upset was particularly brutal; passengers were thrown about, with one woman hitting the fascia above her so hard that it left a dent, before falling back with a bleeding head wound. Most of the injured were those who had left their seat belts unbuckled. A few whose seat belts were loosely attached, and often near the limit of their extension, claimed they became unclasped as they were thrown upward—it seems the clasp got snagged under the arm rest. Those standing were injured, as were the two in the toilet—one seriously, the other less so.

Altogether, one member of the cabin crew and 11 passengers suffered serious injuries. Eight members of the cabin crew and 95 passengers received minor injuries, leaving 200 persons (3 crew and 197 passengers) relatively unscathed, though the roller-coaster experience must have been quite traumatic. As these were the officially reported figures, they may not give the full picture, as others could have been treated privately.

On the Flight Deck

At 12:40:28, and a couple of minutes before the first upset, the autopilot disconnected. There followed a series of warnings that this or that system had failed.

At 12:42:27, and with the pilots having had to take control while still trying to evaluate the confusing situation, the nose of the aircraft pitched downward without any such input command from the pilots. The maximum pitch-down angle was about 8.4 degrees, with the aircraft dropping 650 feet. The crew brought the aircraft back to its cruising height of 37,000 feet, all the time trying to deal with the cascade of failure messages. At 12:45:08 the aircraft made an uncommanded pitch down again. However, this second time the pitch-down angle only attained about 3.5 degrees, with the drop in height less too, at just 400 feet.

The aircrew made a "Pan, Pan, Pan" call, used instead of the well-known Mayday call to signify difficulties rather than a full emergency, and diverted to Learmonth, a backup Australian Air Force base also used by civilian aircraft. When informed by the cabin crew of the gravity of some of the injuries, the pilots made a Mayday call.

Because they were unable to enter a GPS approach into the flight management computer, they made a straight-in visual approach from about fifteen nautical miles, acquiring the precision approach path indicator (PAPI) at about ten nautical miles. They touched down safely at Learmonth at 13:50, with the passengers somewhat apprehensive due to the time taken making orbits before landing.

The Investigation

Though there was much information available from the pilots, the digital flight data recorder (DFDR), the cockpit voice recorder (CVR), the quick access data recorder (QAR), the onboard fault-monitoring systems, and data stored in the various units, it was difficult to determine every detail of what happened and why. Adding to the problem is that it was not quite possible to say why the number one ADIRU seemed to have malfunctioned in the way it did, with the investigations even looking at possible extraneous causes, such as special military radio transmissions. Another aspect was that how the computers react to the—possibly false—data they receive from a failed ADIRU depends on how Airbus programmed them.

To see the great lengths to which the investigators have already gone and to learn much about the A330 fly-by-wire control systems, the author suggests the reader download the actual preliminary report, which, as mentioned, is more for the technically minded. As it gives the facts and details of the work done without conclusions, one cannot sum it up, and to do so would risk distortion.

Other Incidents

Virtually any device can fail, and this is why essential ones, such as ADIRUs, are in triplicate, so that the computers can identify the faulty one as the odd one out. The report said the manufacturer in question stated the mean time between failures (MTBF) for the ADIRU model used on the aircraft in question was about 17,500 flight hours. The chance of two failing together would therefore be minimal. very low figure. Even if they did, that should not affect the flight controls in such as way. Also, Airbus has apparently stated it knows of no such case where the ailerons have been affected, though it seems this aspect was a matter of programming and a rare data input combination.

The ATSB investigated an in-flight upset occurrence related to an ADIRU failure on a Boeing 777-200 aircraft, which occurred 240 kilometers northwest of Perth on August 1, 2005. The ADIRU on that aircraft was made by a different manufacturer and of a different type to that on QF72. Further details of that investigation can be found on the ATSB website.

The ATSB investigated other incidents where the ADIRUs had problems on Airbus A330s, with one occurring geographically rather near and the other somewhat less near the location where this upset happened, leading to speculation that there might be some other factor, such as electromagnetic waves, at play. However, this is thought to be most unlikely.

The ATSB's final report, dated December 19, 2011, attributed the incident to the combination of a design limitation in the flight control primary computer (FCPC) software of the Airbus A330/A340 and a failure mode affecting one of the aircraft's three ADIRUs. This meant that, in a very rare and specific situation, multiple spikes in angle of attack (AOA) data from one of the ADIRUs occurring more than 1.2 seconds apart (that is, longer than the system would retain historic data on the assumption a spike was abnormal) could result in the FCPC commanding the aircraft to pitch down.

Lawyers representing relatives of victims of the Air France 447 crash (see next narrative) attempted to draw parallels with this incident. However, in that case problems with the pitot probes seem to have been a factor, not mentioned at all in this case.

AF447 Pilots Turned Glitch into Disaster (South Atlantic, 2009)

Recorders surprisingly found two years later explained how, but not why.

> Without being able to get into the head of pilot flying, one cannot know
> precisely why he did what he did.
>
> *Air France Flight 447, June 1, 2009*

The arrivals boards at Paris's Charles de Gaulle Airport indicated Air France Flight 447 from Rio de Janeiro as "Delayed," but with the delay becoming interminable it became obvious that something was seriously wrong. Finally, the airline had to admit the aircraft with 216 passengers and 12 crew had disappeared between Brazil and Africa.

With no Mayday call, and the Airbus A330 lost in an area not covered by radar, the only real clue as to what had happened was a short stream of cryptic ACARS maintenance messages sent from it by satellite to Air France in Paris. These showed the computers had switched to alternate law meaning the protections preventing the pilots flying outside the safe operating envelope were disabled.

A few days later the Brazilians began recovering small amounts of wreckage, including the vertical stabilizer (tail fin), from the South Atlantic Ocean surface. They also retrieved some fifty bodies, including that of the captain.

A French submarine and vessels trawling with sideways-scanning detectors began searching for the locating signal of the digital flight data recorder (DFDR) and cockpit voice recorder (CVR). They continued until the batteries in the locators would have run down a month later. It was claimed this search was unsuccessful because the water was almost two and half miles (four kilometers) deep and the ocean floor very uneven in places.

Neither the wreckage nor the bodies showed any traces of explosives, so a bomb was ruled out. The aircraft had clearly hit the water with considerable force, but with so little evidence the investigation was in a limbo. Speculation was rife. Did the A330 have some inherent defect?

Two years later, when the search was about to be abandoned, the organizers sought the help of US experts using mathematical probability theories to analyze searches where the object has initially been missed. Using their input, searchers rescanned areas where the wreckage might have been and soon found it thirteen thousand feet (four thousand meters) below the surface of the ocean.

The cockpit voice and flight data recorders were recovered intact, and to Airbus's relief revealed there was nothing intrinsically wrong with the aircraft. Though only revealed much later in the course of the judicial inquiry, the CVR had recorded the captain telling the junior first officer he had only had an hour's sleep the night prior to the flight and admitting it was "not enough."

Recap of Technical Terms:

"Normal law" simply refers to the usual flying mode where the computers controlling the aircraft are receiving reliable data and fly it optimally with "protections" preventing the pilots doing something stupid such as stalling it.

"Alternate law" refers to the mode where the computers controlling (flying) the aircraft, having determined that the data, such as airspeed, they are receiving must be unreliable, switch to a simplified operating law, where many of the protections, such as preventing the pilot stalling the aircraft or doing something that could otherwise endanger it, are suspended. On switching to alternate law, autopilot and autothrust disengage, and control is passed to the pilots, who are alerted by a "cavalry charge" audible warning.

"Pitch" is the geometric attitude of the aircraft relative to the earth's surface in the longitudinal plane. (Degree of bank is the attitude in the lateral plane.)

The **"angle of attack,"** called **"Alpha" or "α"** by Airbus, is the angle of the airflow striking the wing. If the aircraft is cruising normally at an appropriate speed, it is often not so very different from the pitch. However, if the aircraft is sinking, say in a stall, where the wings are not providing enough lift to keep it up, the pitch can, say, be five degrees and the angle of attack (the angle of the airflow to the wings) perhaps forty degrees as the aircraft "flops" downward.

The **"flight director"** (FD) bars on the primary flight display (PFD) show what the flight management computer (FMC) thinks the pilot and/or the autopilot should do, but in the absence of valid airspeed data may just try to keep vertical g-forces constant, even when climbing or descending, and are not a valid guide when the aircraft is dropping like a brick in a stall and not actually "flying." They possibly indicated the pilots should climb to cruising height. Usually, the pilots tell the FMC what they intend—for example, at what heights they want to fly at various stages of the flight and the waypoints they want to pass over, and so on.

The **"electronic centralized aircraft monitor"** (ECAM) is a screen indicating detailing problems, faults, and advising on the action to take.

The Pilots

The eleven hours' duration of the flight back to Paris meant there were three pilots so that they could take turns to rest.

1. **Marc Dubois**, the fifty-eight-year-old captain with approximately 10,988 flying hours of which 1,700 had been on the Airbus A330. He had made 16 south America rotations;
2. Thirty-seven-year-old **David Robert**, the most senior of the two first officers with 6,547 flying hours and 4,479 hours on the A330, including South America rotations.
   A graduate of Frances elite ENAC (aviation college), he had been elevated to a management position and was making the trip to keep his pilot's license current. Talking to the captain in Robert's absence the junior first officer referred to him as "he" rather than by his first name suggesting he was viewed as an outsider, perhaps being part of the elite;
3. Thirty-two-year-old junior First Officer **Pierre Bonin** with 2,900 hours and only 807 on the A330 and rated for it six months earlier. He had only made five A330/A340 South America rotations.

Though the aircraft took off three hours six minutes before the crash, the cockpit voice recording starts about two hours before it was deactivated by the impact with the sea. This was because to comply with pilots' unions' insistence that the cockpit voice recorder not divulge private conversations that could be exploited by management this one worked on a two-hour loop, meaning the pilots could talk freely mid-flight on the assumption what was said would be overwritten before arrival in Paris.

Time Line of Flight in Hours, Minutes, and Seconds

−02:00:51 Two hours, and fifty-one seconds before disengagement of autopilot
Captain Marc Dubois and the junior first officer, Pierre-Cédric Bonin, are alone on the flight deck. Senior First Officer David Robert is resting in the rest station, just behind the cockpit. [We have omitted what is on this early part of the CVR except for the following exchange where the captain says they would not change course to avoid the cumulonimbus clouds (storms) ahead and suggests Bonin, the junior first officer, must be tired.]

−01:55:05 Captain:
> We're not going to let some cu-nimbies (cumulonimbus) mess us around.
The French is more vulgar, but our translation conveys the intended meaning.

−01:11:58 Captain suggests junior FO gets some rest:
> Try maybe to sleep twenty minutes when he [the senior FO] comes back or before if you want.

Junior FO:
> Okay, that's kind. For the moment, I don't feel like it, but if I do feel like it, I will.

Captain:

> *It'll be a lot for you.*

The captain was obviously concerned about the junior FO getting some rest, but it seems odd that the senior FO is allotted proper rest time in the rest station and not the junior FO. When the captain said, "It'll be a lot for you", he would seem to have been thinking of the total flight time of some eleven hours, not to mention the time on duty before takeoff.

–00:18:07 Captain comments on weather conditions ahead:

> *It's going to be turbulent [bumpy] when I go for my rest*

–00:14:08 Captain notes it is time to wake up the senior FO:

> *Well, right, we just have to wake him up. That's all, eh?*

–00:13:59
Sound of high/low cabin chime (–00:13:55) plus there's a noise like a knock on the partition of the rest station

–00:13:49 Captain asks junior FO about who is landing the aircraft:

> *Er, who's doing the landing? Is it you? Well, right, he's going to take my place.*

The junior FO does not reply.

–00:13:45 Captain confirms FO is a PL [fully licensed pilot], plus there's a change in the background noise:

> *You're a PL, aren't you?*

–00:13:44 Junior FO replies in affirmative:

> *Yeah.*

It was quite logical for the junior FO to perform the landing on their return to Paris, as the senior FO had done so on the way out to maintain his currency and keep his license valid while working in management. However, it seems strange the captain is questioning Bonin's qualifications, something he should have noted prior to departure, and even before the outbound flight.

–00:11:36 to –00:11:34 Private conversation

The BEA has omitted this exchange considering it to be of a private nature. However, one wonders whether it was not just the type of conversation that might have thrown useful light on Bonin's feelings concerning Robert.

–00:11:06 Captain colloquially emphasizes his stint is over—sound chair being adjusted:

> *That's it.*

–00:10:39
[Sound of cockpit door opening.]

–00:10:33 Captain to senior FO:

Okay?

–00:10:29 Junior FO to senior FO:

Did you sleep?

–00:10:27 Senior FO replies:

So-so.

–00:10:21 Captain:

You didn't sleep, then.

–00:10:18 Junior FO answers for senior FO:

He, he said so-so. . . so-so.

–0:09:57 Captain announces he is going for his rest:

Well, I'm out of here. ["I'm scarpering" is a more approximate translation, since the French is very colloquial.]

–00:09:46 Senior FO replies to junior FO's comment:

I . . . I was dozing in fact.
Senior FO asks junior FO whether he is okay.

–0:09:29 Junior FO replies:

Okay.

–0:09:32 Junior FO briefs senior FO on situation in presence of captain

Well, the little bit of turbulence that you just saw we should find the same ahead. We're in the cloud layer. Unfortunately, we can't climb much for the moment because the temperature is falling more slowly than forecast. So, what we have is some REC MAX [recommended maximum height given by computer for the conditions].

–0:08:34 Captain tells junior FO which frequencies to use

Er, sixty-six, forty-nine, fifty-five, sixty-five, and after it's sixty-five, thirty-five.

–00:08:19

[Noises in cockpit. Captain leaving.]

–01:08:05 to –00:06:28 The FOs discuss diversion airports available
–0:06:21 Junior FO:

The intertropical convergence there. Well, we're in it between SALPU and TASIL.

–0:04:10 Junior FO:

We'll call them in the back to tell them anyway because . . .

–0:04:00 Junior FO talks to cabin crew

Yes, it's in front. Tell me in two minutes. There we ought to be in an area where it will start moving about a bit more than now. You'll have to watch out there.

–00:03:52 to 00:03:46 Dialog between senior cabin crew member probably in the front cabin and junior FO:

All right, are we to sit down?

Well, I think that it might be a good idea to tell your . . .

Yeah, okay. I'll call the back (flight attendants at the back); thanks a lot. Thanks. I'll call you when we're out of it.

–00:03:04 Junior FO:

See, we're really on the edge of the layer [and under the squall].

–00:02:02 Senior FO:

Don't you, maybe want to go to the left a bit?

–00:01:58 Senior FO:

You can possibly go a bit to the left. I agree that we're not in manual, eh?

The senior FO suggested changing course a little to the left to avoid turbulence they could see on their weather radar, and they altered the heading to a setting minus twelve degrees off their preset route.

–00:01:24 Junior FO:

What's that smell now?

–00:01:22 Senior FO:

It's . . . it's ozone.

–00:01:21 Junior FO:

It's ozone—that's it. We're all right.

–00:00:19

[Background noise increases]

Other A330-340 pilots identified it as typical noise of impact of ice crystals. Neither pilot alluded to it. The BEA noted that at the time pilots had little knowledge regarding the phenomenon. Nevertheless, Robert (the PNF) took it upon himself to reduce the airspeed down toward Mach 0.8 and switched on the engine anti-icing, no doubt in response to the indication of ice shown by the instruments.

–0:00:02 to 00:00

[Sound of cavalry charge warning that autopilot is disengaging]

The "Glitch"
(00:00:00) [02:10:05 UTC]
AUTOPILOT DISENGAGES

00:00:01 Junior FO Bonin who is the pilot flying (PF):

I have the controls.

00:00:03 Senior FO Robert who is the pilot not flying (PNF):

All right.

Unlike Boeing airliners, which have traditional control columns that move in unison on both sides of the cockpit, Airbuses have sidesticks with no mechanical or simulated interconnection. In addition, as these are located on the coming on the far side of the pilots, it is very difficult for the other pilot to see what his colleague is doing, and particularly so in relative darkness. Apparently, Robert had his seat set right back making it even more difficult to see what Bonin was doing with his stick.

   Thus, the senior FO did not see that the PF had pulled back his stick and put the aircraft into a rapid climb. Without application of more power, the aircraft could only achieve this climb by using up kinetic energy in the form of airspeed and soon lost so much that it was in danger of stalling.

00:00:05 Two SV (synthetic voice) stall warnings (BEA calls it Stall 1 warning)

Stall.

Stall.

00:00:06 Senior FO:

What is that?

Even so, the junior FO was still for some reason keeping the nose up, thus preventing the aircraft from diving and regaining essential airspeed.

00:00:08 Two SV stall warnings, last one truncated

Stall, s—

00:00:09 Junior FO:

We haven't got a good display . . .

00:00:10 Junior FO:

We haven't got a good display . . . . . . of speed.

00:00:11 Senior FO:

We've lost the speeds . . .

00:00:17 Senior FO:

Alternate law, protections law/Low/Lo.

The CVR transcriber could not tell what Robert had said, so it is unlikely Bonin had. In fact, what the ECAM had spelt out was "LOST." Over such a crucial point Robert should have insisted on confirmation that Bonin understood protections had been lost. Worse, while Robert was speaking, Bonin was trying to make a point about the thrust levers. Bad CRM, two talking at the same time.

00:00:19 to 00:00:23 Senior FO:

Wait, we're losing . . .Wing anti-ice. Watch your speed. Watch your speed.

00:00:23 Junior FO:

Okay, Okay, Okay. I'm going back down.

00:00:26 Senior FO:

Go back down.

00:00:27 Senior FO:

According to that we're going up. According to all three you're going up, so go back down.

00:00:30 Junior FO:

Okay.

00:00:31 Senior FO:

You're at . . . Go back down.

00:00:37 Junior FO:

We're in a . . . We're in a climb.

00:00:45 Senior FO:

* * *, where is he?

It seems the senior FO had already called the captain. Anyway, from this time onward, he tried several times to call him back, and this must have drawn his attention away from what the junior FO was doing at this most critical time.

00:00:46

[Almost-continuous stall warnings start] (BEA calls them Stall 2 warnings)

00:01:01 Senior FO:

* * *, where is he?

Bonin (the PF) made slight nose-down inputs alternately to the right and to the left. The rate of climb, which was astounding for that great height with thin air had reached seven thousand feet a minute, had dropped to seven hundred feet a minute, and the roll varied between twelve degrees to the right and ten degrees to the left. The airspeed shown on the left side increased suddenly to 215 knots (Mach 0.68), while the speed displayed on the left primary flight display (PFD) remained invalid for twenty-nine seconds.

By then the aircraft was at an altitude of about 37,500 feet and the recorded angle of attack around four degrees. Stall 2 warnings continue.

00:01:05 Aircraft ultimately reaches a maximum height of 37,924 feet

00:01:20 Senior FO:

Do you understand what's happening or not?

00:01:28 Junior FO:

I don't have control of the airplane anymore now.

00:01:30 Junior FO:

I don't have control of the airplane at all.
Repetitive Stall 2 warnings continue.

00:01:37 Return of the Captain

[Noise of cockpit door opening.]
The captain's return to the cockpit coincided with the aircraft mushing downwards through 35,000 feet, the altitude it had been when he had departed a little earlier. Apart from the alarms and the senior FO's expression of concern, a quick glance at the altimeter might have led him to believe that not much had changed. He was not to know that the A330 had climbed almost 3,000 feet without added power and could well be stalling.

Captain:

Er, what are you [doing]?

00:01:39 Senior FO:

What's happening? I don't know, I don't know what's happening.

00:01:42 Junior FO:

We're losing control of the aircraft there.

00:01:39 Senior FO:

We lost all control of the aircraft. We don't understand anything. We've tried everything.

00:01:48 Captain:

So take that.

00:01:50 Senior FO:

Take that, take that.

00:01:53 Junior FO:

I have a problem; it's that I don't have vertical speed indication.
In fact, the aircraft was dropping so fast the indicator could not show the actual sink rate, though there was a colored mark indicating it was "off the dial."

00:01:59 Junior FO:

I have the impression that we have some crazy speed, no? What do you think? The cacophony generated by the aircraft mushing downwards, with the air howling as it splayed off the wings, may have given the FO the impression they were traveling too fast, and needed the air brakes.

00:01:42 Senior FO:

No, above all don't extend [the air brakes].

00:02:09 Senior FO:

What do you think about it? What do you think? What do we need to do?

00:02:10 Captain:

I don't know, it's going down. [Impossible to translate colloquial "there" that the captain added, and the junior FO starts using too. Some exchanges omitted.]

00:02:22 Senior FO (with stall alarm still sounding):

You're climbing.

00:02:23 Senior FO (stall alarm still sounding):

You're going down, down, down.

Captain:

* * * [Going down]

00:02:25 Junior FO:

Am I going down now?

00:02:27 Senior FO:

Go down.

Captain:

No, you climb there.

00:02:34 Junior FO:

Okay, we're in TOGA. TOGA (takeoff/go-around) is a button that sets the engines to high power, and the control surface configuration, so that when in dense air at low altitude the aircraft can immediately climb out and away. It can also stop the aircraft stalling. However, at high altitude where the air is thin, the situation is quite different. Maximum thrust has hardly any effect, and with plenty of height to play with, the first requirement is to get the nose down so the aircraft dives and gains airspeed.

00:02:37 Junior FO:

On alti what do we have here?

00:02:39 Stall warnings ongoing.
Captain:

> ***, it's impossible.

The junior FO again asked about altitude, without getting a reply, and in the following twenty seconds the captain's main concern was to tell him to get the wings horizontal, as the synthetic voice stall warnings continued.

00:03:13 Junior FO:

> We're there, we're there, we're passing level one hundred [ten thousand feet].

00:03:20 Junior FO:

> What is . . .? How come we're continuing to go right down now?

The stall warnings ceased when the airspeed fell below sixty knots because the creators of the software wanted to avoid it giving spurious stall warnings when landing and taking off. They could never imagine pilots would ever allow airspeed to fall so low in normal flight. The odd result in this context was that when the pilots did something right, such as raising the airspeed above sixty knots, the warnings would restart, possibly making them think their action might be wrong. (Even when the airspeed was somewhat above sixty knots, the angle of attack meant the airspeed registered by the sensors would be less.)

00:03:35 Junior FO:

> But I've been in maxi nose-up for a while.

This remark makes the captain finally realize what the junior FO has been doing wrong all along.

00:03:38 Captain:

> No, no, no, don't climb.

00:03:40 Senior FO:

> So, give me the controls; the controls to me; the controls to me.

00:03:41 Junior FO:

> Go ahead. You have the controls. We are still in TOGA, eh.

00:03:50

With the senior FO taking control, the stall warnings recommenced even though the situation was improving, as explained above. With them dropping at around ten thousand feet per minute and little height remaining, it was too late anyway.

00:04:00 Captain:

> Watch out, you're pitching up there.

00:04:02 Junior FO:

> Well, we need to; we are at four thousand feet.

00:04:11 SV (synthetic voice) sink rate warning

[The "Pull up!" warning continues until the aircraft hits the sea.]

00:04:21 Impact with the Sea

The recordings stopped at two hours, fourteen minutes, and twenty-eight seconds UTC, that is four minutes and twenty-one seconds after the autopilot disconnect.

The last recorded values were:

Vertical speed: –10,912 feet a minute
Ground speed: 107 knots
Pitch attitude: 16.2 degrees nose-up
Roll angle: 5.3 degrees left
Magnetic heading: 270 degrees.

Though some of the passengers, not forgetting the cabin crew, would have been aware something was wrong, the fact that the nose was pointing upwards may have reassured them. The deceleration on hitting the sea, with the aircraft dropping at 10,912 feet a minute, would have been so great that they would have lost consciousness instantaneously.

THE NINE FACETS OF THE DISASTER

Facet 1: Junior First Officer Bonin (PF)

The BEA investigators tentatively attributed the abruptness of Bonin's stick input to the "startle effect" of suddenly having to take over and lack of experience/training for manual flying at that great height.

They also suggested he might have climbed because the icing-up of the pitot tubes had caused the altimeter to erroneously show a drop of a few hundred feet below its programmed thirty-five-thousand-foot cruising height, a deviation which would be indicated by an audible alert. Also, they said earlier repeated exchanges with the captain showed Bonin was fixated on the need to fly higher, and the sight of clear sky above might have reinforced this.

What little practice Bonin had of flying manually most likely involved following the flight director. Later, when the flight director bars reappeared, the investigators thought Bonin might out of habit blindly followed them, although with it simply reacting to changes in g and believing the steep climb and precipitous drop were intended, it was not telling them how to recover.

Perhaps the most plausible reason for Bonin's unbelievably steep climb, at least initially, and one highlighted by psychologists using sophisticated terms such as "cognitive tunneling," was that, not practiced in keeping the wings level under manual control in alternate law at high altitude, he became so mentally consumed by that delicate task that he simply did not realize he was climbing so steeply. Bear in mind that when the captain did eventually arrive, he too was primarily concerned with Bonin keeping the wings level and not the pitch.

As William Langewiesche so rightly points out in a piece for the October 2014 edition of *Vanity Fair*, the French pilots' unions take a very hard stand defending colleagues killed in the performance—or one might say—in the nonperformance of their duties. They say publishing what is on a cockpit voice recorder is an infringement of their deceased colleagues' human rights, though the pilots in question may well have brought about their own deaths and those of hundreds of passengers and crew. Though ostensibly noble, the unions' thereby cow the French investigators into not revealing general poor conduct or incompetency. The BEA perhaps for that reason tell us little about Bonin, or the other two aircrew for that matter. Some on the pprune.org website have said that as a former cabin attendant, like the captain, Bonin had not had to take the usual psychological tests prior to being taken on as a potential pilot.

Before going for his rest, the captain handed over authority to Bonin in a most perfunctory manner, even, unbelievably, having to confirm as an afterthought that the junior first officer was qualified (to make the landing in Paris, or perhaps take command). Admittedly, the Air France protocol (since changed) for the handover was not well-thought-out and implied the pilot flying was also in charge, even though he might be the less experienced of the two.

The BEA said this resulted in the perverse situation where on one level Bonin deferred to Robert, without Robert having executive authority on another level. One commentator has even suggested that while Bonin was vocally acknowledging Robert's recommendations, his actions bore no relation to them—perhaps akin to someone being nagged by a spouse and saying "Yes, yes, dear."

The BEA investigators did not pick up on Bonin not wanting to follow the captain's suggestion that he rest or catnap at his seat, since keeping awake all the way to Paris "would be hard on him." Robert, too, seemed concerned about Bonin's condition, which suggests there must have been some good reason for that. Could some other factor, say excess caffeine, with its five-hour half-life, found in some soft drinks in Brazil, have made him highly strung and twitchy? One can never know.

Finally, Bonin alluded at the beginning and toward the end to them being in TOGA (takeoff go-around mode), evidently believing it would obviate a stall, which was not the case at high altitude. As already said, Robert had also unforgivingly not made clear to him normal protections were lost.

Facet 2: Senior First Officer Robert (PNF)

As said, Robert was unable to assume command because Air France handover protocols at the time inferred the pilot flying (PF) was in charge. Also, he may have been undermined by the cozy relationship established between the captain and Bonin following the helicopter tour of Rio the two had taken with their women on the day of departure. Had he not only taken command, but also taken over the controls it is unlikely he would have made that disastrous climb. There would have been no stall, and as the aircraft moved into more clement conditions the ice in the pitot tubes would have melted and everything returned to normal.

Robert could see that the aircraft was climbing in the half minute following Bonin's assumption of manual control, telling him several times, without significant effect, to go down. Thirty-one seconds into the incident, seeing they were still climbing, he reiterated his wish that Bonin go down, but instead of making sure Bonin did so, he diverted his attention to calling the captain.

Showing his frustration at the captain's tardiness, Robert exclaimed, "Where the heck is he?" at +00:46 and +01:01 against the background of the repetitive stall warnings, to which he otherwise might have paid more attention. This was the critical moment, wasted, when the situation was about to radically worsen. Maybe he felt only the captain could deal with Bonin.

Four seconds later the aircraft reached its apogee, and having run out of kinetic energy, its airspeed decayed. Hanging there, it began to truly stall despite a couple of desultory nose-down inputs from Bonin.

The lack of airspeed also meant the A330 did not easily respond to stick inputs, so when Robert asked Bonin at +01:20, one minute twenty seconds after the autopilot disengagement, whether he knew what was happening, Bonin answered he no longer had control of the aircraft, repeating himself ten seconds later.

How could Robert, after an almost 3,000-foot climb in thin air, not have even given some credence to the stall warnings we shall never know. Furthermore, he had his seat set far back, as if he were just a casual observer, and his seat, later found in that position, stayed back. It made it more difficult for him to see what Bonin was doing. Was it so the captain could more easily take his place? Was he longsighted?

Though Robert, as a graduate of the ENAC (National School of Civil Aviation), was considered part of the elite and assumed to be supremely qualified by commentators, his course there had been in the field of air traffic control. In consequence, though an experienced pilot, he had no special knowledge concerning the operation of computer-controlled aircraft. Prior to the trip to Rio, he had not even landed an A330 for almost three months, and with piloting not being his role at Air France, he might not have been assiduously trying to learn everything possible in that domain.

There is something to be said for the "captain" not taking over and concentrating on the thinking, but in this case, Robert was not only distracted from that by repeatedly calling the captain, but any conclusion was negated by Bonin not following his advice, or worse, making him believe he was doing so when doing the opposite.

Facet 3: Captain Dubois

By deciding early in the flight not to alter course to avoid the storms, the captain in a sense set up the scenario that produced the glitch. Admittedly, it was allegedly Air France policy not to load significant extra fuel for major diversions to avoid storms on those Rio flights.

According to a mutual woman friend of both the captain and his veteran flight attendant-opera singer companion, who was to have joined them on the jaunt to Rio but missed the outbound flight because of indisposition, he was greatly looking forward to the three-day shared layover as a release from the stress at home related to his divorce. One can surmise that on the return flight, in addition to being extremely tired he would have been euphoric and more likely than usual to "not let some cumulonimbus mess them around," whereas other aircraft that night had changed course to avoid such storms.

According to a medical expert, his lack of sleep would have put him at times in a soporific state, with his faculties diminished like a drunk, especially on returning to the cockpit after relaxing and going into a quasi-sleep. Bill Palmer, author of *Understanding Air France 447*, thinks the captain was not in the rest station and returned to the cockpit because of the strange behavior of the aircraft, for instead of asking what was happening, he had said, "Err, what are you [doing]?"

If true—the French investigators did not belabor this point—this means his return was delayed because he was most likely with his companion. Admittedly, any pilot might go back to chat with the flight attendants in the galley before taking his or her rest. Nonetheless, it was perhaps one of many factors contributing to the disaster, in that when he did finally return to the cockpit, the aircraft was mushing downward through the altitude it had been when he departed. For a moment, he might have thought nothing much had really happened in the interim. Had he returned some thirty seconds earlier, the aircraft would have been at its apogee, making him perhaps realize what was happening and the stall warnings were continuously sounding for good reason. Even fifteen seconds earlier could have made a significant difference.

Only at the very end, when Bonin told him he had been in "maxi nose-up for a while," did the captain realize too late what had been happening and tell him not to climb. Considering the captain's sleep deficit, even that was quite a feat. To be fair to him, it would be difficult for anyone to properly assess the situation without being told they had just made what was at that altitude a ridiculously steep climb. Even if he had, was he in good enough shape to take over aggressively and appropriately?

A retired British Airways captain who used to fly that route from Rio de Janeiro has said he would never have gone for his rest before entering the potentially dangerous intertropical convergence zone, something that was accepted practice for Air France captains.

Had the captain been in his bunk next to the cockpit and returned sooner, Robert would not have diverted his attention for so long away from monitoring the aircraft and perhaps would have made sure Bonin was complying with his advice.

Facet 4: Cockpit Resource Management (CRM)—Culture of the airline

After the overrun of an Air France A340 at Montreal in 2005, the freelance French aviation writer François Hénin said the airline was making serious efforts to rectify the easygoing attitude of its aircrew. William Langewiesche, in the perceptive *Vanity Fair* article mentioned above, inferred that at Air France the egalitarian discipline of CRM had devolved into a self-indulgent style of flying. Though confirming actions with one's colleague may seem a waste of time in normal circumstances, getting into the habit of doing so means one is much more likely to do so in a crisis. Some pilots at Air France may have been somewhat lax in this regard and lost the habit.

Although at the beginning cognitive tunneling may have meant Bonin was too preoccupied with keeping the wings level to consider anything else, he should later have made it clear to Robert he was pulling back on his stick, and, as said, only mentioned that to the captain, right at the end. That was too late but indicated he was aware of what he had been doing.

As said, Robert did not even confirm Bonin understood they had lost the usual protections such as stall protection, and in fact did not finish his phrase.

Adding to the confusion was the fact that once the aircraft lost airspeed and ceased responding immediately to sidestick inputs, statements or interrogative remarks by Robert, such as "You are going up/down," did not correlate with movements of Bonin's sidestick. For Bonin to push the nose down, he would have had to push his sidestick forward and maintain it there for some time, perhaps reducing engine thrust to stop it levering the nose up.

Facet 5: Seventy-Five Synthetic Voice Stall Warnings Not Heeded

The crash occurred because the aircraft stalled, but at no point did any of the pilots mention the word "stall," though at one stage Robert said, "What's that?" when the warning sounded without making clear to what he was referring. These stall warnings went on for a very long time, and the captain must have heard them even though they stopped soon after his arrival when the airspeed fell below sixty knots, as explained.

The BEA and others point out that when a stall warning sounds in normal flying conditions, it is usually of little concern to the pilots, as the computer—under normal law—is always there to protect them. Also, it is very possible Bonin thought the stall situation was covered, as he twice says, as if reassuring himself, that they are in the TOGA (takeoff/go-around) configuration, which at low altitude (but not high) would normally get them out of a stall. As mentioned elsewhere, it has been suggested that sounds tend to get mentally filtered out in crises. However, the stall warnings did sound seventy-five times in all so could hardly have passed unnoticed.

Facet 6: Pilot-Machine Interface

In its conclusions and recommendations, the BEA report euphemistically refers to the pilot-machine interface as ergonomics, when in fact they give it a much broader meaning.

For anyone particularly interested in the pilot-machine interface facet of this incident, we must again recommend *Understanding Air France 447*, by Bill Palmer. Particularly interesting is the way he explains the great gulf between flying an Airbus A330 in normal and alternate law.

Sullenberger and others, and the BEA, have mentioned possible improvements, not all really coming under ergonomics:

1. There should be a clear indication of the reason that the autopilot disengages of its own accord. This should be shown on the ECAM, as should stall warnings—in addition to the audible ones.
   This would have meant the pilots would not only have faced less uncertainty but, perhaps more importantly, would have meant the pilot not flying (PNF) could have concentrated on what the aircraft was doing rather than wasting time trying to fathom what was happening and reading out irrelevant messages appearing on the ECAM.
2. In a crisis, humans pay more attention to visual cues than to aural cues, which are often lost among a cacophony of other aural warnings, and there should be a visual cue in addition to the one on the airspeed indicator tape, which was not valid anyway.

3. Following on from the previous point, airliners should be equipped with an instrument showing the angle of attack, which would confirm the stall and give pilots a better indication of what is happening. Its installation on certain military aircraft apparently significantly reduced accidents. However, getting the angle of attack indicator to provide useful information it not simple as stalling speed depends very much on height.

4. As already mentioned, the sidesticks do not move in unison, making it difficult if not impossible for the PNF to see what the PF is doing, and notably something silly. Some means of indicating this should not be difficult. Likewise, the throttles do not move to remind pilots of what the engines have been told to do. Eliminating such features has some pluses, but the case needs to be argued.

   Interestingly, though Boeing are resolutely sticking to their control column concept, more users, military, high-end business jet users, and now civil, are favoring the sidestick, and notably "active Sidesticks." These not only enable the pilot flying to better feel what he is doing, but also allow the other pilot to see or feel what he or she is doing and might well have enabled Robert to prevent Bonin making that disastrous climb in the first place.

Airbus ironically may be right in that the pilot-machine improvements proposed, though in theory desirable, may not after proper training be needed to prevent an absurd repeat.

Facet 7: Air France Training

The BEA very much exculpates the pilots because Air France had not trained them to cope with such a situation at high altitude. Post disaster, Air France, not to mention many other airlines, have been training their pilots to handle high-altitude upsets and stalls and how to avoid them in the first place.

Facet 8: Pitot probes and choice thereof

The icing-up of the pitot tubes supplied by the French conglomerate Thales—intent on ousting the US makers of such equipment who had gained a stranglehold after World War 2—also provided the trigger for the disaster, and Air France should have replaced them, and were in the process of doing so.

However, the disaster resulted from the way the two first officers mishandled the situation. Other instances where the pitot tubes had iced up had been handled competently, though in many of those cases it was daylight and the pilots had more visual references. Anyway, in the year or so prior to the disaster we are discussing, Air France had nine instances of problems with the A330 and A340 pitot probes and was changing them.

Facet 9: Investigative and supervisory bodies—BEA, DGAC, and EASA

When the BEA announced that while the pitot tubes may well have been a factor in the crash, but were surely not the only factor, the Air France pilots' union said it was only trying to cover up its own failures and those of the DGAC (French Directorate General of Civil Aviation). The argument was that it had failed to pick up on problems with the Thales pitot tubes, and the European Air Safety Agency had not been diligent in applying the latest lessons in certifying them.

The BEA report said the A330s had originally been fitted with Goodrich probes, but the DGAC, on finding some problems, ordered that they be fitted with a more advanced Goodrich probe or with the Thales probe that iced up on AF447. Thus, the French DGAC opened the door for the Thales probe, but Air France was still responsible for going through with it, perhaps with Thales being very persuasive.

Mutual Dislike of The Two First Officers

Because we think they had a material bearing on the disaster, we have alluded to private matters one would not normally divulge. Firstly, the ones leading to the captain's sleep deficit, and even his very likely euphoric and soporific frame of mind when early in the flight he just might have bothered to adjust course to avoid the storms, as did the captains of other airliners that night.

Bill Palmer's fascinating book, *Understanding Air France 447*, dismissed our point regarding the captain's supposed fatigue as unproven and high-mindedly focused on the technical aspects, all the while suggesting fatigue on the part of the two first officers may have been a factor, which of course could be even more pertinent. However, Bill Palmer was possibly not aware at the time of writing that the French investigators had not revealed that the captain himself had said on the CVR that he had only had an hour's sleep and it was not enough. One can understand them expunging him pronouncing they were going to die, but apparently the language overall on the CVR was more vulgar and ultimately panic-stricken than the BEA made public. But surely the mention of lack of sleep cannot be regarded as a private matter.

Secondly, for what it's worth, an aviation expert familiar with Air France culture and much else told the author that Robert and Bonin hated each other intensely, and that this is even evident from nuances in the "sanitized" version of the CVR. When talking to the captain, Bonin referred to Robert (who was having his rest) as "he" rather than using his first name, at the very least suggesting he was an outsider. The awkward atmosphere could have prevented Robert relaxing properly during that rest and led Bonin to spurn his advice and even do the opposite. Robert was from the academic elite and part of management, in contrast to Bonin, whose engagement as a pilot had been via a humbler route.

While the captain's questionable comportment underlay the disaster, it was the two first officers at the coalface who turned what was a mere glitch into a disaster. Their relationship is therefore very relevant.

Interestingly, Bill Palmer makes the point that yaw damping, still functioning in alternate law, probably prevented the aircraft flipping over under Bonin's control.

Unlikely to Recur

Lessons learnt should mean that never again will there be such a stall following a high-altitude autopilot disengagement due a loss of credible airspeed data. Pilots will be reminded that all they need do—apart from keeping the wings level (not easy without the computer) and continuing to fly level at the already proven power setting—is *nothing*.

They are also being taught how to recover from a high-altitude stall. Besides reducing power to prevent the nose being pushed up in a stall it might also be necessary to remove the resulting upward trim on the horizontal stabilizer in the most unlikely event of someone Bonin-like having been continuously "pulling back" on his stick. Unfortunately, when Bonin announced that was what he had been doing all along, they were dropping at ten thousand feet a minute, with only half that height remaining. It was quite hopeless, for even after putting the nose down and having regained airspeed, considerable time, or rather height, would be still needed to break that vertiginous fall.

The Future—Artificial Intelligence (AI)

Research is being done on an autopilot featuring artificial intelligence that can, for instance, guesstimate airspeed from more data sources, such as GPS and inertial guidance systems, before "giving up" and disengaging. The team developing it found it could even learn from experience gained on different models of aircraft. That would apply to cases where pilots had flown on manually with no problems after the autopilot disengaged due to lack of airspeed data.

Such an autopilot could continue, as the lawyers say, "under advisement" and avoid the pilots being thrown in at the deep end, where the startle effect and lack of manual flying dexterity are liable to make them do something untoward, as tragically happened here.

The Causes were Complex

Frequently cited as the ultimate example of how automation has made pilots lose their flying skills and crash, this disaster was also due to many other incidental human and technical factors. As Sullenberger so wisely says, "Bad outcomes are almost never the result of a single fault, a single error, a single failure—instead they are the end result of a causal chain of events."

Rather than an example of pilots losing their flying skills this could also be a prime example of how complex underlying situations can be.

By "Pull Down!" Captain Did Not Mean Pull Stick (Java Sea, 2014)

Copilot pulled back his sidestick, putting aircraft into a steeper climb and stall

> An intermittent fault of no great consequence had meant pilots had to
> repeatedly reset a system. On the fourth time that flight, the captain tried
> a novel technique, not realizing the consequences.
>
> *AirAsia Flight 8501, December 28, 2014*

The A320 on the AirAsia two-hour flight from Juanda International Airport, Surabaya, Indonesia to Singapore had 166 people, 155 passengers and 7 crew on board when it took off virtually on schedule at 5:35 in the morning. The captain was fifty-three-year-old Iriyanto, a former Indonesian Air Force officer with a total of 20,537 flying hours, 6,100 of which were with Indonesia AirAsia on the Airbus A320. The copilot was forty-six-year-old First Officer Rémi Emmanuel Plesel, a French national born in Martinique. He had a total of 2,247 flying hours with Indonesia AirAsia.

The aircraft seemed to be in good order, except that for almost a year the rudder travel limit unit (RTLU) had an intermittent fault that had manifested itself twenty-seven times, both in the air and on the ground. The RTLU is a unit that prevents the rudder being moved too far in flight and perhaps breaking off, as happened at Queens, New York, when the pilot made excessive rudder movement while caught up in wake turbulence.

Instead of repairing the RTLU fault properly—it was due to a badly soldered connection in a circuit board—the system had simply been reset, at least twenty-three times in all. Just three days earlier, the captain had seen how, when it failed on the ground, the mechanic first rebooted it, and when that did not work, he tripped and reset the circuit breakers for the main computer. When the captain asked the maintenance personnel whether it would be okay to do that in-flight if the problem came back, they assured him it would. While this was okay on the ground, it was not in-flight, as it would lead to a series of computers being disabled, including those for autopilot and autothrottle, with the aircraft passing into alternate law without most of the safety protections.

At 6:12, with the aircraft flying at their allotted cruise altitude of thirty-two thousand feet over the Java Sea, the pilots saw a storm ahead and requested permission to climb to thirty-eight thousand feet. This request was denied because of other aircraft in the vicinity, although they were allowed to alter course to the left.

An alarm then sounded indicating failure of the RTLU. The first officer being the pilot flying told the captain to see what ECAM action should be taken. The ECAM (electronic centralized aircraft monitor) is a monitor (screen) like a super-checklist that tells the pilots what to do in various situations, prioritizing problems according to urgency. In this case, it told the pilots to reboot two of the flight augmentation computers (FACs). The warning disappeared after the captain had done that.

A minute or so later, the same thing happened, and the captain again followed the identical ECAM instructions, with the warning disappearing. They contacted air traffic control at 6:14 and were given authorization to climb to thirty-four thousand feet, which they did not acknowledge, perhaps because the alarm had gone off again, requiring the captain to go through the same sequence of actions.

When they might have been about to confirm their climb authorization, the alarm went off a fourth time.

"Enough with this," said the captain.

When the first officer said, "Follow the ECAM," as before, the captain stated he had a better idea and tripped the two circuit breakers. This stopped the warning, as it did on the ground, but not only that, for the autopilot disengaged, and the aircraft passed into alternate law. This means that the normal law protections, such as those preventing stalling, were lost.

The A320 rolled to the left.

The captain, who had been standing in order to operate the circuit breakers, returned to his seat and in doing so noticed the aircraft rolling.

"Level, level, level," he said.

The first officer, having inexplicably allowed the roll to continue for nine seconds without intervening, turned to the right, and overbanked to the right.

He then overbanked to the left, putting the aircraft into a steep climb of six thousand feet per minute—twice what would be normal. This seemed to be similar to what Bonin did in AF447, described in the previous piece, in that so preoccupied was he with managing the lateral stability that he inadvertently climbed more than he should.

"Pull down, pull down, pull down!" exclaimed the captain, no doubt meaning pull the nose down.

The first officer evidently thought he meant he should pull back on his sidestick. The aircraft climbed five thousand feet in fifty-four seconds to an altitude of thirty-seven thousand feet, the opposite of what the captain intended, for the flight data recorder later showed the captain had pushed his sidestick forward but forgotten to push his priority button, with the result that their inputs had canceled each other out and the aircraft continued to climb.

The aircraft finally stalled at that 37,000-foot apogee, banked deeply to the left, and dropped.

The captain again said, "Pull!"

Whatever that meant, pulling on the sidesticks perhaps was not the way to recover.

The captain could have had full control himself and not concerned himself with the first officer simply by pressing the red priority button on his sidestick. Though it has been said the priority button has to be held down for forty seconds, that is only to latch it, meaning it no longer needs to be pressed to maintain priority. Incidentally, the priority button can also be used to take control from the autopilot.

Disappearing from Jakarta radar at twenty-four thousand feet, the A320 plunged into the Java Sea, with no possibility of survivors, though of course controllers could not be sure of that at the time.

A massive search was undertaken using aircraft and ships from many nations. Fishermen had seen some wreckage, but it was only on returning to shore the next morning that they were able to tell the authorities where to look. Search aircraft sighted wreckage on December 30 very close to where the Jakarta radar had shown they had come down.

Over the following days, wreckage and bodies were retrieved from the surface, and on January 3 wreckage was found on the seafloor. Within ten days the flight data recorder and cockpit voice recorder had both been recovered.

Because of the bad weather at the time of the crash, and the controllers' refusal to allow the A320 to climb over it due to congestion, everyone was assuming stormy weather was the cause, with even the sudden climb attributed to an incredible updraft. After all, it was, like AF447, in the intertropical convergence zone, renowned for fickle weather. The recorders were able to show that was not the case and that the accident happened as described above.

The Investigation

As it was not a US-built aircraft, the NTSB, who had disagreed with the Indonesians over the SilkAir 737 crash, were not part of the investigation, which was led by Indonesia's National Transportation Safety Committee (NTSC), with the assistance of Australia, France, Malaysia, and Singapore.

Although Singapore had granted AirAsia the right to fly that route on a daily basis, the Indonesians had only granted rights for fewer days and not for Sundays. The flight had, from their point of view, been operating illegally.

The flight data recorder had shown the weather was not the cause. Regarding the actions of the crew, the investigators were surprised that following the disengagement of the autopilot it took nine seconds for the first officer, who was the pilot flying, to correct the yaw and bank caused by the rudder being no longer under control by the computer. Letting the situation get so extreme, a bank angle of fifty-four degrees, might have contributed to his sense of panic and led him to overcompensate and forget he was climbing excessively.

The crux was miscommunication between the two pilots caused by the captain saying, "Pull down" when he must have meant pull the nose down, not pull the sidestick back. The report recommended that work should be done on establishing standard terminology for such things—something that could have made a difference in the Air France AF447 crash, where the non-flying pilot was saying "You are going down," without it being clear whether he was referring to the aircraft or the sidestick.

Conclusions

The accident sequence was triggered by the repeated failure of the rudder travel limit unit (RTLU), which made the captain experiment. But as in the case of Air France AF447, triggered by the freezing of the pitot tubes, it was the pilots' erroneous response to what should have been merely a glitch that caused the crash.

Had the computer not been disabled, it would have prevented the copilot doing the wrong thing.

Airbuses are designed to prevent pilots even getting themselves into dangerous predicaments outside the flight envelope. While the controls may not be as good as they could be outside that envelope, the fact that they, in most cases, stop pilots ever getting there may have saved many lives. How many is difficult to know, especially as there might be no reason to report such instances.

Chapter 18
SINGULAR CASES

Plane Fell from the Sky, with Pilots' Reputation (Saginaw, 1979)
The captain was regarded as a hero, but that was soon to change.

> Because it was only an "incident," with no loss of life, and the aircraft was
> immediately repaired, thus destroying the evidence, it was not
> investigated as rigorously as it might have been.
>
> *TWA Flight 841, April 4, 1979*

The fourteen-year-old TWA Boeing 727 trijet took off from New York's JFK airport at 8:25 p.m. bound for Minneapolis–Saint Paul with eighty-two passengers and seven crew. They did not have far to go, but a very strong headwind meant it would take longer than usual. To save fuel and have a smoother flight, the pilots decided to climb from thirty-five thousand to thirty-nine thousand feet, though flying at that great height was more delicate.

They and the passengers found the flight much smoother; moreover, the headwind had fallen from above one hundred to eighty-five knots.

There were three people in the cockpit. Forty-four-year-old Captain Harvey Gibson was nicknamed "Hoot" because he had once played the part of an owl in a school play, which involved running across the stage shouting, "Hoot! Hoot! Hoot!" Before joining TWA, he had been in the Marine Corps for two years to support a young wife and children, been an air traffic controller at Chicago Midway Airport, and had spent a year as a pilot at a company flying Hugh Hefner and Playboy bunnies around the US.

He was qualified to fly not only the Boeing 727 but also the DC-9, Lockheed 1011 Tristar, and the Boeing 747, having in fact been flying the jumbo as first officer for the experience and to enjoy flying the long-haul, and more exciting, routes. He had been off work for three months due to a broken ankle and had just returned as a 727 captain after three weeks of ground training and route checks. He had allegedly told his two colleagues to keep an eye on him, as it was his first time back on the 727 as captain, which would suggest it would be most unlikely he would risk doing anything unconventional, especially at very high altitude, where margins between going too fast and too slow are slim.

His copilot was First Officer Kennedy, who had served for eight years as a flight engineer on joining TWA eleven years earlier, before becoming a first officer pilot. He had a total of 10,336 flying hours, of which 8,348 had been on the 727, and was no novice.

The flight engineer, which any sizable airliner had to have in those days, was thirty-seven-year-old Second Officer Banks, who, though a qualified commercial pilot, had worked as a flight engineer for nine years. He had also served in the military.

The Upset According to the Pilots

With the autopilot set to hold as they cruised at thirty-nine thousand feet, Hoot was rummaging in his flight bag for the charts for their next stop when there was a vibrating sensation in the rudder pedals, followed by buffeting that could be felt through the airframe. The instruments showed the aircraft was still flying with the wings level, but the autopilot had moved the control wheel twenty to thirty degrees to the left. This had not been noticed by his two colleagues, who were busy.

Taking over manual control, Hoot found he, too, had to maintain the control wheel over to the left. The aircraft then suddenly yawed to the right, then even more, until it developed into a roll and then a pitch downward, which Hoot tried to correct by applying full left aileron and pulling back on his stick, but to no apparent effect.

Application of full left rudder slowed the roll momentarily, but then, as the aircraft banked forty-five to sixty degrees to the right and seemed about to turn over, Hoot pulled the thrust levers back again, to no avail. The aircraft then turned over and pitched downward.

It plunged on and on down, and it seemed they were doomed. As a last resort, Hoot got his copilot to deploy the landing gear, even though that meant the gear doors would be ripped off in the slipstream. Whether it was because of the landing gear drag slowing the aircraft, or the rupture of hydraulic lines controlling a misbehaving rudder, or the greater controllability in thicker air, or a combination of all three, Hoot was finally able to pull the aircraft out of its dive, with it at moments having had to withstand 5 g without the wings breaking off.

How close to the ground the aircraft came is difficult to say. Officially, it was five thousand feet. Unofficially, it was much closer, even hundreds of feet—after landing Hoot found a tree branch in the undercarriage, but that might have been picked up afterward.

The pilots had not only survived a precipitous dive but had to bring the damaged aircraft back to an airport and land safely. The one they chose was a compromise. The nearest, Saginaw, only had short runways and limited facilities, and the favorite, Chicago, was too far away to risk going for in an aircraft that might become unflyable. They therefore opted for nearer Detroit, despite the weather not being so good there.

They did a low-level flyby, with rescue personnel shining searchlights on the landing gear, and were told by the tower that both were down, but the one on the right looked misaligned. Hoot came in fast at 205 knots and touched down perfectly on the left undercarriage, keeping the right one off the ground as long as possible. It held until they came to a stop but collapsed when a tractor began pulling the aircraft away from the runway, showing how close a shave it had been.

On being told fuel was leaking from the wings and that passengers should evacuate, the pilots decided to have them leave via the easily deployable stairs at the rear—an almost unique feature of the 727—without risking injury on the slides.

Only eight people on board were injured during the upset, partly because Hoot, knowing passengers would not have their seat belts fastened, had pulled back on his control column to maintain some positive g. In fact, when calm was restored a passenger gulped down the cocktail he had ordered, indicating that although the passengers had endured a terrifying experience, it was more like a roller coaster ride, and only the pilots knew how close to death they had been.

A passenger looked into the cockpit to thank the aircrew before exiting and noticed two of them staring ahead as if they were in shock, which might explain why they did not remember whether they had erased the cockpit voice recording (CVR). They were initially seen as heroes.

No NTSB Go-team

Because there was no crash site and no need for deployment of the always on standby" go-team" consisting of NTSB staff with appropriate qualifications for specific roles, the incident was not properly investigated at the outset.

Officials working for the FAA and the NTSB then surreptitiously revealed to journalists that the CVR had been erased. Since the recorder ran on a half-hour loop to protect pilot privacy, and longer than that had been taken to get back to an airport and park, the CVR would not have included the upset or any of the time leading up to it. However, there remained the possibility they erased it to delete discussion of a cover-up, even though there would hardly have been time for that when bringing back a crippled aircraft with doubts as to whether the undercarriage would hold up. Nevertheless, the press felt they were onto a good story.

Whether it was increasing pressure from the press, or out of a desire to stop pilots routinely erasing the CVRs, or a combination of both, the chairman of the NTSB took responsibility for the investigation away from their local technical man and put in charge an investigator whose qualifications were principally legal, ordering him to hold the initial formal deposition by the pilots in the presence of the press, something unheard of, as the media would normally only be present at the end of an investigation prior to issuance of the final report.

On the day of the deposition, a week after the incident, Hoot, copilot Kennedy, and First Officer Banks were not in good condition, having suffered from lack of sleep, the trauma of the near-death experience, and many difficult interviews. Though the crew explained to those attending—NTSB and FAA staff, representatives from the pilots' union, TWA, and Boeing—what had happened, the focus switched to the question of the deletion of the DVR, with Hoot saying that he could not remember having done so, though he usually did. This led to the man from the FAA whose usual role was pursuing pilots for not following regulations obsessing on a line of questioning about whether erasing the CVR was permissible.

As a result, instead of the press properly reporting what pilots had said about what had happened, they focused their stories on the erasure of the CVR and what it might imply.

The head investigator then latched on to a rumor that to save fuel, pilots of the 727 were flying it with two degrees of flap, having prevented the slats deploying by disengaging their circuit breakers. This scenario was given legs by a passenger way back in coach saying she had seen the flight engineer come into the cabin and hand food trays from the cockpit to a steward just before the incident. This led the investigators to conclude that in the engineer's absence, the pilots had deployed two degrees of flap, but on his return, he had unsuspectingly reset the circuit breakers. When the pilots then retracted the flaps to retract the slats, the misaligned number seven slat failed to come back in, with the result the upset occurred.

The following comment, posted on the pprune.org pilots website in September 2016 by a contributor called Hipexec, is relevant:

> The amazing thing about the Boeing suggestion about pulling circuit breakers and extending two degrees of trailing-edge flaps in cruise was I never—I mean never—discussed, much less contemplated, it during any one of my ten thousand hours on the 727. We constantly discussed ways of saving fuel but never talked about extending flaps in cruise. Where Boeing ever dug that one up is beyond me. If Boeing knew such a procedure would save fuel, why didn't they design a procedure so we could use it? It sounds to me like a big, fat CYA red herring to distract any litigation away from Boeing.

No one who had used the technique was ever found, and in 1980, flight tests by Boeing showed that rather than reducing fuel consumption the supposedly widely used technique increased it. The reason the rumor spread as it did may have been that "hero pilots" or prima donnas are not always popular with colleagues or even flight attendants, with some pilots who have skillfully saved their aircraft and passengers even preferring not to be officially commended in the knowledge that others have suffered or had their careers torpedoed after their feat became publicized.

Thereafter, the investigation locked on to this "Boeing Scenario," excluded other possibilities, and, unbelievably, did not question the pilots again, perhaps in the belief they would be lying. They did not even do simple things like interview the cabin crew to check the facts regarding the supposed handing over of the food trays by the flight engineer just prior to the upset. Neither did they ask them about the vibration and the yaw to the right mentioned by the pilots.

Although it was claimed wiring to the CVR had been checked to show only the pilots could have erased it, this was untrue. The flight data recorder only recorded a few parameters onto aluminum foil. When it did not seem to confirm the Boeing scenario, interpretations were altered. The FDR was anyway far too primitive to properly record the motions of an aircraft rolling over and did not record the positions of the controls or airfoils.

The investigators became locked into the idea that the pilots were guilty—otherwise, the 727 might have to be grounded—and focused on evidence suggesting that. The final report, with one member of the board reticent, declared the pilots must have been at fault.

The chief investigator was later promoted to administrative posts in the NTSB but never again served as an investigator. His appeal against a later demotion was denied on the grounds that "his performance did not demonstrate the excellence required to meet the goals of the special executive service of civilian government."

Justice Delayed, Justice Denied

Emilio Corsetti III has probably written the definitive book on the incident. Called *Scapegoat: A Flight Crew's Journey from Heroes to Villains to Redemption*, it looks in painstaking detail at every aspect: the pilots and the passengers as individuals; the 727 as one of the most successful airliners, whose grounding would have been disastrous for the manufacturer and airlines; Boeing; TWA; the media; the pilots' union; the NTSB itself and its investigators; the FAA; and the more probable causes of the upset and the resetting of the CVR.

He says the pilots were vindicated in the end, but this was only morally, for the long fight for justice had taken its physical and mental toll and was never legally won.

He seems to prove that the NTSB, after being sidetracked by the erasure of the CVR, the loss of the number seven slat, and the Boeing scenario, failed to consider the most likely causes, such as yaw damper and rudder problems.

In that respect, the case has some parallels with the longest-running NTSB investigation ever, that of the quirk of the rudder of the Boeing 737, described in Chapter 6, where the NTSB did not jump to conclusions—quite the opposite.

Not All Pilots Can Conduct Air Tests (Off French Med. Coast, 2008)
What was expected to be a mere formality turned to disaster

> This highlights the danger of pilots who are not test pilots carrying out tests without defining what constitutes a pass or fail, and the problems a pilot can face on taking over manual control from the computer.
>
> *XL Airways Flight 888T, November 27, 2008*

Faced with a refusal by the air traffic controller at Bordeaux on the French Atlantic coast to allow them to engage in test flights in general air traffic, the pilots of the Airbus A320 had cut short their "acceptance flight" and turned back to Perpignan Airport on the Mediterranean. In just under an hour—half the time intended—they had surreptitiously managed to fit all but one of the tests into their flight plan without air traffic control objecting.

The pilot flying (PF) was XL Airways 51-year old Captain Kaepell with 12,709 flying hours, of which 7,038 were on that type of aircraft. His copilot was 58 with 5,525 hours on type. Based in Frankfurt, they would have been tired, having taken a taxi at 4:30 that morning to catch the flight to Montpellier, 80 miles by car from Perpignan Airport where the A320 had been sent for final maintenance and paintwork before XL Airways returned it to Air New Zealand at the end of its two-year lease. Sitting behind them was Air New Zealand's Captain Horrell.

At 15:43:37, on their descent towards Perpignan Airport, Captain Kaepell disengaged the autopilot, saying in somewhat broken English:

Down below the clouds so you want what?

Captain Horrell replied:

To go slower you mean?

Kaepell reduced the thrust to idle, all the while keeping the nose down to maintain airspeed in accordance with the Perpignan controller's instructions. Their altitude at that point was 4,080 feet and their airspeed 166 knots.

The two pilots were evidently treating this more as a demonstration than a test for they were also busy explaining to the Perpignan controller that they did not want to do a full-stop landing but a go-around and proceed to Frankfurt.

Little did they know the demonstration could never work for in hosing down the fuselage mechanics had failed to cover the angle of attack sensors on the side. This had allowed water to penetrate deep inside the mechanism, that had frozen when the aircraft reached 31,000 feet locking two of the three sensors for the rest of the flight. Besides making it impossible for the stall protections in normal law to work, it also meant $V_{\alpha PROT}$ and $V_{\alpha MAX}$ on the primary flight display speed strip were underestimated. "α" is the abbreviation Airbus use for angle of attack and $V_{\alpha PROT}$ is the airspeed at which the stall protections should come into play.

With full flap set and gear down, the A320 descended through 3,100 feet. Kaepell allowed the nose to rise, and the airspeed fell to 136 knots, then to 107 knots. At 15:45:05 the stall warning triggered with the pitch +18.6 degrees and airspeed 99 knots and Kaepell's sidestick still held slightly back. Treating it as an approach to stall, Kaepell applied go-around thrust and pushed his stick forward, but due to the load factor, not enough for the elevators to go beyond the neutral position to the stops and remove the extreme nose-up trim which unfortunately the computer had just set to enable the aircraft to fly so slowly.

Kaepell overcorrected a roll to the left producing a stalling sideslip to the right that engendered discrepancies in the air data. Not able to make sense of the situation, the fly by wire system switched to direct law. The investigators believe the pilots missed the aural warning of that because it was masked by the stall warning. Neither pilot seemed to have noticed "USE MANUAL PITCH TRIM" that also appeared in amber on the primary flight display.

The elevators (operated via the sidestick) could not overcome the combined effect of the pitch trim set at its maximum nose-up and the go-around thrust of engines set low down under the wings also levering the nose up.

We do not have space to give details of subsequent pilot inputs and the resultant behavior of the aircraft, only that by 15:45:42, after rising enough for the stall warning to stop momentarily, the airspeed had fallen to 40 knots. Two seconds later the maximum values recorded were a pitch of +57 degrees (i.e. exceedingly nose up), and an altitude of 3,788 feet, at which point the aircraft lurched over and careered down into the sea. All on board, including three Air New Zealand engineers and a New Zealand Civil Aviation Authority official in the passenger cabin, were killed.

Only sixty-two seconds had elapsed between the time the stall warning first triggered and the moment the recordings stopped!

The Investigation

This crash was very troubling because a month earlier another Airbus, a Qantas A330, had behaved very bizarrely (see Chapter 17) off the coast of Australia, albeit with the pilot managing to recover. After first suspecting the Air Data Inertial Reference Units as the Qantas A330 investigators had done, the French investigators were able to show the cause was the freezing of the water in the mechanisms of the two angle of attack sensors on the left of the fuselage. A third sensor still working was ignored by the system on the "odd man out" principle.

The investigators thought it likely that Horrell by saying "alpha floor, we're in manual" likely thought that the alpha floor function had triggered, and "manual" signified the autopilot had disconnected. They also note that the aircraft's lack of reaction to the nose-down control input did not draw the pilots' attention to the position of the horizontal stabilizer and the loss of the auto-trim function.

The pilots did not understand what was happening but had little time in which to do so. Earlier, as the airspeed dropped they had not noticed $V_{\alpha PROT}$ and $V_{\alpha MAX}$ on the primary flight display speed band were unusual and low.

Other Factors:

1. The test should have been carried out at 14,000 feet.

2. What was a pass? What was a fail? If instead of relying solely on the calculations of the aircraft's computer, the lowest airspeed for initiation of the stall protection $V_{\alpha PROT}$ had been determined independently and had been kept in mind, the test would have been aborted in time.

3. Neither airline deemed it necessary to roster check captains for the task, let alone properly trained test pilots who always take precautions and for instance would never do such a test at low altitude.

Captain SandL on pprune.org sums up the situation very aptly:

I know that the major leasing companies ask for a "2-hour demonstration flight" but they do not specify what is included. Part of the reason may be that not all airlines have crew who are able to conduct a full air test.

Valuable lessons for Airbus and others

The official BEA (French Aviation Investigation and Analysis Bureau) final report did not analyze the fly-by-wire reactions—no G-trace. Nor did it speculate on how the pilots might have recovered had the test been done at 14,000 feet in accordance with the original plan. Though the last recorded values at 15:46:06.8, namely a pitch of -14 degrees, a bank angle of 15 degrees to the right, and an airspeed of 263 knots (and altitude of 340 ft) suggest that with a little more height to play with they could have easily recovered, the elevators were surprisingly in the nose-down position even though Kaepell was pulling back on his stick to raise the nose! The report says this was probably because a load factor higher than the value commanded leads to a nose-down movement of the elevator and a rapid rise in pitch is offset by the flight control law.

Appendix 10 of the BEA report, however, looks at how the computer systems functioned in flight and what data they considered or dismissed. It is extremely complicated and may not be worth studying at length as programming will surely have since been modified in the light of the lessons learned.

Designers and programmers of fly-by-wire systems must envisage all possibilities, however improbable. Though this impromptu low-speed test proved fatal for those on board, it provided lessons for pilots, Airbus and others that may well have saved lives on revenue flights.

Landing Without Monitoring Airspeed (San Francisco, 2013)

Asiana Airlines 777 was flying too slowly to climb over airport seawall

> This incident highlights not only the absolute no-no of allowing airspeed to fall too low but also the problems an evacuation can pose.
>
> *Asiana Airlines Flight 214, July 6, 2013*

Pilots had been notified in advance that the instrument landing system (ILS) for San Francisco's 28L left runway would be out of service. Without the ILS, pilots use the runway's visual precision approach path indicator (PAPI) to help them keep to the five-degree glide slope.

PAPI is a simple but reliable system whereby four lights set beside the runway threshold appear to change color according to the angle of approach. If the pilots are too high, there are more white lights than red; if they are on the glide path, there is an equal number of each; and if they are too low, there are more red than white.

All four lights were white, indicating the Boeing 777 from Inchon, South Korea, was too high as it came over the sea from afar to land on that runway on recovered land, with a facing of boulders capped by a low seawall. Without ILS, the Asiana Airlines pilots could not, as they invariably did even in good visibility, use the Autoland system to keep them on the glide path. They were nevertheless using the autopilot to make the aircraft do what they wanted in terms of airspeed and rate of descent, but not altitude, which they had just set to 3,000 feet, as that would be the altitude for a go-around should they have to abort the landing.

About three nautical miles from the runway, with them still too high, the pilot flying for some reason changed the autopilot mode to "Flight Level-Change." This made the autopilot latch on to the 3,000 feet that had been programmed for an eventual go-around and initiate a climb, not at all what the pilot wanted. To counter this, the PF disconnected the autopilot and pulled the thrust levers back to idle, putting the autopilot, unusually, in "Hold" mode, meaning it no longer controlled airspeed via the autothrottles.

With the engines idling, the aircraft's glide slope steepened, and about 1.4 nautical miles from the runway they found themselves precisely on the glide path and by chance virtually at the 137-knot target speed they had selected for the actual landing. Unknown to the pilots, the autothrottle was not responsible for that, as it was in "Hold" mode. Apart from that perhaps easy misunderstanding, their sink rate was about 1,200 feet per minute, when it should have been 700. The observer pilot had a few seconds earlier politely mentioned the sink rate but did not repeat the warning when, with them so near the ground, it was even more pertinent. One can only assume that all three pilots were so fixated on the PAPI lights turning to red that no one noticed the falling airspeed, though of course the pilot flying was pulling back on his control column, desperate to climb and get some white lights, but with less and less effect due to the low airspeed.

Eleven seconds before reaching the seawall, there was an audible warning that their airspeed was too low. Four seconds later, with the airspeed at about 108 knots, the pilot monitoring pushed the throttles forward.

The blast from the engines spewed water up from the sea, but unlike internal combustion engines in motor vehicles, jet engines take time to spool up, and with the raised nose causing extra drag, the aircraft further slowed to 103 knots as it continued dropping. The control sticks then began shaking to warn of an imminent stall.

In the seven seconds remaining, the increasing thrust raised the airspeed by a mere three knots, not nearly enough to produce the lift required to fully arrest their descent let alone enable them to climb over the seawall, which the main landing gear struck violently, separating, as designed, from the aircraft without causing structural damage. The tip of the tail, even lower due to the nose-up attitude, struck the wall, too, with the tail cone shearing off along the line of the aft pressure bulkhead, just behind the four flight attendants sitting by the galley.

Because the aircraft had been just high enough for the landing gear to suffer and temper the initial impact, the aircraft hurtled on relatively intact for quite some distance before cartwheeling and coming to a stop with the wings still attached. The gyrations subjected the fuselage to extreme lateral stresses, causing two evacuation slides on the right side to inflate inside the cabin without the hatches being opened. In doing so, they trapped the two flight attendants seated at those exits. One was about to suffocate when a member of the flight crew came with an axe from the cockpit to deflate the one trapping her, while a member of the cabin crew deflated the other with a knife from the galley.

Flight Attendants and Passengers Ejected, Then Fire

Three of the four flight attendants sitting by the galley right at the rear just in front of the aft pressure bulkhead were ejected still strapped in their seats and survived, albeit seriously injured. Two teenage girls sitting at the back whose seat belts were not attached were not so lucky on being ejected. One, thrown out early on, probably died on impact with the ground, while the other, who was ejected just before the aircraft came to a stop, was twice run over by a fire appliance while lying covered by fire-retardant foam. There were two inquests, with the first finding she was still alive when run over, and the other not. This cast a shadow over the heroic efforts of the firefighters and notably those who climbed up the slides into the cabin to rescue five passengers unable to escape, one of whom later died.

Evacuation

Ordering an evacuation is a decision not easily taken. Pilots sitting right at the front can see ahead but have little idea of what is going on behind them. Unless told by their instruments, they do not even know whether an engine is on fire.

In nose-up crash landings, where the tail suffers the initial impact, they can even believe it was a relatively smooth landing, when those right at the back are injured or killed. Most evacuations expose the occupants, and especially the older ones, to injury. Following protocol, the pilots liaised with the controllers in the tower, who had a clear view of the aircraft, and waited for communications from the cabin crew. Passengers were told to remain seated.

Though the engines had separated, one had ended up lying beside the fuselage by the right wing with its fuel tanks. Oil from a tank on it caught fire, and when the lead cabin attendant was informed of this, she ordered an evacuation—ninety seconds after the aircraft had come to a halt, ironically the time it should theoretically take for an airliner to be evacuated with only half the exits usable.

Only two out of four slides on the left deployed, but passengers were able to evacuate not only via them but also through the gap at the back where the tail cone had been. The lead flight attendant was praised by the rescuers for her professionalism and for having ensured everyone got out. What with three cabin crew at the back having been ejected and others injured, only five were able to assist with the evacuation and help and, perhaps more importantly, control the passengers, some of whom, as they so often do, collected their luggage.

The fire finally penetrated the cabin, but firefighters who came in before it did much damage noted the seats at the front were pristine, while those right at the back were contorted.

Surprising Number of Survivors

On board there had been 12 crew and 291 passengers. Many of the passengers were Chinese, who had difficulty understanding what was happening. They included a group of teenagers on their way to a summer camp in the US who had the misfortune to be sitting at the back.

Surprisingly, 99 percent of the passengers survived, and only one passenger would have been killed—by a dislodged rear door—had all had their seat belts been attached. Altogether, 187 people were injured, of whom 49 were injured seriously. This is difficult to believe when one sees the videos taken from the control tower.

The Investigation

The NTSB's chair, Deborah Hersman, gave press conferences immediately after the event, leading to objections from ALPA (Air Line Pilots Association) that the release of information was leading to rampant speculation as to the cause.

Interestingly, she pointed out that exchanges between the pilots and the controllers are in the public domain, while material on the cockpit voice recorder is confidential and can only be made public as a transcript. In the author's view, this is a pity, as the tone of voice can reveal the atmosphere in the cockpit and the relationship between the pilots.

The accident happened essentially because the pilots failed to monitor the airspeed in the erroneous belief the autothrottle would ensure it was maintained at the target landing speed they had set.

However, this came about for a whole variety of reasons:

1. The pilots, in line with company policy, relied on automation whenever possible, and the pilot flying had never made a landing, even in visual conditions, without the help of the Autoland/ILS to keep him on the glide path. He was nervous, especially as he was being checked out and had performed less than perfectly on a previous check flight.

2. The Boeing documentation did not make it clear that the autothrottle did not control airspeed in "Hold" mode and that the clever safety feature whereby it would come to life in an emergency, such as a much-too-low airspeed, did not operate if one pilot's flight director was on and the other pilot's was off. The PF had previously flown Airbuses, which had somewhat different philosophies.

3. The check captain was performing that task for the first time and may not have had the confidence to order a go-around earlier.

4. Crew resource management was poor: the PF did not call out what he was doing. The pilot monitoring could perhaps have repeated his warning about the sink rate and should have concentrated on the instruments rather than, presumably, on the all-red PAPI lights outside.

Aftermath

The airline revised its training techniques to ensure pilots could cope with flying manually. Boeing reviewed documentation and training so that the various autothrottle modes are better understood and followed NTSB advice to make low-airspeed warnings, for example, kick in earlier when the aircraft is near the ground and the time left to save the situation is limited. Makers of escape slides strengthened their mechanisms to lessen the chance of them opening prematurely. Firefighters took measures to better manage accident zones and lessen the chance of injuring escaping passengers inadvertently.

Was Disappearance of MH370 Almost the Perfect Crime? (2014)
The greatest mystery since that of Amelia Earhart

> The Malaysian authorities' handling of the aftermath was so baffling that
> it is difficult to know whether it signified ineptitude or a cover-up.
> *Malaysian Airlines Flight 370, March 8, 2014*

MH370 was the Malaysian Airlines overnight flight from Kuala Lumpur to Beijing. The twelve-year-old Boeing 777 had a total of 53,460 hours but had been used on relatively long-haul routes and so only had 7,525 cycles, that is takeoffs and landings, which are what determines the age an aircraft in terms of metal fatigue. Bearing that in mind and that it had not been involved in an incident like a tail strike entailing major repairs, catastrophic failure of the fuselage was most unlikely.

Taking off from Kuala Lumpur's new airport at 00:41 Malaysia time with 12 crew and 267 passengers, it gradually gained height as it flew over the Malay Peninsula to the northeast. At 01:06, just before reaching the coast, its ACARS data link automatically transmitted an update of its status, including the fact that 96,600 pounds (43,800 kilos) of fuel remained and that there were no particular faults. Out over the South China Sea, it reached its thirty-five-thousand-feet cruising height.

At 01:19, Kuala Lumpur air traffic control instructed them to switch frequency to that of the Vietnamese Ho Chi Minh air traffic control. The captain of MH370 acknowledged and signed off by simply saying:

Good night, Malaysian Three-Seven-Zero.

These were the haunting last words heard from the pilots.

The Vietnamese controllers would not have been following their blip closely, as not having checked in there was nothing to say. Not having been listening in to the exchanges between MH370 and the Malaysian controllers they would not know exactly when to expect the call. Finally, when they thought it was taking far longer than expected, they could find no echo for MH370 on their screens.

Military radar, called primary radar, is designed to detect enemies from feeble waves reflected from the aircraft's surface. However, civilian air traffic control radar, called secondary radar, relies on a transponder on the aircraft to send back a much stronger echo with the identity, altitude, and airspeed of the aircraft. It also includes the four-figure squawk code assigned to the flight, which the pilot can change to, say, 7500 to report a hijacking, 7700 to indicate an emergency, or even 7600 for failure of the radio.

There was no blip for MH370 on the Vietnamese air traffic control radar.

There was not even the ACARS data link.

ACARS (aircraft addressing and reporting system) is a data link that regularly sends information required for maintenance. Pilots can also use it for weather updates or even for requests, such as when medical assistance is needed for a disembarking passenger. It can send the data cheaply via VHR radio and via satellite, which is much more expensive, and for which Malaysian Airlines had not signed up. Malaysian Airlines had not paid for satellite transmission, the satellite would at intervals "shake hands" with the onboard device if it had power.

It was not until 1:38, nineteen minutes after MH370 should have checked in, that Ho Chi Minh control called Kuala Lumpur ATC to ask what was happening, only to be informed MH370 had been told to contact them and signed off.

At 1:46, the Vietnamese again queried KL ATC, and they were told there had been radar contact but no verbal contact at the IGARI waypoint. Ho Chi Minh replied that the radar blip had disappeared at BITOD, a waypoint slightly farther on.

At 1:50, KL ATC asked HCM ATC to confirm they had finally made contact and were told they had not. At 1:57, HCM informed KL ATC that they were declaring officially that there had never been any contact with MH370, despite direct calls on various frequencies and indirect attempts to contact them via other aircraft in the vicinity.

It was then that Malaysian Airlines Operations confused everyone by saying the aircraft was in Cambodian airspace, information that KL ATC passed on to HCM at 02:03. Fifteen minutes later, at 02:18, a mystified KL controller asked HCM whether the flight plan did indeed include a route through Cambodian airspace and was told it was solely through Vietnamese airspace, and that HCM had checked with the Cambodians, only to be told they, too, had had no radar contact or communication with MH370.

At 3:30 Malaysia Airlines Operations astoundingly admitted that the coordinates they had given for MH370's location in Cambodia were merely based on projections and were not reliable for flight positioning, and KL duly informed HCM of this. At 4:25, HCM, obviously greatly concerned as MH370 should have passed through their patch, questioned KL as to whether they had heard anything from Hong Kong or Beijing!

At 4:30, KL ATC finally alerted search and rescue, which then, and for days, concentrated on the South China Sea off Vietnam and China, on the aircraft's scheduled route.

Where Had It Gone?

MH370, with the transponder not working or disabled, had immediately executed a U-turn and had long left the Vietnamese air traffic control area which it had in fact hardly entered. Thus, even search by primary radar would not have picked it up because they would be looking in the wrong place—in fact looking further along its route on the flight plan in the direction of Beijing.

Later analysis of Malaysian military radar showed it had flown back over the Malaysian Peninsula and not to Kuala Lumpur but to Penang, farther north, then turned right to fly northwest up the Strait of Malacca beyond the tip of Sumatra where there was no military radar coverage either from Malaysia or Indonesia. Several days later, experts from the satellite telecommunications company Inmarsat were able to show it must have turned left there and flown in a southerly direction, to come down hours later somewhere in the southern Indian Ocean perhaps two thousand miles off Perth, Australia.

Even when it became known the aircraft had not crashed in the South China Sea the Malaysian authorities persisted with the search there not only wasting the time and money of countries deploying their aircraft and ships but delaying the search in more likely places.

What Had Happened?

The briefings by the Malaysian authorities were a disaster, perhaps partly because knowing little and understanding little they gave premature answers to unbelieving Chinese relatives. Usually when there is a crash, people know where it was and roughly what happened, giving relatives something to latch onto and allowing them to accept the assertion, justified or not, that the investigators cannot say much. Here there was nothing to investigate.

The Malaysian military were accused of being slack, though again, to be fair, one must admit that relatively junior officers would understandably hesitate to send up fighters to investigate nonthreatening blips in the middle of the night without their superiors' authorization, and particularly so if the blip was exiting the country and obviously not a threat. Reports their radar operators had seen the aircraft climbing to a great height—supposedly to incapacitate the passengers—and descending and climbing steeply were later discounted, since doing so would have used so much fuel that that the 777 could not have flown for as long as it did.

Also, it was alleged the military in neighboring countries were keeping quiet so as not to divulge what they could see, or as others said, more likely not to reveal how little they could see.

The Crew and Passengers

The captain was fifty-three-year-old Zaharie Ahmad Shah from Penang. The fact that the aircraft flew over Penang on its way out to the Indian Ocean led to suggestions that he was committing suicide and wanted a last look at the place where he had been brought up. Penang also had a long runway that would have made it a good place to head for to make an emergency landing.

Zaharie had 18,365 flying hours on various aircraft, including many on the 777. He loved flying and had a flight simulator at home. However, one must be wary of taking comments by family members at face value, as it is in their interest, even financial, to show their relative must be innocent.

The copilot was twenty-seven-year-old Fariq Ab Hamid, from Kuala Lumpur. Though it was only his sixth flight in the 777 and the first time without being supervised by a check pilot, he had 2,763 flying hours, having joined the airline in 2007.

Nothing suggested he was a man going to commit suicide. He was about to wed his wonderful fiancée, a captain flying for the low-cost airline AirAsia, whom he had known ever since they had been students at Langkawi Flying School. He was a relaxed fun-loving type if one is to believe the photos given to the media by two South African teenage girls showing them sitting with him in the cockpit and them all laughing and smoking away.

Military Radar—Track Back over the Malaysian Peninsula

Although it was not known or revealed until later, the Malaysian military had seen the aircraft, as mentioned, turn left after passing the IGARI waypoint, which is near a point where three air traffic control regions—those of Malaysia, Vietnam, and Singapore—meet, so that the aircraft is in a zone where it is effectively under the control of the center actually handling it. Also, VHF radio, used both for voice and ACARS communications, can be intermittent, which may in part make controllers wait for communications to improve before worrying or ask other aircraft to make contact for them.

It flew back over the Malaysian Peninsula, over Penang, before turning almost forty-five degrees to the right to go up the Strait of Malacca without passing over land. It then went out of radar range.

Satellite Handshakes

Malaysian Airlines had not paid for ACARS data to be transmitted via satellite, but the satellites nevertheless performed handshakes at regular intervals to confirm the system was working should it be required, whereby the satellite transmits a signal (a ping) and waits for a reply (a pong). Following the disappearance of Air France flight AF447 over the South Atlantic, where the company was paying for satellite transmission of ACARS data, Inmarsat, the British company providing the service, found there was not enough information about the aircraft's location and incorporated a timing feature in the handshake that could give some general indication of the location.

This new feature provided the startling news that MH370 had flown on for almost six hours after being lost from view on radar near the northern tip of Indonesia. The satellite, hovering at thirty-five thousand kilometers above the ocean, showed the aircraft must have come down on one of two arcs, one to the north, the other to the south. The northern one was discounted because wide-ranging military radar coverage there was such that it could never have been missed for so long.

The conclusion was that it had come down off Perth, Australia, in an isolated area with terrible sea conditions known as the Roaring Forties. Much time had been lost looking in the South China Sea, and the locator beacons, which only last thirty days, were almost out of power when the proper search began. There were moments when it was thought signals had been picked up, but these proved to be red herrings.

Aircraft from various nations scoured the surface of the sea for debris without finding any that could ultimately be proved to be from the aircraft. Had the search concentrated on that area earlier, there would of course have been a better chance of finding something. Recently, some French satellite images have appeared showing items in the water that could be man-made, but definition was too poor to draw definitive conclusions as to what they might be.

Debris—Flaperon with Trailing Edge Broken Off

On July 29, 2015, sixteen months after MH370's disappearance, a flaperon from the aircraft was found on a beach in Saint-André, on Réunion, off Madagascar. Reference numbers made its origin virtually certain and dashed the hopes of relatives that their loved ones might be still alive.

The trailing edge of the flaperon had clearly broken off, something requiring considerable force, making some conclude the flaps had been deployed by a human so the aircraft would come down relatively gently under control and not break up into thousands of pieces. Indeed, a part of a flap was later found in the same general area with the trailing edge similarly broken off.

A small amount of other debris was found, all of which could have been brought there by currents from where the aircraft supposedly went down.

The flaperon was sent to France, and the French were keeping it on the grounds that it is evidence in a judicial inquiry and that La Réunion is French territory.

The Underwater Searches

The location of the presumed final resting place meant that Malaysia, with its limited resources, was in charge of the investigation, and though an expensive underwater search, managed by Australia, was undertaken with vessels and equipment mainly from China and Australia, Malaysia was obviously a partner. After two years' searching in extremely difficult conditions, it was called off, just as revised wreckage-drift analysis and rethinks regarding the satellite data and the possible point where the turn to the south Indian Ocean was made suggest more certainty as to the likely resting place of sunk wreckage from the flight on the ocean floor.

Hopes were high when on January 10, 2018, Malaysia's government announced it would pay Texas-based Ocean Infinity up to $70 million if it could find the wreckage or black boxes of Malaysia Airlines Flight 370 within three months. Unfortunately, despite the deployment of eight underwater drones able to stay under water for forty-eight hours at a time—which unlike those used in previous searches are not tethered to the surface ship by cable and reduce noise to yield data of better quality—the search was unsuccessful.

For the relatives, finding the aircraft and their loved ones, and hopefully determining who might be responsible for such a dastardly thing, is of prime importance. For others, and from an air safety point of view, it may be less so, as the aircraft flew on for hours with changes of course unlike Air France Flight 447 lost in the South Atlantic after dropping out of the sky while still on course and with the ACARS link working. There were fears and accusations that there was something inherently wrong with the Airbus A330, making it of the utmost interest for the manufacturer and the French government to prove otherwise. In the end, discovery of the wreckage and recovery of the recorders after two years proved the pilots had turned a glitch into a disaster.

Even recovering MH370's cockpit voice recorder would not reveal how the incident was orchestrated, for to protect pilot privacy, CVRs overwrite themselves at half- or in this case two-hour intervals. Passengers' mobile phones might provide answers, but not if they died through lack of oxygen before there was anything of note to see.

Searching for the Truth

Analyzing the satellite data regarding the location of descent into the sea and associated evidence is extremely complex, and a subject for specialists.

Though interpretation of the satellite data gives some idea of along what arc MH370 came down, that is a line thousands of miles long. Where on that line, and how far away from it to search, was based on how far the aircraft could go with the fuel it had and whether someone might have flown it away from that line after the fuel ran out. Complicating the issue is the fact that after going northwest up the Strait of Malacca, MH370 may have continued much farther than thought before turning south. An extra fifteen minutes would mean almost half an hour's difference in flying time to the south, meaning the expensive underwater search had been much too far south.

Also, if the pilot had recovered the aircraft after it had dived after running out of fuel he (or even she) could have glided a fair distance to either side of the arc, making the search area much vaster. More recent drift analysis based on the few pieces of wreckage found suggest the aircraft had come down further north on the arc than previously assumed.

Anyone fascinated and wanting to keep up to date technically might well visit Victor Iannello's http://mh370.radiantphysics.com/ website. An American engineer living in Virginia, Iannello has been passionately following the affair. His blogs are not for the faint-hearted and much too technical for most, but some of the comments make fascinating points, besides revealing how many different points of view there are, with people getting quite worked up, and even angry.

Interestingly, data released almost three years after the event to one of the next of kin contained all of the information transmitted via an Inmarsat satellite link on the previous flight by the jet, from Beijing to Kuala Lumpur, picking up features that were common to both flights and not anomalies unique to MH370.

Tentative Conclusion (Summary)
Taken individually, many of the twists and turns are explicable, but taken as a whole, the coincidences, leading to suspicion it was an intentional act, are just too many:

1. It all started seemingly all too conveniently just after 01:19 Malaysia time at the point where the Malaysian and Vietnamese flight information regions (FIRs) intersect. Having just left Malaysian control, MH370 disappeared from civilian radar without having signed on with the Vietnamese. Not only had the transponder stopped working, the ACARS data link had, too. Though they could have reversed course to get back to Malaysia to make an emergency landing, it also meant the Vietnamese controllers would be looking for them, possibly with primary military radar, in quite the wrong place.

2. Because they turned left some 150 degrees to Penang, rather than 180 degrees back to Kuala Lumpur, the possibility of the unidentified aircraft being suspected as having been hijacked and about to be flown into the Petronas twin towers, 9/11-style, was avoided.

3. Turning to the right after Penang to go north up the Strait of Malacca avoided drawing the attention of the Indonesians.

4. Malaysian military radar tracked them going up the strait until 02:22, after which they turned left on reaching some point above the northern tip of Sumatra, to fly south over the ocean to one of the most remote parts of the world, where, without the satellite handshake data, no one would ever dream of looking. Last satellite contact was at 08:19, seven hours after "Goodnight, Malaysian Three-Seven-Zero."

Was someone trying to prove how clever they were by creating what Victor Iannello's blog calls "the world's greatest aviation mystery"? If one rules out the about-to-be-married first officer, who had shown a love for the good things in life, that only leaves the technically minded captain, unless there was a rogue passenger or someone hiding on board, even in the avionics bay. If he was genius-like going to create the greatest-ever aviation mystery he surely would be clever enough to hide his anger from his spouse.

Since the cockpit voice recorder records on a two-hour loop, if it is recovered it will probably only reveal whether someone was breathing, praying, or ranting at the very end. On the other hand, that person, at the very least, might be identifiable.

ATSB Final Report

On October 3, 2017, the Australian National Transport Safety Bureau (ATSB) issued its lengthy final report on MH370, or rather, on its role relating to the unsuccessful 1,046-day underwater search off Australia. We were lucky to be able to use the information in the report in writing this piece.

It was not the FBI, as rumored, but a Malaysian contractor that the Malaysian authorities engaged as part of a police investigation, the details of which were never made public, that analyzed the captain's home computer and found traces of a simulated maneuver far south in the Indian Ocean.

An Emirates airliner going northwest up the Strait of Malacca just thirty-three nautical miles behind MH370 had the same track on the military radar, confirming the latter was also following the waypoints like a normal flight.

If the intention was to commit suicide creating the greatest mystery in aviation, one must wonder how someone reasonably clever could do it at the expense of so many.

Final Report

The 400-page final official report revealed on July 30, 2018 said little of note other than that the initial 180-degree turn was made manually because the autopilot could not have done that in 2 minutes 10 seconds and that the transponder stopped transmitting, "whether with intent or otherwise." Adding "Although it cannot be conclusively ruled out that an aircraft or system malfunction was a cause, based on the limited evidence available, it is more likely that the loss of communication prior to the diversion is due to the systems being manually turned off or power interrupted to them."

The report did not throw any new light on the mystery. In fact, it deepened it by virtually dismissing the idea that the pilots could be responsible, saying that there was no sign of stress in their radio communications and investigators had found no evidence of anything suspicious regarding them.

Pilot suicides we have described in this book show pilots pushed to the edge with passenger's lives secondary. Might passengers, as a group, or as an individual or family, have been primary in this case? Just a troubling thought. [See UPDATES for latest info]

Conclusion

The lessons learned from the incidents described and advances in technology have made flying so safe that some experts say innovative ways have to be found to prevent the few accidents that there are. Apart from terrorist acts, the greatest risk seems to be unforeseen clear air turbulence (CAT), which is increasing in frequency and severity. Even that challenge is perhaps on the way to being solved with Lidar, a form of radar using light from lasers.

Particularly in the United States, we are entering a new era of air traffic control, called NextGen, based on GPS, with aircraft transponders within a wide radius automatically communicating with each other as well as with air traffic control. Besides the traditional flight number, squawk code, airspeed, and altitude, the content of the exchange includes the aircraft's precise location, determined by GPS, and even maneuvers in the course of being executed—in the past, controllers could only see an aircraft was changing course after it had flown some distance on that heading.

The system will allow aircraft to look after themselves without so much intervention by the controllers and fly closer together. Potential conflicts will be recognized much earlier, and aircraft will be able to fly more direct routes instead of following lines between waypoints.

Long gone are the days when internationally regulated airfares allowed some pilots to have extravagant salaries. Now, conversely, there is the danger of pilots being undervalued and airlines having insufficient applicants with the ideal combination of physical and mental qualities, including self-discipline, good judgment, and technical intelligence, with the ability to apply it in the unlikely event of being suddenly faced with the unexpected.

There has recently been talk of having just on just one, or even no pilot in the cockpit with the aircraft controlled from the ground. As things stand, passengers would not accept having no pilot. However, having just one would mean he or she could be paid more and be better trained with, paradoxically, more extensive manual flying experience. Cabin crew and others with aptitude could be trained to be their assistants cum emotional hand-holders and be quite capable of landing the aircraft with the help of automation should the rostered pilot become indisposed.

Could having one good pilot be better than having two questionable ones, even captains about whom, as happens in some crashes, the airline has reservations? Not having a second pilot's opinion would have drawbacks but having conscientious "assistants" (like nurses in a hospital) might result in better monitoring, including that of airspeed. It could open up a rewarding career of pilot-flight attendant, with perhaps several taking turns, thus making the monitoring more effective.

There have been black swan events where the presence of even more than the complement of two pilots proved vital, namely the amazing landing of an uncontrollable DC-10 using engine power alone at Sioux City, and the bringing back to Singapore of a Qantas A380 superjumbo with most of its control systems damaged. However, those additional pilots on board just happened to be the best of the best. We are talking about ensuring there is at least one pilot of unquestioned ability flying the aircraft by using them more efficiently. Lest we forget, the flight engineer was once thought indispensable.

Updates

January 2018—AF447
According to news agencies and services, preliminary findings of the judicial inquiry into AF447 lay the blame largely on the pilots with Air France objecting and the association representing the families of the victims saying they "disgusted" by the conclusion absolving Airbus. The parties have two months to comment.

The conclusion of the judicial inquiry seems to concord with what we wrote in Chapter 17.

August 2018—Mystery of The Disappearance of MH370
In response to the final official report by the Malaysian investigators on July 30, 2018, the research section of the French Air Transport Police announced it would conduct its own review of the satellite data (pings from the aircraft) as there had been four French citizens on board, and verifying it was necessary for their investigation. This implied they were not satisfied.

Publication of the report was also an occasion for the world's media to bring up the many theories regarding what might have happened, including the suggestion a stowaway might have been hiding in the avionics bay (accessible from the ground and via a hatch behind the cockpit), and the possibility that the aircraft could have been taken over electronically by hackers. About the latter, Boeing said that although it had patented a remote-control system as a possible way to take control away from hijackers such as those in 9/11, it had not installed such a system on any commercial aircraft.

The Malaysian report said the first officer's mobile phone had been on and had made a handshake at Penang with no call, and that tests had been made showing that was possible. Some experts were surprised no attempts were made to see whether any of the Malaysian passengers' phones had similarly tried to log on to a Malaysian mobile phone network as at least one or two passengers tend to leave their phones on.

It also noted "no traces of explosion were found" on two pieces of debris that are almost certainly came from the cabin interior.

Malaysia's civil aviation chief resigned after the report which found failures by Malaysian air traffic control, notably lapses by controllers in relation to emergency procedures. The aircraft had been missing, the report said, for about 20 minutes before authorities sent an alert. As he was close to retirement anyway, his resignation was not so meaningful.

The mystery is as great, if not greater than ever.

As we said in the piece on MH370, the best link, though rather technical, for keeping up to date on the latest MH370 developments may be http://mh370.radiantphysics.com/.

Index

**Page numbers refer
to the printed version**

Acknowledgments

Many official reports, excellent articles, and books, not to mention films of varying accuracy, have contributed to this book's realization. Twenty thousand or more items available on the Internet provided background knowledge. In addition, I am sometimes citing from memory a detail or point made in now irretrievable coverage in the local media in the countries where the incidents occurred, and I happened to be at the time.

Though it may seem ghoulish, reviewing cockpit voice recorder (CVR) transcriptions included in official reports often seemed to give a real feel to what was happening in the cockpit and a good idea of the cascading dilemmas facing the pilots.

I wasted much time, without coming to definitive conclusions, in some cases where official reports were contested as conspiratorial, as can happen when all parties—airline, manufacturer, and investigating authority, and even victims— are from the same country. Once someone claims officials have tampered with evidence, have shredded it, and are lying to boot, there is no definitive answer. From a safety point of view, this may not matter overly, as the conspirators, if any, know the truth, and stealthily, but eventually, take remedial measures.

In preparing the material, I was fortunate to come across *Air Disaster*, by the late Macarthur Job, which reviews in several volumes the most notable jet-age airliner disasters up to 1994. Job succeeds in putting them in perspective, using a wealth of technical and human detail. Those captivating volumes would be a good starting point for anyone wishing to read about a number of the pre-1995 accidents in much more detail than is possible here. Although now out of print, those books recently became available on Kindle.

Similarly, *Disasters in the Air: The Truth about Mysterious Air Disasters*, by Jan Bartelski, was of great value in making me look again at the assumptions that had generally been made regarding some notorious disasters, and notably the worst-ever multi-aircraft disaster, in which two 747 jumbos collided on the ground in fog at Tenerife in 1977.

Some of the episodes of the excellent TV series *Air Crash Investigation*, called *Mayday* in some countries and broadcast on the National Geographic Channel among others, yielded useful new insights. Besides bringing events to life much as we have tried to do here, the highly successful series has aroused the interest of many. The producers obviously devote considerable effort, and not least money, to interviewing passengers in addition to aircrew, investigators, and experts to produce such captivating and informative videos.

At the other end of the spectrum in that it is a textbook stuffed with concentrated information regarding all aspects of aviation safety was *Commercial Aviation Safety*, by Alexander T. Wells. Its depth of detail is such that it even mentions how cockpit noise affects pilots, and in so doing helps put other safety aspects in perspective.

A Human Error Approach to Aviation Accident Analysis, by Douglas A. Wiegmann and Scott A. Shappell, was a valuable introduction to the academic work being done on accidents. Although purchased too late to be much help, I should mention *Aviation Disasters*, by David Gero, covering major civil airliner crashes since 1950. Much less discursive and opinionated than this book, it is a good tool for specialists wanting the cold bare facts concerning virtually every disaster.

Other books consulted are cited where relevant in the accounts.

While too technical for many, the Professional Pilots Network on http://pprune.org has some very interesting threads on specific topics.

Australian aviation journalist Ben Sandiland's *Plane Talking* blog provided leads and helpful insights on many topics, including MH370. Even though critically ill in hospital, he unstintingly gave advice for the second edition of this book. His outspoken and knowledgeable commentary on aviation and even astronomy will be sorely missed by many.

Also, the official reports by air accident investigative bodies, such as the NTSB, AAIB, ATSB, BFU, and BEA, often made fascinating reading besides providing factual details.

On a personal level, I would particularly like to pay tribute to the late John Hawkins for encouraging me. His experience both as a metallurgist and as managing director and chairman of numerous companies in the British Alcan Group, developing high-temperature alloys for aircraft, including the supersonic Concorde, meant he was able to advise me about the evolution of these materials. He also informed me about some aviation incidents that although in the public domain did not make the headlines.

Others kindly read the original manuscript and gave valuable advice. They include Keith Lakin, Adrian Wojcieh, James Denny, Mike Pegler, Gerald Burke, Jonathan Evans, Hélène Bartlett, and Go Sugimoto.

Finally, I must thank Marcus Trower for copyediting this complex manuscript, and Nikita Wood for kindly doing a final check.

Christopher Bartlett

Author

Christopher Bartlett initially trained as a mining engineer, a field where ensuring compliance with safety standards is of prime importance. His passion, however, has been flying, and notably air safety.

This was engendered as an air cadet during his youth and as a member of the British Interplanetary Society, as well as during sessions on fighter simulators at the Air Ministry. He completed his two years' military service in the British Royal Air Force.

After taking a degree in Modern Chinese and Japanese at the School of Oriental and African Studies, London University, he became, among other things, a professional translator of Japanese scientific and technical material. This included Japanese rocket tests. He also wrote for magazines in the Far East.

His fluency and understanding of English, French, and Japanese enabled him to undertake research based on material published in its original format and note opinions and facts that were not widely publicized. In addition, his coincidental residence in countries when and where headline air crashes occurred has enabled him to add local color and extra details to several of these accounts, and notably the worst-ever single aircraft crash, JL123, in Japan.

Online Shop (5x copies for training or gifts)

Our website https://chrisbart.com includes an online shop where the book can be purchased in multiples of five copies for shipping only to addresses in the UK, Australia, and US (mainland) where the books are also printed by Lightning Source.

Made in the USA
Monee, IL
14 December 2023

49288787R00234